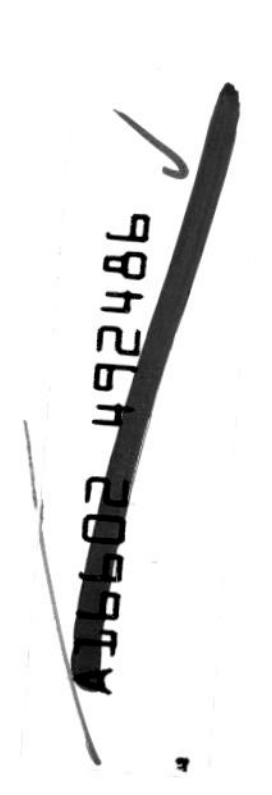

Introduction to
STATISTICS

McGraw-Hill Series in Probability and Statistics

David Blackwell and Herbert Solomon, Consulting Editors

BHARUCHA-REID Elements of the Theory of Markov Processes and Their Applications
DE GROOT Optimal Statistical Decisions
DRAKE Fundamentals of Applied Probability Theory
EHRENFELD AND LITTAUER Introduction to Statistical Methods
GIBBONS Nonparametric Statistical Inference
GRAYBILL Introduction to Linear Statistical Models
HODGES, KRECH, AND CRUTCHFIELD StatLab: An Empirical Introduction to Statistics
LI Introduction to Experimental Statistics
MOOD, GRAYBILL, AND BOES Introduction to the Theory of Statistics
MORRISON Multivariate Statistical Methods
RAJ The Design of Sample Surveys
RAJ Sampling Theory
STEEL AND TORRIE Introduction to Statistics
THOMASIAN The Structure of Probability Theory with Applications
WADSWORTH AND BRYAN Applications of Probability and Random Variables
WASAN Parametric Estimation
WOLF Elements of Probability and Statistics

Introduction to STATISTICS

Robert G. D. Steel
Professor of Experimental Statistics
School of Physical Sciences
North Carolina State University

James H. Torrie
Professor of Agronomy
College of Agriculture
University of Wisconsin, Madison

McGraw-Hill Book Company
New York St. Louis San Francisco Auckland Düsseldorf Johannesburg Kuala Lumpur London Mexico Montreal New Delhi Panama Paris São Paulo Singapore Sydney Tokyo Toronto

INTRODUCTION TO STATISTICS

1234567890DODO79876

This book was set in Times Roman.
The editors were A. Anthony Arthur, Alice Macnow, and Shelly Levine Langman;
the designer was Scott Chelius;
the production supervisor was Angela Kardovich.
The drawings were done by J & R Services, Inc.
R. R. Donnelley & Sons Company was printer and binder.

Library of Congress Cataloging in Publication Data

Steel, Robert George Douglas, date
Introduction to statistics.

(McGraw-Hill series in probability and statistics)
Bibliography: p.
Includes index.
1. Mathematical statistics. I. Torrie, James Hiram, date joint author. II. Title.
QA276.S79 519.5 75-22235
ISBN 0-07-060918-7

In Memoriam
Jonathan Chester Steel
1954–1975

Contents

Preface

This is an introductory text, presenting some of the basic concepts and techniques of statistical inference. These concepts and techniques have played an important role in a long list of very diverse fields, in fact, wherever one finds observational or experimental data, whether it be counts or measurements. It is because of the chance element in the origin and nature of data that statistics has become such an important tool, since statistics depends upon probability, which is concerned with chance.

The first four chapters of the book deal primarily with the presentation and summarization of large amounts of data; they indicate the need to distinguish between samples and populations and to consider the possibilities of drawing inferences about the latter from the former. At the same time, some of the procedures given are regularly applied to small samples.

The introductory chapters are followed by probability theory. It is here that the notions of sample space, random variable, and probability function are introduced. Several frequently encountered distributions are considered at some length.

The general concepts of statistical inference, namely the estimation of parameters and testing of hypotheses, are next discussed, with special reference to and examples for the binomial and normal distributions. However, each aspect of inference once applied to a specific distribution is then looked at from a nonparametric point of view. Consequently, the students' understanding of the topic as originally introduced is reinforced.

The analysis of variance and regression analysis are included, as are nonparametric alternatives to these techniques.

At no time should one find a need for more than a working knowledge of high school algebra. Even when such algebra is used, as in finding the expectation of a binomial sum, one can generally go straight to the results. For those who enjoy algebra, there are a number of problems with algebraic content.

The book was written to provide sufficient material for a two-semester course. However, Chapters 1 to 13, with some topics deleted, together with topics selected from later chapters, perhaps 15 and 17 in particular, can serve as a one-semester course. For those instructors with an interest in nonparametric statistics, Chapters 14 and 18 can substitute for 15 and 17. Section titles should help with one's choice of topics to be deleted or included.

We are indebted to the literary executor of the late Sir Ronald A. Fisher, F.R.S., to Dr. Frank Yates, F.R.S., and to Longman Group Ltd., London, for permission to reprint Table III from their book "Statistical Tables for Biological, Agricultural and Medical Research."

We are also indebted to E. S. Pearson and H. O. Hartley, editors of "Biometrika Tables for Statisticians," vol. I, and to the *Biometrika* trustees for permission to make use of Tables 8 and 18; to John W. Tukey and to Thomas E. Kurtz and Prentice-Hall, Inc., for permission to reprint Tables B-4 and B-5; to the late Frank Wilcoxon, Roberta A. Wilcox, and the American Cyanamid Company for permission to reprint Table B-6; and to Dr. James H. Goodnight for producing Table B-8 for us.

In addition, we are indebted to many authors, journal editors, and publishers for their generous permission to use the data that appear in the text and exercises. Their names are listed in the references.

Finally, we are grateful to a number of people for their contributions during the preparation of the manuscript, in particular to Mrs. Dorothy Green for her careful typing, and to Marcia and Jonathan Steel for their helpful suggestions toward a final draft.

ROBERT G. D. STEEL
JAMES H. TORRIE

Chapter 1
INTRODUCTION

1-1 YOU AND STATISTICS

Statistics plays a part in our daily lives in numerous ways—ways we may not always appreciate. This is a relatively new phenomenon, and is a consequence of the many problems that science and technology have chosen to study and deal with using statistical methods. Notice that the word "statistics" is used in the singular. The term no longer refers simply to the collection and presentation of data in tables, graphs, and descriptive figures. Today, statistics is a subject or discipline. As such, it is very much involved in the physical and social sciences, in business and industry, and indeed in most human activities that require decisions involving uncertain outcomes.

Statistics is used extensively in the production of many of our foodstuffs. The flour you use is most likely the result of breeding experiments with wheat, resulting in the selection of varieties or cultivars possessing high quality, and of baking tests requiring careful measurements, at various stages, of the appropriate properties. Both the breeding techniques and the baking tests are based on sound statistical procedures. The drinking water we use so casually for so many purposes is almost certainly under continuous statistical quality control. For much of our clothing, rigorous statistically analyzed trials have determined what mixtures of fibers and what insulating materials provide a reasonable compromise between maximum comfort and minimum care.

The paint on your house may have been selected for the market as a consequence of a weathering trial conducted according to statistical guidelines at a forest products or industrial chemistry research laboratory. The building code of your community may have been changed recently on the basis of a statistical comparison of a long-used building material with one developed since the code was adopted.

Consumer protective agencies evaluate statistical evidence that may lead to the removal from the market of clothing that could prove to be a fire hazard, food additives that could cause genetic damage, and drugs with undesirable side-effects.

Unfortunately, statistics in the form of numerical "facts" is sometimes used to confound us. For example, we may be told that a gasoline with a special additive has given up to 3 miles per gallon more than the average of a number of leading brands without the additive. The advertisers have not said that their average mileage is any better than the average given by the leading brands, though they almost certainly hope we will conclude so. Sometimes our would-be confounders compare wage rates or crime rates where definitions have changed over time or place. They may compare sales for consecutive time periods when the only significant comparison would involve comparing the same period in consecutive years, and they may quote percentages without saying whether the figures were based on a small or a large number of observations.

As a consequence of the ever-present role that statistics plays in our lives, it is a virtual requirement that we attain some degree of statistical literacy. Without such knowledge, we will be unable to interpret satisfactorily a reasonable number of the statements and situations with statistical content with which we come in contact.

1-2 WHAT IS STATISTICS?

Already we have implied that statistics is concerned with knowledge in the form of numbers relevant to various questions. The numbers initially collected are organized, processed arithmetically, interpreted, and finally presented in tabular or graphic form or as summary numbers such as averages or percentages. Usually interpretation involves a conclusion or inference.

The originally collected numbers or *observations* are called, collectively, *data*. Clearly they are almost invariably a fraction or *sample* of a larger set of possible observations. For example, only a small amount of flour will be used to provide an observation or *sample value* in a baking test. The same is true in measuring water quality, house-paint weathering, or gasoline mileage.

The set of observations which constitute all possible measurements is called a *population*.

In the cases mentioned, it is easy to visualize the population as consisting of more observations than we can number, that is, as an *infinite population*. However, if we are sampling United States income tax returns for 1974, say for the purpose of auditing them, we can number them all in sequence and so have a *finite population*. Note that the term population applies to the data rather than to the people or objects which are observed to provide the data.

If a census or complete enumeration of the observations in a finite population exists, one can find the true average for that population. This is a constant and not subject to any variation. Thus, if we have complete census data on ages in the American population, the computed average age will be the true average for the time of the census. In the case of an infinite population, we cannot have the numbers to do the arithmetic of finding a true average, but we can conceive of such a value which, like that for the finite population, is a constant.

Notice that for a population to exist at all, it must first be capable of definition. Definition of any population is a matter for careful thought. It involves providing a rule that will identify every unit on which an observation might be made. For example, what will be our rule to identify "farmer" for a study of farm incomes in a county? How do we identify each and every dwelling unit in a city block for a study of housing density? Answers, of course, depend on the aims of the investigation.

Any inference drawn from a sample must, of course, be one concerning the population sampled. Thus, if we want a sample of registered voters, we go to voter rolls rather than to telephone listings. Since we are reasoning from the part or sample to the whole or population, the end result is an *inductive* or *uncertain inference*. Because of sample-to-sample variation, even within the same population, it is evident that our inferences cannot be completely dependable; they must be qualified in some manner. To make sure that most of our inferences about descriptive measures like averages will be valid, we must have good samples, samples resulting from sound sampling schemes and experimental designs.

What is a good sample? We would certainly like any sample to be typical or representative of the population. However, there is no way to determine what typifies a population unless we can examine all of it, a course of action not available to us. Consequently, we need a technique to tell us, at least, how well our sampling and inferential procedures work. When the laws of probability theory apply, this theory provides such a technique.

Random sampling is done when each possible sample has the same probability of being selected or when its probability of being selected is known. As a consequence, the laws of probability can be applied.

Some random samples lead to valid inferences while others do not. However, probability theory is able to provide a measure of how well we are doing, on the average, with inferences from random samples. Also, random sampling assures us that we will not obtain samples and, in turn, inferences that are dependent on our personal biases if such should exist. Consequently, we will be concerned only with random sampling.

Statistics provides guidance in designing random-sampling schemes for efficient studies and applies probability theory in evaluating the precision achieved. Although statistics is different things to different people, and a formal definition covering all its aspects seems impossible to achieve, an appreciation of some of its elements can help us interpret our environment and react to it in a knowledgeable way.

1-3 WHAT DOES A STATISTICIAN DO?

Some statisticians are concerned with the strength of highways as affected by the materials and methods used in their construction. They work with highway engineers, helping to plan the differently constructed sections to be included in their studies, selecting places to make test observations, and analyzing the results. Other statisticians observe incoming election returns and use these, with the experience and knowledge gained from past elections, to predict or forecast the eventual outcome. Still others help design clinical trials to evaluate the performance of new drugs relative to standard ones.

All these statisticians analyze data and draw inferences. They serve as consultants or partners in an enterprise where their statistical knowledge is required. Most of them are involved with several projects simultaneously, although, on occasion, a specific project may require a statistician's undivided attention for some time.

Other statisticians lecture, conduct research, and direct students working on research theses. Statisticians may teach the application of statistical methods and data analysis or the mathematics needed for statistical theory. Research may be on statistical problems raised by geneticists and physicists, involve attempts to provide mathematical descriptions of social and biological phenomena, or have no immediately obvious application.

A volume of interesting statistical and probabilistic applications from many fields, entitled "Statistics: A Guide to the Unknown," has been edited by Judith M. Tanur, et al. (1972).

1-4 WHERE DO STATISTICIANS WORK?

According to the second edition of "Careers in Statistics," published in 1966 for the Committee of Presidents of Statistical Societies, about 40 percent of the more than 20,000 people in the United States who call themselves statisticians are in business or industry. They are associated with research departments working on new products or improving old ones, comparing performance on the basis of samples. They may be in a quality control section and watch over industrial output. Some will be a part of management and deal with problems of inventory control, auditing and accounting procedures, or the location of new plants and offices.

Another 25 to 30 percent of the statisticians in this country are in the federal government, working with birth rates and migration rates, cost of living and unemployment indices, crop sampling, and the census.

Approximately 20 percent are in academic life or research centers. They teach, conduct research, and consult with research workers in the many disciplines of the social, biological, and physical sciences. One somewhat unusual statistical investigation concerned inference applied to the disputed authorship of 12 "Federalist" papers ascribed to both Alexander Hamilton and James Madison.

Many of the remaining 10 percent are private consultants, often working with a number of companies and research institutes. Presumably, these statisticians enjoy a certain freedom of choice regarding what studies they will undertake.

Certainly an increasing number of people every year call themselves statisticians, and with science and technology creating and collecting more and more data, the demand for statisticians can only continue to increase.

1-5 AIM OF THE TEXT

Our aim is to present enough theory and applications to illustrate some of the more general problems statisticians deal with, the type of training they require, and the tools they use. We do not wish to make professional statisticians of all our readers but to help them attain some appreciation of what statistics encompasses and at the same time provide a measure of statistical literacy.

In our presentation, we have adhered, where possible, to the use of statistical symbols and notation as recommended in *The American Statistician*, June, 1965.

Chapter 2

COLLECTION, ORGANIZATION, AND PRESENTATION OF DATA

2-1 INTRODUCTION

When an investigation of a fact of scientific interest is conducted, whether it be of the reading ability of sixth-grade children across the nation or the effect of a fungicide on seed germination, it is necessary to observe and record some property of the entities studied. Such recorded observations are the raw material of the research worker, census taker, and survey sampler. When large amounts of numerical data are collected, they must be especially carefully organized for presentation, often in summary form, before inferences can be drawn and decisions made.

This chapter is concerned with the need to collect, organize, and informatively present appropriate data.

2-2 VARIABLES

Observations are made on some characteristic or property which distinguishes or identifies an individual. Thus we say, "Joe's batting average is .312" or "Jack is the fat brother," and we have something to help us identify Joe and Jack. We know that other batters have other averages and that Jack's brothers appear to be slimmer than Jack is. We are dealing with attributes that vary.

A *variable* is an observable characteristic or property associated with an individual and differing among individuals. *Observations* are specific values of a variable.

We observe the age of a person, the weight of a package, the cost of a standard basket of groceries, or the percentage of soluble solids in grape juice. These are variables. Variation is universal and, without it, we would be unable to describe any object.

A variable may be qualitative or quantitative.

A *qualitative variable* is one that leads us to assign an individual to one of a set of mutually exclusive or nonoverlapping categories because of a particular quality or attribute. For example, we say a person is a male and so assign him to a particular category because he possesses that particular attribute; he is not a female. A college student may be assigned to the category freshman, sophomore, junior, or senior according to the number of hours successfully completed. Here, an underlying quantity has led us to set up categories which, in turn, have made the variable a qualitative one. Data on a qualitative variable will usually be presented as counts of individuals assigned to the various categories.

A *quantitative variable* is one where the observations are basically numerical to begin with. Such a variable may be *continuous*, as when heights or weights are measured, or *discrete*, as when we count the insects on a plant or the fruit on a tree limb. With continuous variables, all values in some range are possible, though we may be limited by our measuring device or as a matter of convenience.

A qualitative variable may be quantified if it serves some useful purpose. For example, if to the qualitative variable "year in college," with categories, freshman, sophomore, junior, and senior, we assign the discrete values 1, 2, 3, and 4, respectively, we have quantified the variable and can now compute the average number of hours successfully completed by this set of students.

EXERCISE

2-2-1 Classify the following variables into as many categories as apply. The categories: qualitative, quantitative, continuous, discrete. The variables: hair color, yearly incomes of high school teachers, lengths of 500 nails produced by a factory, number of automobile accidents on a highway, weights of male students at a university, daily Dow-Jones stock averages, number of fish caught daily in a park.

2-3 FREQUENCY DISTRIBUTIONS

Data consisting of a large number of observations, whether a sample or a population, may be summarized by the use of categories or classes. For example, student weights, a continuous quantitative variable ordinarily measured in pounds, might be assigned to 10-pound weight classes. In the case of a qualitative variable with many categories, the original number of categories can be reduced. The summarized data in a tabular arrangement, that is, a table, then give the number of individuals in each class. Such a table is called a *frequency distribution* or *frequency table.* Now, the raw data have been reduced to more manageable form, providing a basis for graphical presentation, or indication of trends.

Classes to be used for a frequency table are best chosen prior to data collection with the nature of the data and the study objectives clearly in mind. All data must be includable and ambiguities avoided. In the case of qualitative data, it may be difficult to foresee all eventualities and an "other" class may be needed. This category may also be used to combine several intended classes when each turns out to have only a few observations.

Consider the amount of money invested in ordinary life insurance in 1962. This is essentially a continuous quantitative variable. Table 2-3-1 shows a possible set of classes not determined by the variable itself. Five categories result in a reasonably manageable summary. We can easily observe facts such as that investments in whole life insurance were about five times those in term insurance at that time, and can quickly estimate the percentage of each type. Some forms of graphical presentation follow naturally, and if several years' data were tabulated, trends might be apparent.

TABLE 2-3-1 / Money Invested in Ordinary Life Insurance, United States, 1962

Type of Insurance	Millions of Dollars
Whole life	212,600
Combination	97,600
Term	42,100
Endowment	27,500
Retirement	9,400
Total ordinary life	389,200

Source: Data courtesy of Institute of Life Insurance, "Life Insurance Fact Book," 1964.

If a finer classification system is desired, each category can be further partitioned. For example, the original data allow us to present the whole-life class as follows:

Straight life : 146,800
Limited-payment life : 48,900
Paid-up life : 16,900

Term insurance could be subdivided as regular, decreasing, and extended.

Ambiguities occur with overlapping categories, and consequently these should be avoided. If automobiles are classified by body type as sedans, station wagons, hardtops, and four-doors, certain cars in the first three categories also qualify for the fourth. More specific definitions are required.

For a discrete quantitative variable, for example, the number of heads observed in the toss of five coins, the classes should be obvious. Table 2-3-2 is such a frequency table.

TABLE 2-3-2 / Heads from 5 Coins Tossed 100 Times

Number of Heads	Frequency
0	5
1	16
2	31
3	28
4	16
5	4
	100

Sometimes the number of classes will be large. Thus, if 20 coins are tossed, there are 21 possible classes. If such a large number does not summarize sufficiently, then we group a number of classes into a single class. Here, we might assign the 0, 1, and 2 heads classes to the first class; 3, 4, and 5 to the second; and so on.

For a continuous variable, when classes are to be on the scale of the variable, classes are chosen somewhat arbitrarily. However, the choice is in part dependent on factors such as total number of observations, range of observations, precision required for statistical calculations, and extent of summarization desired. The last two factors are sometimes in opposition. That is, a large number of classes implies greater precision for subsequent calculations, but if the number is too great, it may not be possible to summarize sufficiently and the many small irregularities found in the data may obscure general trends.

The number of classes in a frequency table is usually between 5 and 20. Table 2-3-3 consists of *raw data*, that is, data as originally collected. Here, there are numerical grades for 126 students in a statistics course. Table 2-3-4 shows these data assigned to 11 classes. Tally marks, used in constructing the table and explained later, are not ordinarily included.

TABLE 2-3-3 / Numerical Grades of 126 Students in a Statistics Course

90	65	85	71	80	82	77	81	90	83	81	90	88
79	68	64	75	68	80	82	80	77	86	84	80	57
92	94	79	98	83	79	82	82	80	91	88	60	51
89	72	84	80	68	68	73	73	81	89	88	55	87
79	76	85	83	88	79	89	88	67	82	82	81	63
86	84	71	59	76	74	78	69	87	72	89	85	72
51	81	57	82	92	80	78	73	81	81	72	94	
90	80	83	88	74	76	93	83	88	85	86	75	
60	81	80	93	85	72	86	90	76	90	56	48	
83	50	83	79	77	87	86	55	47	88	89	68	

TABLE 2-3-4 / Distribution of Grades in a Statistics Course of 126 Students

Grade	Tally	Number of Students
46–50	///	3
51–55	////	4
56–60	~~////~~ /	6
61–65	///	3
66–70	~~////~~ //	7
71–75	~~////~~ ~~////~~ ////	14
76–80	~~////~~ ~~////~~ ~~////~~ ~~////~~ ////	24
81–85	~~////~~ ~~////~~ ~~////~~ ~~////~~ ~~////~~ ~~////~~	30
86–90	~~////~~ ~~////~~ ~~////~~ ~~////~~ ~~////~~ //	27
91–95	~~////~~ //	7
96–100	/	1

True class boundaries are values dividing the total range so as to permit the assignment of every value to a particular class. Consequently, they must be values that are impossible as observations. In the illustration, 45.5 and 50.5 are appropriate for the first class.

True class intervals are determined by the true class boundaries. Thus, the first true class interval for Table 2-3-4 is from 45.5 to 50.5.

The *class range* is the difference between the true boundaries of any class. For Table 2-3-4, the class range is 50.5 − 45.5 = 5 grade points.

The *class mark* or *class midpoint* is at the center of the class interval. For the first class, this is (46 + 50)/2 or (45.5 + 50.5)/2, namely 48. In computations using a frequency distribution, all observations in a class are given the value of the class midpoint, here 48.

To understand the preparation of a frequency distribution, consider the statistics grades in Table 2-3-3. Grades range from a low of 47 to a high of 98. If we use a class interval of 5, we will have 11 classes. The odd number, 5, has the possible convenience that all class midpoints can be observed values.

In general, when setting up classes, it is desirable to make the lower limit of the first class slightly less than the smallest value in the sample; here the lowest value is 47, and the lower boundary of the first class is at 45.5. The remaining classes must not overlap and must accommodate all observations. This is the case for Table 2-3-4.

All class marks in Table 2-3-4 coincide with potential grades, and there is no reason to expect grades to concentrate at specific values. When such concentration is to be expected—for example, food prices may tend to cluster around 19 cents, 29 cents, and so on—it is desirable to have midpoints coincide with these values because they are a truer representation of what the class contains. In other words, it is desirable to have the class midpoints at most frequent values when the data tend to cluster at specific values. For such food prices, boundaries could be at 14, 24, and so on.

When boundary points and class intervals have been determined, the data are sorted into classes, ordinarily by one of three commonly used methods. The first is the familiar tally method seen in Table 2-3-4. The observations are gone through in the sequence observed and each is recorded as a stroke or tally in the appropriate class. For convenience in counting, each fifth stroke is placed through the preceding four. Unfortunately when there is a lack of agreement in checking frequencies, the source of error can be difficult to find.

The second and perhaps safer method of sorting is to write each observation on a card and order them from smallest to largest. The result of such an ordering is called an *array*. This is very convenient but entries must be carefully checked. As an alternative to ordering the cards, class ranges may be written on other cards which are then arranged in order on a desk and the observations sorted to the appropriate range cards. Frequencies are determined by counting.

The third method requires card-punching and -sorting equipment. Each observation is punched on a card and the cards are mechanically sorted into classes. Careful checking of cards is necessary.

The assigning of values to classes is simple for Table 2-3-4. However, if grades were averages of several tests and reported to the nearest first decimal point, then the proper assignment of any observation that coincided with a class boundary would present a difficulty. Several solutions are available. It has already been pointed out that values impossible as observations may be used. This would call for an extra decimal place in selecting class boundaries. If boundary points are to be possible observations also, then when an even number of observations falls on a boundary, assign half to each of the adjacent classes; for an odd number, assign one by the toss of a coin and the rest as for an even number. Finally, and more commonly, we use *intervals open on the right* (or on the left). For example, an interval open on the right would include all observations beginning with 45.5 but less than 50.5, beginning with 50.5 but less than 55.5, and so on.

Closed classes are most common when the data are the result of an enumeration and consist of counts. *Open classes* are frequently used when the data are observations on a continuous variable.

With tabulations derived from *asymmetrical* or *skewed* distributions, there is a tendency to have sequences of classes with few or no observations, especially if procedures outlined earlier are used. To avoid this unsatisfactory state, class intervals of unequal length are sometimes employed. Open-ended classes, as in Table 2-3-5, offer another alternative. These permit the inclusion of a wide range of values without specifying the extremes. If desired, extremes can be given in a footnote. Both open-ended and unequal intervals present difficulties when computing descriptive statistics and in graphing.

TABLE 2-3-5 / Age of Licensed Drivers in United States, 1962

Age	Number to Nearest 1,000
Under 20	6,450
20–29	20,300
30–39	21,850
40–49	18,400
50–59	13,200
60–69	7,650
70 and over	3,150

Source: Data courtesy of Automobile Manufacturers Association, "Automobile Facts and Figures," 1963.

For *relative-frequency distributions*, divide the observed class frequencies by the total number of observations; multiply by 100 if a *percentage frequency* is desired. For example, for the 71 to 75 class of student grades, $14/126 = .1111$ is the relative frequency and 11 percent is the closest precentage frequency. Table 2-3-6 shows both distributions for the grade data of Table 2-3-3.

Cumulative-frequency distributions are used to show how many or what percentage of observations are "less than" or "not more than" a specified value, the second expression allowing us to include values at an endpoint if this is desired.

TABLE 2-3-6 / The Percentage Distribution of Grades in a Statistics Course of 126 Students

Grade	Frequency	Relative Frequency	Percentage Frequency
46–50	3	.0238	2
51–55	4	.0317	3
56–60	6	.0476	5
61–65	3	.0238	2
66–70	7	.0556	6
71–75	14	.1111	11
76–80	24	.1905	19
81–85	30	.2381	24
86–90	27	.2143	21
91–95	7	.0556	6
96–100	1	.0079	1
Total	126	1.0000	100

We may also have a "greater than" or "not less than" cumulative-frequency distribution. Thus for Table 2-3-4, there were no grades less than 46, 126 grades were 46 or more, 3 grades were less than 51, 123 grades were 51 or more, and so on. Table 2-3-7 is the completed table for this approach. Cumulative-frequency distributions for relative or percentage frequencies are similarly prepared.

TABLE 2-3-7 / Cumulative-frequency Distribution of Grades of 126 Students in a Statistics Course

Grade	Number of Students	Grade	Number of Students
Less than 46	0	46 or more	126
Less than 51	3	51 or more	123
Less than 56	7	56 or more	119
Less than 61	13	61 or more	113
Less than 66	16	66 or more	110
Less than 71	23	71 or more	103
Less than 76	37	76 or more	89
Less than 81	61	81 or more	65
Less than 86	91	86 or more	35
Less than 91	118	91 or more	8
Less than 96	125	96 or more	1
Less than 101	126	101 or more	0

EXERCISES

2-3-1 Arrange the following numbers in an array: 21, 5, 17, 81, 4, 51, 16, 43, 32, 9.

2-3-2 Heights of school girls measured to the nearest inch ranged from 48 to 72 inches. How many classes would be appropriate to use in a frequency distribution if the number of girls measured was 50? 200? 1,000? For each set, give the class limits and midpoints as if you were to construct a frequency distribution.

2-3-3 Toss 10 coins 150 times. Record the number of heads appearing in each toss. Construct a frequency table. Prepare a table giving the relative-frequency and percentage-frequency distributions.

2-3-4 The accompanying data are the weights in grams of the livers of 80 hens.

26.9	26.0	39.6	32.3	30.4	35.6	29.0	34.9	29.4	26.3
29.5	41.2	29.8	32.4	25.9	24.9	24.1	22.8	23.4	28.3
34.1	41.5	25.8	32.5	29.5	30.3	27.3	35.0	27.3	31.7
23.8	29.8	24.7	28.3	26.6	23.4	25.5	29.7	28.2	32.8
26.3	23.1	29.3	31.6	31.2	27.0	34.6	27.4	30.0	26.1
33.3	25.9	26.0	25.2	27.3	32.2	37.7	27.5	36.2	26.7
30.6	27.7	39.1	25.6	22.6	28.3	29.3	28.9	30.6	40.3
44.7	26.4	28.2	22.9	31.0	30.9	21.5	32.3	32.2	31.7

Source: Hopkins and Biely (1935). Reproduced by permission of the National Research Council of Canada, *Can. J. Res.*, **12**: 651–656 (1935).

Construct a frequency distribution. Prepare relative-frequency and percentage-frequency distributions.

2-3-5 The accompanying data are the ages of pitchers on the National Baseball League spring roster, 1965.

30	44	30	28	27	26	37	20	23	26	23	25	27	22	39	28	32	33
23	24	24	22	28	36	38	23	24	27	22	27	23	26	25	24	36	22
35	23	23	21	36	24	29	19	34	19	20	24	21	28	25	19	20	23
27	25	21	28	25	38	31	24	28	25	21	25	28	33	36	25	31	24
19	27	28	23	29	20	29	25	26	27	21	34	21	23	34	31	29	19
21	30	22	30	36	30	24	31	20	21	22	26	19	29	23	23	31	18
27	24	26	27	28	19	29	36	30	27	26	25	29	21	25	19	19	28
23	37	24	29	24	25	20	24	23	26	24	26	20	32	30	31	21	19
30	35	26	21	34	21	26	22	23	36	25	22	30	30	20	26	28	26
22	22	22	27	26	24	34	27	25	32	29	32	26	37	23	21	22	20
24																	

Source: Street and Smith (1965).

Construct a frequency distribution and percentage-frequency distribution.

2-3-6 The accompanying data are the average June temperatures in degrees Fahrenheit for Ohio during the period 1890–1939:

73.3	72.0	69.8	69.2	65.9	66.8	68.9	73.0	70.0	67.4
71.0	69.5	70.9	69.8	70.9	64.7	73.4	65.9	70.8	70.3
73.0	68.1	66.9	65.6	66.6	66.9	70.9	64.6	71.0	69.7
70.6	71.9	64.4	69.2	69.8	68.8	71.0	65.0	74.4	68.5
71.3	71.5	68.4	70.1	71.1	74.2	68.0	67.1	76.0	72.5

Source: Hendricks and Scholl (1943).

Construct a frequency distribution using a class interval of 1 and the midpoint of the first class as 64.5.

2-3-7 In the game of bridge, four hands of 13 cards each are dealt. Before bidding, the players will generally compute their high-card point counts by assigning 4 to an ace 3 to a king, 2 to a queen, and 1 to a jack. The accompanying data are the high-card point counts in 200 bridge hands.

8	8	9	13	19	8	13	17	11	9	13	17	10	2	2	10
10	12	12	12	6	12	3	9	9	9	10	21	18	15	13	5
21	11	17	12	13	20	8	10	14	11	18	4	7	7	8	12
5	6	9	24	10	15	11	9	5	13	13	14	7	11	7	14
6	8	5	19	16	10	6	15	10	12	8	13	16	4	12	16
11	15	19	12	14	17	12	9	16	11	5	13	14	10	21	12
17	10	16	11	14	21	5	10	9	10	12	8	5	13	15	17
5	13	8	10	17	15	9	5	11	21	11	9	11	5	16	16
10	9	9	16	16	6	10	19	8	8	12	11	8	12	12	13
14	17	17	11	6	16	15	5	10	15	11	8	12	8	21	17
5	19	9	5	14	11	13	16	7	5	5	9	5	11	10	19
18	8	14	13	15	12	12	16	21	11	13	15	12	16	15	5
7	14	13	11	3	4	15	13								

Use class intervals of 2 and prepare a frequency and a percentage-frequency distribution.

2-3-8 The accompanying data are the vertical loads at ultimate strength of concrete corbels.

619	954	491	782	606	563	583	491	672	961
602	1,090	544	542	729	778	370	436	392	659
497	754	563	594	457	885	622	420	339	570
397	699	922	777	434	691	959	581	347	519
599	543	888	849	556	1,010	822	932	374	481
641	495	981	713	606	859	804	814	301	418
749	878	1,300	1,090	660	715	809	811	435	668
474	804	1,110	1,090	625	648	664	718	435	621
517	774	787	1,650	572	640	674	716	424	458
515	681	1,150	898	533	844	675	477	620	493
546	967	1,340							

Source: Kriz and Raths (1965).

Prepare a frequency distribution. Do you think this distribution warrants the use of open classes or classes of unequal size? If so, illustrate the modifications necessary in the preparation of the frequency distribution.

2-4 GRAPHIC PRESENTATION

Graphic presentation of data is intended to set forth the essential characteristics of a frequency distribution in a readily comprehensible manner. Such devices tend to give less information than tables but attract more readers.

Pie charts are often used to present qualitative or categorical data, particularly percentage distributions. It is best to keep the number of components under six. Figure 2-4-1 is a pie chart of the data in Table 2-3-1. Construction is simple.

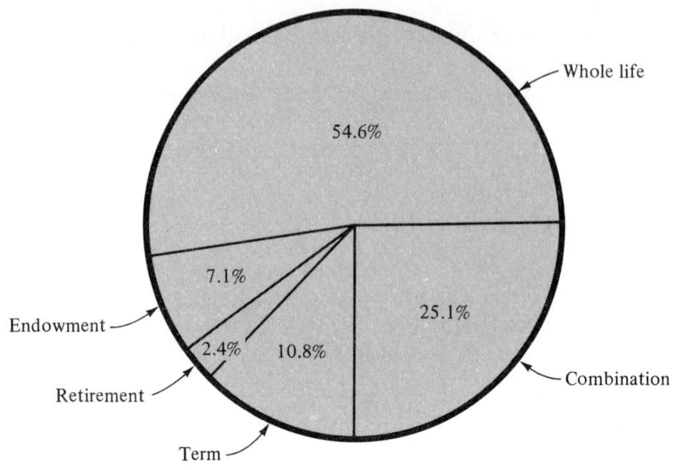

FIGURE 2-4-1 Types of Ordinary Life Insurance, United States, 1962. (Data courtesy of Institute of Life Insurance, 1964 Insurance Fact Book.)

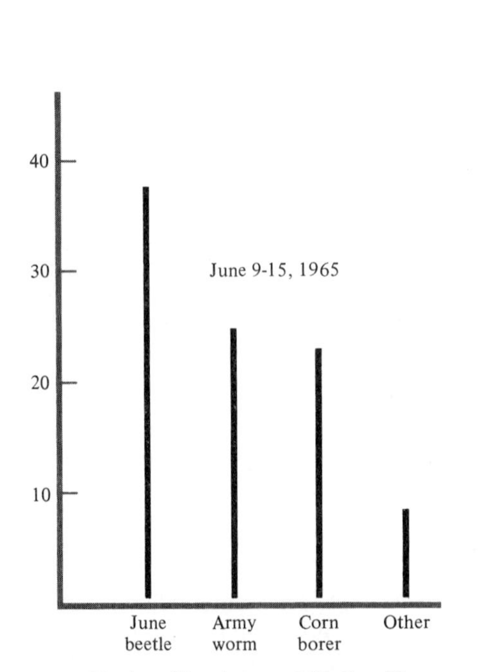

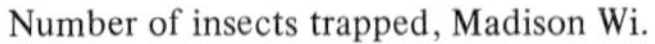
Number of insects trapped, Madison Wi.

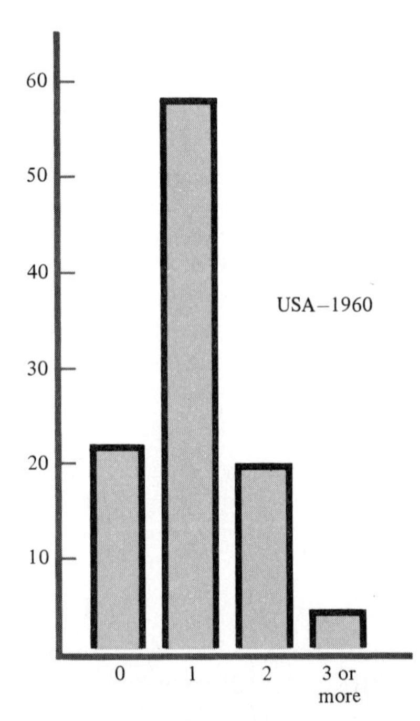

Source: Data courtesy of Automob. Manuf. Assoc., "Automobile Facts and Figures," 1963

FIGURE 2-4-2 Presentation of Discrete or Categorical Data.

The entire circle represents 100 percent, and the sum of all central angles is 360°. Thus 1 percent is represented by a central angle of 3.6 degrees. Term insurance requires (42,100/389,200)100 = 10.8 percent of the pie and so is represented by 10.8 × 3.6 = 38.9 degrees.

The use of vertical *line charts* and *bar charts* or graphs with heights equal to or proportional to frequencies is illustrated in Figure 2-4-2. For the insects, we are able to read the actual numbers. If no vertical scale or only a percentage scale is provided, the reader sees relative frequencies and is unaware of the sample size unless this is specified. A vertical scale can supply both absolute and relative frequencies; in fact, two scales may be used.

Histograms and *frequency polygons* are often used to present data summarized in frequency distributions.

Histograms are similar to bar charts except that the sides of the bars are moved horizontally until those of adjacent bars touch. The resulting figure tends to indicate that we are dealing with measurement data, that is, from a continuous variable, whereas bar charts tend to indicate that categorical or count data, that is, from a discrete variable, are being considered. In the histogram, common sides may be deleted. Figure 2-4-3 is a histogram of the grade data in Tables 2-3-3 and 2-3-4.

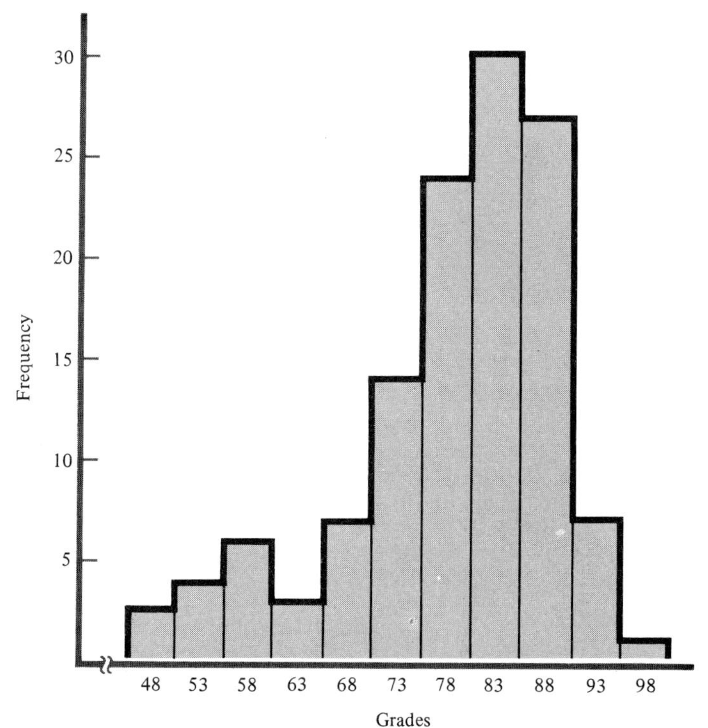

FIGURE 2-4-3 Histogram of Grades in a Statistics Course.

When prepared from distributions with classes of unequal size, histograms can be misleading if heights are proportional to frequencies; suitable adjustments are needed. To illustrate, combine the first four classes of Table 2-3-4 to give Table 2-4-1. This might well be done if it was deemed necessary to arrange tutorials for those with test grades of 65 or less. Figure 2-4-4 is a histogram for Table 2-4-1, with the heights proportional to the frequencies, as one familiar with bar charts would expect to find. This leaves the false impression that about one-third of the students had grades of 65 or less, because particular areas catch our eye. Figure 2-4-5 is a truer representation of the data because areas are now proportional to frequencies. Obviously the better approach to constructing histograms is to make

TABLE 2-4-1 / Distribution of Grades in a Statistics Course with the First Four Classes Combined

Grade	Number of Students
46–65	16
66–70	7
71–75	14
76–80	24
81–85	30
86–90	27
91–95	7
96–100	1

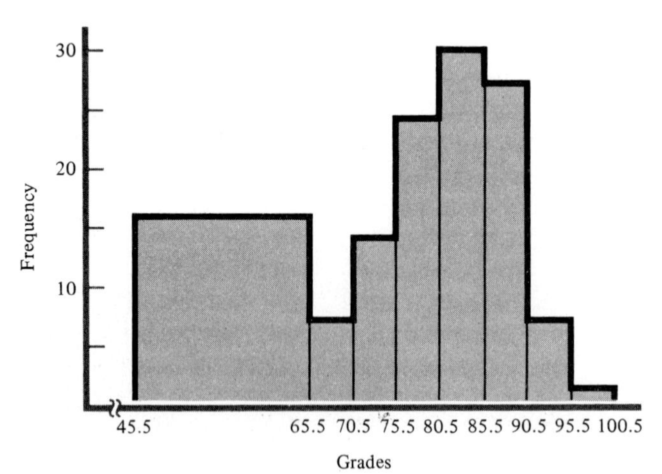

FIGURE 2-4-4 Incorrectly Drawn Histogram for the Grade Data of Table 2-4-1.

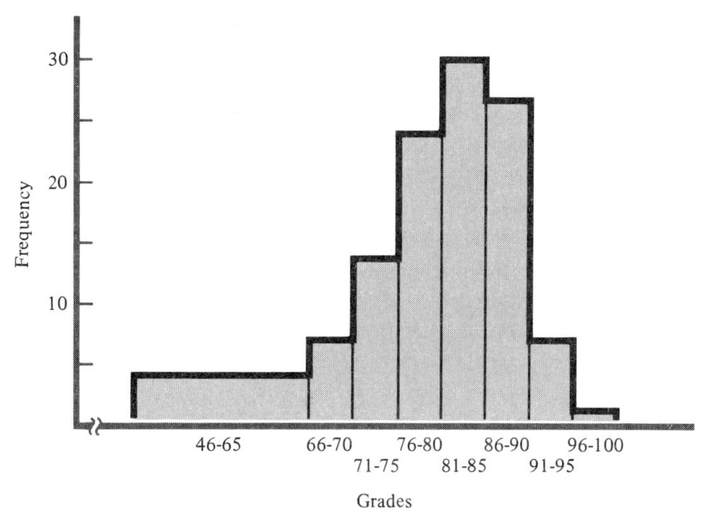

FIGURE 2-4-5 Correctly Drawn Histogram for the Grade Data of Table 2-4-1.

areas, rather than heights, proportional to frequencies. Thus if one class interval is twice as long as the others, make the height of its rectangle one-half of its frequency; if it is three times as long, make the height one-third of its frequency; and so on.

Three methods, all illustrated here and all in common use, serve to indicate the classes of a histogram. Midpoints of classes are used in Figure 2-4-3, class boundaries in Figure 2-4-4, and class ranges in Figure 2-4-5.

The frequency polygon for graphically representing a frequency distribution is prepared by connecting midpoints of the tops of histogram rectangles by straight lines, as in Figure 2-4-6, which is a frequency polygon for the data of Table 2-3-4. The rectangles themselves are not shown. A class with zero frequency is usually added at each end so that the polygon rests on the horizontal axis. Like histograms, frequency polygons cannot be used for distributions with open-ended classes. For distributions with unequal class intervals, adjustment is made as for histograms.

An *ogive* is a graphic representation of a cumulative frequency distribution such as Table 2-3-7. To prepare an ogive, plot a point with height representing cumulative frequency above the upper limit of each class. Join these points, as in Figure 2-4-7 for the data in Table 2-3-7. Note that 46, the lower class limit of the first nonempty class of the "less than" ogive, is used as a starting point with zero height. An ogive of "or more" type is prepared similarly but starting from the other end. Here 101, the upper limit of the last nonempty class, has zero height.

The histograms, polygons, and ogives shown here are all based on absolute or observed frequencies. Similar graphs can be prepared using relative frequencies or percentages.

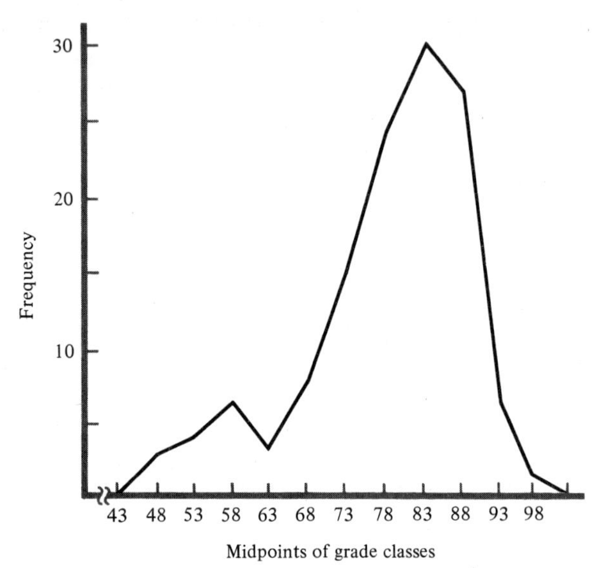

FIGURE 2-4-6 Frequency Polygon of Grades in a Statistics Course.

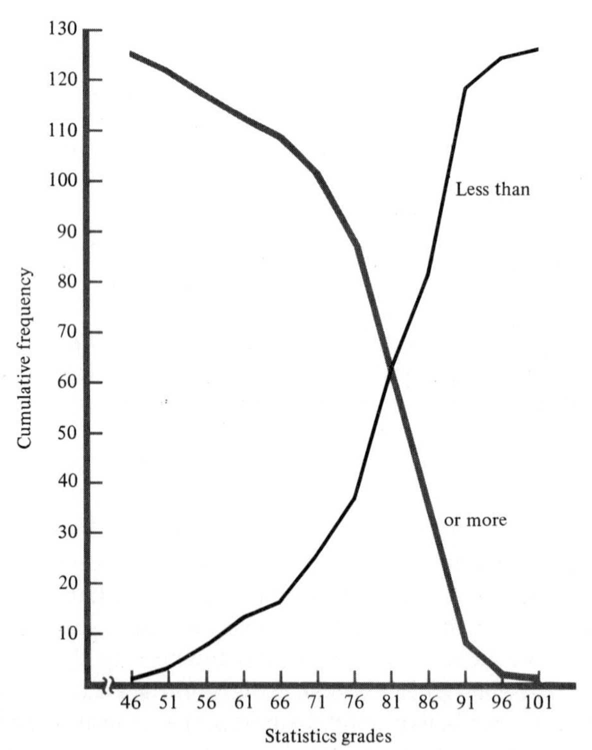

FIGURE 2-4-7 Ogives for the Cumulative Distribution of the Grades in a Statistics Course.

Histograms and polygons can have almost any shape. However, there are standard types with names that fit most of the distributions observed in practice. The more important are shown in Figure 2-4-8. The most common is the *symmetric* or *bell-shaped distribution.* This bell-shaped distribution plays an important role in the theory of statistics. The data of Table 2-3-2 giving the number of heads observed from tossing 5 coins 100 times illustrate a symmetrical distribution. There is, of course, some variation due to chance in tossing the coins.

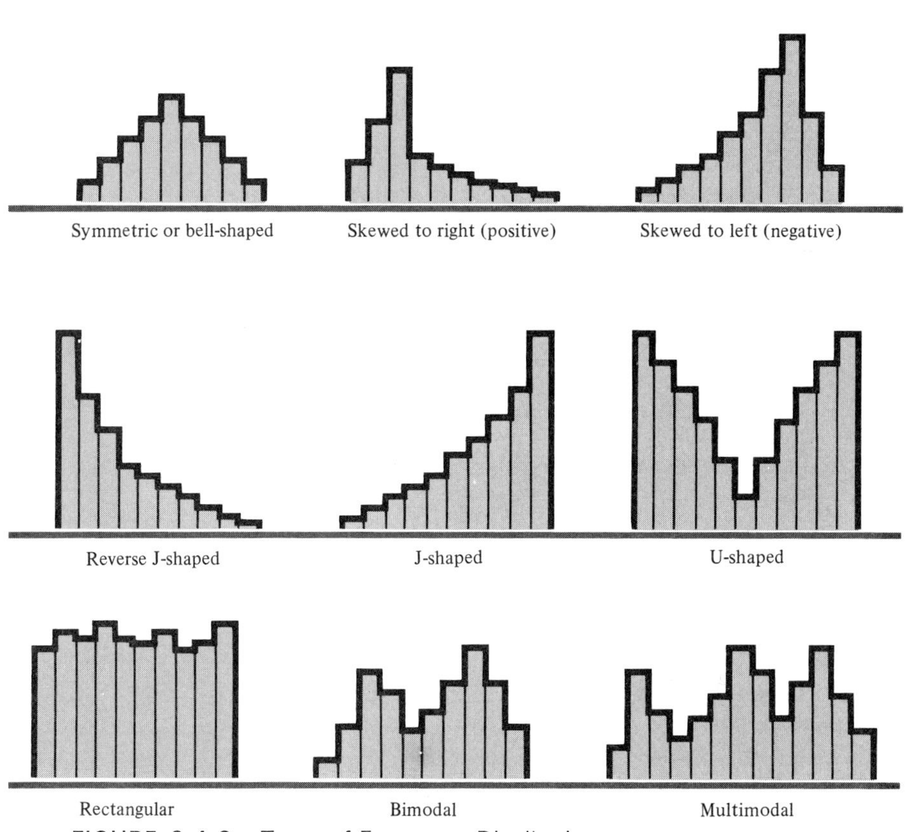

FIGURE 2-4-8 Types of Frequency Distributions.

The grade data in Figure 2-4-3 illustrate a distribution with *negative skewness*, one said to be *skewed to the left* where the long tail occurs. The class of greatest frequency is somewhere to the right of the middle class. A positively skewed distribution is skewed to the right where the long tail occurs, and the class of greatest frequency is to the left of the middle class.

J-shaped and *reverse J-shaped* distributions are skewed with a maximum at one end. Table 2-3-1 is a J-shaped distribution of the amount of money invested in different categories of life insurance. This could be considered to be artificial since there is really no horizontal scale.

U-shaped distributions have a maximum at each end. An example is the data of Exercise 2-4-6, the frequencies of estimated intensities of cloudiness at Greenwich during July.

Rectangular distributions have approximately the same observed frequency in each class, that is, the same within random-sampling fluctuations. The data of Exercise 2-4-7 are frequencies of the digits 0, 1,. . . , 9 in a table of random numbers and illustrate a rectangular distribution.

For any frequency distribution, the *mode* is the class of greatest frequency. Thus for Table 2-3-2 the mode is at 2; for Table 2-3-4 the mode is the 81 to 85 class. These distributions are said to be *unimodal.* A mode may also be considered as relative to nearby classes. This leads to a consideration of *bimodal* distributions where there are two relative maxima or modes and to *multimodal* distributions with even more modes. Such distributions are also illustrated in Figure 2-4-8.

EXERCISES

2-4-1 Prepare a pie chart of the types of life insurance in force in the United States in 1963 for both amount and number of policies.

Type	Amount in Millions	Number in Thousands
Ordinary	418,856	101,431
Group	228,540	50,908
Industrial	39,672	93,443
Credit	43,555	52,856

Source: Data courtesy of Institute of Life Insurance, "Life Insurance Handbook," 1964.

2-4-2 Prepare a bar graph of the sources of gross farm income in the United States and Wisconsin (10-year average, 1944–1953) expressed in percentage of total gross income.

Source	U.S.	Wisconsin
Crops	39.9	11.3
Cattle and calves	15.9	12.8
Dairy	12.6	45.5
Hogs	10.8	11.3
Poultry	9.0	8.6
Other	11.8	11.3

Source: Data courtesy of Wisconsin Statistical Reporting Service, *Bulletin* 331.

2-4-3 Graduate enrollments in a southern university for 1953–54 and 1973–74 are given in the accompanying table. These have been categorized by schools. Prepare a pie chart for the 1953–54 data. Prepare a pie chart for the 1973–74 data. If you were required to reduce the number of categories, what suggestions would you make? Prepare new pie charts that incorporate your suggestions.

School	1953–54	1973–74
1 Agriculture and life sciences	157	556
2 Design	. . .	119
3 Education	54	428
4 Engineering	113	346
5 Forest resources	12	63
6 Liberal arts	. . .	204
7 Mathematics and science	. . .	237
8 Textiles	14	65
9 Unclassified	45	340
Totals	395	2,358

2-4-4 Look through a number of newspapers and financial reports for pictorial or graphic presentations not mentioned in the text. List what you have found.

2-4-5 The accompanying table gives the number of passenger cars in the United States by age groups for 1964. Draw the histogram excluding the class "16 and older" which can be given as a footnote. Note the unequal class intervals. Draw or-more and less-than ogives including the last class.

Age in Years	Number in Thousands
Under 1	5,847
1–2	7,348
2–3	6,629
3–4	5,455
4–5	6,134
5–6	5,586
6–7	3,971
7–8	5,124
8–9	4,683
9–10	4,715
10–11	2,634
11–12	2,471
12–14	2,466
14–16	1,882
16 and older	1,054

Source: Data courtesy of Automobile Manufacturers Association, "Automobile Facts and Figures," 1965.

2-4-6 The accompanying table is the distribution of estimated intensities of cloudiness at Greenwich during July of the years 1890–1904 (excluding 1901). Construct a histogram for these data. Prepare the less-than and or-more cumulative frequency and percentage distributions. Construct the less-than and the or-more ogives.

Degree of Cloudiness	Frequency
10	676
9	148
8	90
7	65
6	55
5	45
4	45
3	68
2	74
1	129
0	320

Source: Pearse (1928).

2-4-7 The accompanying table is the frequency of the digits, 0, 1, 2, . . . , 9 found in a table of 10,000 random digits. Construct a histogram. Prepare the or-more and the less-than frequency distributions and corresponding ogives.

Digit	Number	Digit	Number
0	1,015	5	932
1	1,026	6	1,067
2	1,013	7	1,013
3	975	8	1,023
4	976	9	960

2-4-8 Histograms, frequency polygons, and less-than and or-more cumulative distributions and ogives may also be prepared for the data in Exercises 2-3-3 to 2-3-7.

2-4-9 Prepare a histogram and frequency polygon for the data in Exercise 2-3-8. If you have used open classes or classes of unequal size, what modifications have you introduced in preparing the histogram?

Chapter 3
LOCATING A DISTRIBUTION

3-1 INTRODUCTION

Data summaries such as frequency distributions and other graphical devices can point only to outstanding features and general trends. Something more precise and mathematically tractable, yet readily comprehensible, is needed for serious investigations.

Suppose we know the general shape of a graph, for example, a smoothed form of one of those shown in Figure 2-4-8, which will describe a population of data to be sampled. Presumably this shape or form has a general mathematical description or formula and this can serve as a reasonable approximation to the true nature of the data, and we require only one or two numbers, such as the location of a peak or a pair of endpoints, to make the mathematical formula unique. Thus, we are now in a situation where we need to provide measures or formulas that define such features adequately, lend themselves to relatively simple mathematics, and are reasonable comprehensible.

For many distributions, the observations tend to cluster about or be near a central value that can be said to locate the distribution. Distribution-locating values, the formulas that define them, and some of their properties will be the concern of this chapter.

3-2 POPULATIONS AND SAMPLES; PARAMETERS AND STATISTICS

For every collection of data, it must be clearly established whether the data constitute a population or only a part, or sample, from a population.

A *parameter* is a number that summarizes information contained in a population.

Parameters are fixed numbers; that is, they are constants. Symbols for parameters are most often chosen from the Greek alphabet (see Table A-1).

Suppose we have a finite population such as the admitted incomes of all licensed doctors in the United States for a particular year. We can now compute parameters such as the average, minimum, and maximum incomes for this population with complete certainty; there can be no argument, since we have the population. Any census should provide data that constitute a population.

A *location parameter* is one that attempts to locate a population by a single number; that is, it tries to tell, in the simplest terms, where the population is. The observations for many commonly sampled distributions tend to cluster fairly obviously in a region, and a value from such a region has meaning as a central value. Indeed, such a number is often said to be the center. In some sense, then, a central value locates a population, and location parameters are also called *measures of central tendency*. One common measure of central tendency is the arithmetic mean or arithmetic average. This is easily defined for a finite population but less readily defined for an infinite one.

A location parameter is intended to provide information on only one distinguishing attribute of a population, namely its location. Parameters providing information about other aspects are necessary for more comprehensive population summaries.

A *statistic* is a number that summarizes information contained in a sample.

Statistics vary from sample to sample from the same population, and consequently they can be used to draw only uncertain inferences when used to estimate corresponding parameters. Symbols for statistics are chosen from the Roman alphabet.

When statistics are to be obtained, the sampled population must be clearly defined if inferences are to be meaningful, and the sampling procedure must be such that a sound theory of inference can be developed. When the principle of randomization has been operative, probability theory applies. Probability theory allows us to make estimates of the long-run chances for success when using any particular approach to the problem of inference. That is, we can know how frequently our inferences will be correct, on the average. Also note that repeated sampling can generate populations of statistics. Such derived populations are used in the development of a theory of inference.

3-3 NOTATION AND ALGEBRAIC RULES

In order to define parameters and statistics, we need the convenience of symbols. In our discussions, we may not always want to name a variable, such as weight or income, and again may resort to symbols. For sample values and the computations involving them, we need to use general terms since sample values and sample sizes vary. Thus, notation and an algebra are needed to handle numbers in the abstract. We now proceed to the necessary definitions.

A *random variable* is a variable whose individual observed values or outcomes are the result of random sampling, so that chance has played a part in the outcome and, consequently, probability theory can be applied.

Let Y denote a random variable, such as the height of an individual, a measured value, or the number of seeds produced by a plant; in other words, let Y represent all the possible numerical values associated with some observable characteristic. In practice, any capital letter may be used.

A sample of n observations will need to have its members distinguishable, so we add a subscript to our sumbol for a variable and write $Y_1, Y_2, \cdots, Y_n$, with the general or ith observation written as Y_i and read "Y-sub-i." Y_i is also a random variable, representing all possible candidates for the ith observation, that is, all values in the sampled population. If we are to have a random sample of four exam grades, we write Y_1, Y_2, Y_3, Y_4. Here $n = 4$, the sample size.

So far, we have been talking abstractly about samples. Let us say we have a real sample with real numbers as sample observations. Chance has played its part, and the sample values are now nonrandom numbers. These are designated by lowercase letters. Thus, if we draw a random sample of residents in a home for senior citizens and determine their ages, and if the sample observations turn out to be 90, 85, 81, and 94, then we write $y_1 = 90$, $y_2 = 85$, $y_3 = 81$, and $y_4 = 94$.

At first glance, this additional notation may seem unnecessary. However, we often wish to know the probability of obtaining a larger value than some observed value. For example, we may want to know the probability that a fifth sample observation, that is, a value not yet obtained, will be smaller than the smallest already observed. In this case, the smallest sample value is $y_3 = 81$, one about which there is no longer any probability involved. Thus, it is necessary to make a distinction between the observed value, $y_3 = 81$, and the future value, an unknown Y; this is accomplished by using lowercase and capital letters, respectively.

Let $y_1, y_2, \cdots, y_n$ represent the numbers in a sample. Addition of these or of other numbers constructed from them is often required. Thus, if an average of numbers is required, the sum of the observations, namely $y_1 + y_2 + \cdots + y_n$, will be needed. Since n is a letter symbol rather than the integer it represents, we have already had to propose additional notation, namely the three dots, to include values that cannot be specified without knowledge of n. An even more compact notation is given by the expression $\sum_{i=1}^{n} y_i$, which says to take the sum of the y_i's for $i = 1, \cdots, n$. Here $\sum$, the Greek letter capital sigma, is a symbol that tells us to perform the operation of addition. Symbols which tell us to carry out an operation, such as the symbol $\sum$ for addition, are called *operators*. Further, we are told to operate on the y_i's. But i is no longer a specific or fixed value, as when originally used as a subscript for the general observation. It is, rather, a changing one since it represents all consecutive whole-number or integer values between 1 and n, as is indicated below and above $\sum$. Consequently, i is now called an *index of summation.*

In summary, we have the definition of $\sum$ given by Equation 3-3-1.

$$\sum_{i=1}^{n} y_i = y_1 + y_2 + \cdots + y_n \tag{3-3-1}$$

Since we usually add all the sample observations, the symbol may be abbreviated to $\sum_i y_i$ or $\sum y_i$ or even $\sum y$.

This summation notation is much used in statistics. Often we require the sum of the squares of the sample observations.

$$\sum_{i=1}^{n} y_i^2 = y_1^2 + y_2^2 + \cdots + y_n^2$$

For example, if $y_1 = 8$, $y_2 = 12$, $y_3 = 9$, $y_4 = 10$, $y_5 = 8$, $y_6 = 10$, and $y_7 = 8$, then

$$\sum y_i^2 = 64 + 144 + 81 + 100 + 64 + 100 + 64$$

At other times, we multiply sample observations by weights and add these products. Sometimes these products will be of corresponding observations on two different variables measured on the same individual.

$$\sum x_i y_i = x_1 y_1 + x_2 y_2 + \cdots + x_n y_n$$

For example, if we rewrite the numbers of the previous example in increasing order as $y_1 = 8$, $y_2 = 9$, $y_3 = 10$, and $y_4 = 12$ and set $x_1 = 3$, $x_2 = 1$, $x_3 = 2$, and $x_4 = 1$, these representing the number of times the corresponding y_i's occur, then the sum of the numbers in that example is

$$\sum x_i y_i = 3 \times 8 + 1 \times 9 + 2 \times 10 + 1 \times 12$$

In dealing with numbers, as opposed to letters, there are underlying rules that let us validly perform needed arithmetic with relative convenience. We tend to use these rules unconsciously. Thus, if we are required to add a series of mixed positive and negative numbers, we very likely add the positive ones separately from the negative ones and then take the difference.

For handling numbers in the abstract, three much-used algebraic rules are now presented. These, too, should be used unconsciously.

RULE 1:

The summation of a sum or difference of observations on two or more variables associated with an individual is the sum or difference of the individual summations.

$$\sum_{i=1}^{n} (x_i + y_i - z_i) = \sum_{i=1}^{n} x_i + \sum_{i=1}^{n} y_i - \sum_{i=1}^{n} z_i \qquad (3\text{-}3\text{-}2)$$

This rule is a particular case of a general rule that says that the order in which we do our algebraic addition is optional.

For example, let us say that in manufacturing a certain product, we know the cost of the raw materials initially provided and of labor. However, the material is cut and some of the unused part is salvaged. Now, if the costs and salvage return for the ith item are x_i, y_i, and z_i, respectively, then its total cost is $x_i + y_i - z_i$ and the cost of n items is given by Equation 3-3-2. On the left, we compute the cost of the individual items and add to obtain the cost of all items; on the right, we find total costs for material, labor, and salvageable material and use these three to compute the same cost of all items. The arithmetic has simply been rearranged.

RULE 2:

The summation of the products of a constant C by an observation y_i is the product of the constant and the summation of the observations.

$$\sum_{i=1}^{n} Cy_i = C \sum_{i=1}^{n} y_i \qquad (3\text{-}3\text{-}3)$$

Suppose we have a number of observations and their sum. We decide we should have used one-third of each, or 10 times each. Instead of taking one-third of each or multiplying each by 10 and adding again, it is only necessary to take one-third of the total or to multiply it by 10. Here, C is $\frac{1}{3}$ or 10, respectively.

RULE 3:

The summation of a constant is the product of the constant and the number of terms in the summation.

$$\sum_{i=1}^{n} C = nC \qquad (3\text{-}3\text{-}4)$$

The need for Equation 3-3-4 could arise in an experiment such as one where our observations are all temperatures of chemical cells. Suppose these all lie between 65 and 66°C, and suppose that small differences in temperature are important so that the temperatures must be read to two decimal places. A great deal of arithmetic that might be associated with this problem can be done by temporarily ignoring the 65. Thus any temperature can be written as $65 + y_i$, where y_i represents the digits in the first two decimal places. For example, if the temperature is 65.18, then $y_i = .18$. To find the mean of 25 temperature values, we need only find the mean of the y_i's, say .47, and the true mean temperature must be 65.47. However, if we need the sum of the temperatures, then it is

$$\sum_{i=1}^{25} (65 + y_i) = \sum_{i=1}^{25} 65 + \sum_{i=1}^{25} y_i$$

by Equation 3-3-2. To evaluate $\sum_{i=1}^{25} 65$, we must write down 65 for each value of i, that is, 25 times. Hence, $\sum_{i=1}^{25} 65 = 25 \times 65$ as stated in Equation 3-3-4.

Clearly all the symbolization used here applies with a change from lowercase to capital letters. Thus, instead of talking about a particular sample, we might be concerned about the behavior of a sum in repeated sampling, that is, about the random variable $\sum Y_i = Y_1 + \cdots + Y_n$.

We now see the need for algebraic competence in our abstract computations. The accompanying exercises will help measure your algebraic literacy.

EXERCISES

3-3-1 Write in full the following:

(a) $\sum_{i=1}^{6} y_i$ (b) $\sum_{i=1}^{n} y_i$ (c) $\sum_{i=2}^{5} y_i$

(d) $\sum_{i=1}^{3} (y_i - 8)$ (e) $\sum_{i=1}^{4} (x_i + y_i)$ (f) $\sum_{i=1}^{n} K$

(g) $\sum_{i=1}^{n} w_i Y_i$ (h) $\sum_{i=1}^{4} (y_i - 3)^2$ (i) $\sum_{i=1}^{4} (y_i^2 - 3^2)$

3-3-2 Express the following in summation notation:

(a) $y_1^2 + y_2^2 + y_3^2 + \cdots + y_n^2$ (b) $y_2 + y_3 + \cdots + y_{n-1}$

(c) $w_1 y_1^2 + w_2 y_2^2 + \cdots + w_n y_n^2$ (d) $(x_1 + y_1) + (x_2 + y_2) + (x_3 + y_3)$

3-3-3 Demonstrate by example that the following pairs of values are not equal:

(a) $\sum_{i}^{n} y_i^2$ and $\left(\sum_{i}^{n} y_i\right)^2$ (b) $\sum_{i=1}^{n} x_i y_i$ and $\sum_{i=1}^{n} x_i \sum_{i=1}^{n} y_i$

3-3-4 Let

$x_1 = 3$ $y_1 = -2$

$x_2 = 5$ $y_2 = 0$

$x_3 = 7$ and $y_3 = 1$

$x_4 = 8$ $y_4 = 2$

$x_5 = 11$ $y_5 = 5$

Find

(a) $\sum_{i=1}^{5} x_i$ (b) $\sum_{i=2}^{5} x_i$ (c) $\sum_{i=1}^{5} x_i y_i$

(d) $\sum_{i=1}^{5} y_i^2$ (e) $\sum_{i=1}^{5} (x_i - y_i)$ (f) $\left(\sum_{i=1}^{5} y_i\right)^2$

(g) $\sum_{i=1}^{5} x_i \sum_{i=1}^{5} y_i$

Students who enjoy algebra may be interested in doing the following problems and observing the results. They are not necessary to an understanding of what follows. Some additional notation is introduced.

3-3-5 Since examples such as required for Exercise 3-3-3 do not constitute proofs, prove that the paired expressions are unequal.

3-3-6 Write in full the following:

(a) $\sum_{i \neq j} y_i y_j$ where i and j can take on the values 1, 2, 3, and 4, and $i \neq j$ means "*i cannot* equal *j*."

(b) $\sum_{i<j} y_i y_j$ where i and j are as for (a), and $i < j$ means "i is always less than j."

3-3-7 Show that:

(a) $\left(\sum_{i=1}^{4} y_i\right)^2 = \sum_{i=1}^{4} y_i^2 + 2 \sum_{i<j} y_i y_j$

(b) $\sum_{i=1}^{4} x_i \sum_{i=1}^{4} y_i = \sum_{i=1}^{4} x_i y_i + \sum_{i \neq j} x_i y_j$

3-3-8 For the data of Exercise 3-3-4, find the following:

(h) $\sum_{i \neq j} x_i y_j$ (i) $\sum_{i=1}^{5} x_i y_i + \sum_{i \neq j} x_i y_j$

(j) $\sum_{i \neq j} y_i y_j$ (k) $\sum_{i=1}^{5} y_i^2 + \sum_{i \neq j} y_i y_j$

(l) Using the results of Exercise 3-3-4 also, observe that (f) = (d) + (j) = (k); (g) = (c) + (h) = (i).

(m) Provide proofs of the statements made in (l).

3-4 THE ARITHMETIC MEAN

The most common measure of the location of a distribution is the arithmetic mean, the arithmetic average, or simply the mean when there is no ambiguity. It is also the measure generally considered to be the "best" for its purpose.

A population mean is designated by the lowercase Greek letter mu, μ; it is a parameter determined once and for all time and is, thus, a constant. Any mean is in the same units as the observations.

Suppose a finite population consists of the values $y_1, y_1, \ldots, y_N$, with N denoting population size to distinguish it from n for sample size. The arithmetic mean is then defined by Equation 3-4-1.

$$\mu = \frac{\sum_{i=1}^{N} y_i}{N} \tag{3-4-1}$$

When an honest die is tossed, the number of dots on the upturned face will be 1, 2, 3, 4, 5, or 6. The number of dots so appearing is, then, a variable and a random one because of the way chance is involved although this is not pertinent until we deal with sampling. These six numbers, being all possible values of the variable, constitute a population. Thus, from Equation 3-4-1,

$$\mu \doteq \frac{1 + 2 + \cdots + 6}{6} = 3\tfrac{1}{2} \text{ dots}$$

More advanced mathematical techniques are needed to define the mean of a population with a continuous variable so that the computation can be performed. We will have no need to perform such computations, only the need for the notion of such a measure of location and a symbol to represent it.

The sample mean varies from sample to sample and so is a variable with its own frequency distribution. Since it summarizes information from a sample, it is a statistic, and its symbol is taken from the Roman alphabet. It is designated as $\bar{Y}$, read as "*Y*-bar."

For a random sample, one still in the future with values designated as $Y_1, \ldots, Y_n$, the mean is defined by Equation 3-4-2.

$$\bar{Y} = \frac{\sum_{1}^{n} Y_i}{n} \tag{3-4-2}$$

However, if we have at hand a sample that was drawn at random, then we designate its values by $y_1, y_2, \ldots, y_n$ and define the sample mean by Equation 3-4-3.

$$\bar{y} = \frac{\sum y_i}{n} \tag{3-4-3}$$

Now $\bar{y}$ is a particular value of the random variable $\bar{Y}$, and our concern might be with the probability of obtaining a value of $\bar{Y}$ larger than the observed value $\bar{y}$. Thus we might draw a sample of heights of male individuals and find the mean to be 5 feet 10 inches; we write $\bar{y}$ = 5 feet 10 inches, using the lowercase *y* to symbolize a particular value. Or, if a die is tossed four times and the number of dots on the upturned face is 2, 1, 3, and 2, the sample mean is

$$\bar{y} = \frac{2 + 1 + 3 + 2}{4} = 2 \text{ dots}$$

Since all observed values are less than the average, we might wonder if two dots is an unusually small value for the mean when using a fair die and ask the question, "What is the probability that a random sample of four observations made on the toss of a die will have a mean of 2 or fewer dots?" Clearly, we can conveniently make use of the symbol $\bar{Y}$ in referring to future samples of this random variable, and use $\bar{y}$ to refer to the observed mean with value 2. In the case of randomness, as we would expect with a reasonably well-tossed die, probability theory applies and this question has an answer.

In cases where many observations are identical, it may be most convenient to perform the computations by first preparing a frequency distribution. For example, the data giving rise to Table 2-3-2 were once a random ordering of five 0's, sixteen 1's,..., and four 5's; the sample consisted of 100 observations, where each observation involved observing and recording the number of heads that appeared when five coins were tossed. Application of Equation 3-4-3 to the original data gives

$$\bar{y} = \frac{\sum_{i=1}^{100} y_i}{N}$$

After reordering the data, we compute

$$\bar{y} = \frac{(\underbrace{0 + \cdots + 0}_{5 \text{ times}} + \underbrace{1 + \cdots + 1}_{16 \text{ times}} + \cdots + \underbrace{5 + \cdots + 5}_{4 \text{ times}})}{100}$$

But, clearly, we could just as well write

$$\bar{y} = \frac{5(0) + 16(1) + 31(2) + 28(3) + 16(4) + 4(5)}{100}$$

So, in general, for a frequency distribution, we write Equation 3-4-3 as 3-4-4.

$$\begin{aligned} \bar{y} &= \frac{f_1 y_1 + f_2 y_2 + \cdots + f_k y_k}{f_1 + f_2 + \cdots + f_k} \\ &= \frac{\sum f_i y_i}{\sum f_i} \end{aligned} \tag{3-4-4}$$

where y_i is one of the set of k distinct, possible observations, here 0, 1,..., 5, and f_i is the frequency with which it occurs. Note that $\sum f_i = n$; here $k = 6$ and $n = 100$.

We have described the arithmetic mean as the "best" measure of location. Criteria that lead to this sort of statement are discussed toward the end of Section 3-9. Here we will mention some of the other and more obvious properties that make the mean a desirable measure.

The mean is widely used and thus well understood by most people. It can always be computed for numerical data, makes use of all the observations, and is unique. Not every measure of location possesses all these properties. Several means can be combined readily to give other means, either directly or using weights if this seems important. It is analogous to the center of gravity. Thus if we have a lot of very tiny lead balls of the same weight and place these on a weightless axis at points corresponding to observations, then the mean will correspond to the point where a fulcrum can be placed and the axis with its weights will be balanced. This is illustrated in Figure 3-4-1 for the four tosses of a die used as an example earlier in this section. In the case of a frequency distribution, we may visualize a sheet of material of uniform weight throughout cut to the same shape as the corresponding histogram. This sheet will balance on an axis through the center of gravity, which is the mean, and will be vertical to the horizontal axis on which the observation values are recorded.

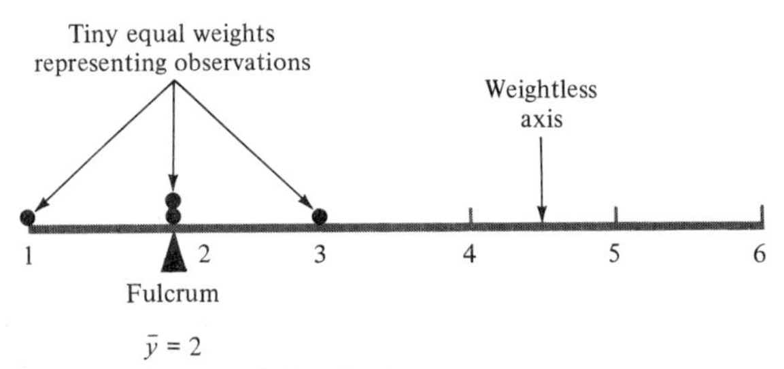

FIGURE 3-4-1 Center of Gravity Is at $\bar{y}$.

One fairly obvious disadvantage of the arithmetic mean is that extreme values readily affect it. For example, suppose a real estate agent reports that sales for the month were five houses at an average price of \$41,000. From these figures, it is easy to visualize sales in a \$30,000 to \$50,000 range. However, further investigation reveals that the actual prices were \$20,000, \$18,000, \$22,000, \$25,000, and \$120,000, representing a much lower range than visualized, except for one very expensive home. Some other measure of location, or an additional descriptive measure, is required to convey this information.

3-5 THE WEIGHTED MEAN

A weighted mean is used when the separate observations have differing amounts of information. Thus, for example, a mean of 100 observations would seem to have 10 times as much information as one based on only 10 observations, and weights should be assigned accordingly.

Consider a baseball player who has played parts of three years in a major league. His batting averages are .412, .250, and .274. If he had been at bat approximately the same number of times each year, we would compute his 3-year average using an unweighted mean, namely (.412 + .250 + .274)/3 = .312. Now suppose the player had been at bat 10, 50, and 200 times, respectively. The unweighted average becomes somewhat suspect. Weights proportional to the number of times at bat suggest themselves, and we compute

$$\frac{10(.412) + 50(.250) + 200(.274)}{10 + 50 + 200} = .275$$

This is a weighted mean, and its computation is seen to be a procedure that might be described as one for retrieving some earlier totals and continuing with these to obtain a new grand total and eventually a mean, where each original observation has the same weight.

To define a weighted mean, suppose $y_1, y_2, \ldots, y_n$ are observed values and $w_1, w_2, \ldots, w_n$, respectively, are appropriate weights. Then the *weighted mean* of the observations is given by Equation 3-5-1.

$$\bar{y} = \frac{\sum w_i y_i}{\sum w_i} \tag{3-5-1}$$

As previously, capital letters would be used if we were talking abstractly about random observations rather than data at hand. Note also that Equation 3-4-4 defines a weighted mean of the k distinct observational values in a sample; the weights are the frequencies with which each distinct value occurred and are given the symbols f_i rather than w_i.

The 126 students of Table 2-3-3 were in four sections. The first section had 29 students with a mean grade of 77.17, the second had 34 with a mean grade of 79.41, the third had 34 with a mean grade of 80.10, and the fourth had 29 with a mean grade of 76.61. The weighted mean of these averages is

$$\bar{y} = \frac{29(77.17) + 34(79.41) + 34(80.10) + 29(76.61)}{29 + 34 + 34 + 29} = 78.44$$

This average first reconstitutes the totals for the four sections, then divides by the total number of students. Thus the weighted mean of these means is an unweighted mean of the grades of the individual students. Actually, Equation 3-4-3 is a special case of 3-5-1 where all the weights are equal and each $w_i = 1$.

Weighted means are much used in business, economics, and government where they are often called *index numbers.* Union contracts may tie wages to the Consumer Price Index or Cost-of-Living Index, farm subsidies depend on the Parity Index, and business decisions are conditioned by the Dow-Jones stock averages and similar indices.

3-6 THE MEDIAN

Recall that when observations are arranged in order of magnitude from smallest to largest, they are in an array.

For an odd number of observations, the *median* is the middle value of their array; for an even number of observations, the median is the average of the two middle values of the array.

For example, the median of the array 2, 4, 12, 16, 40 is 12; and that for 5, 10, 12, 18, 28, 40 is $(12 + 18)/2 = 15$. Half the values are less than or equal to the median and half are greater than or equal to the median in our illustrations. However, this statement is not a satisfactory definition if many observations have the value of the median. For example, the array 1, 2, 3, 3, 3, 4, 5 has median 3, but 5 of these 7 numbers are less than or equal to the median and 5 are greater than or equal to it. Consequently, we find we must say that the median is neither less than nor more than half of the observations.

When observations are arranged in order of magnitude rather than as observed, they may be designated by $y_{(1)}, \cdots, y_{(n)}$. Capital letters would be used if we were talking abstractly about random sampling rather than about a sample already obtained. Thus in an array of the 126 grades in Table 2-3-3, the median would be the average of the 63d and 64th grades, $y_{(63)}$ and $y_{(64)}$. Both these grades are 81, and so the median is 81. If there were only 125 grades, the median would be the 63d grade or $y_{(63)}$. Thus we may conveniently find the subscript or subscripts for the median item by using Equation 3-6-1.

$$s_m = \frac{n + 1}{2} \tag{3-6-1}$$

For $n = 126$, we find $s_m = (126 + 1)/2 = 63\frac{1}{2}$, and so we use both 63 and 64; for $n = 125$, we find $s_m = (125 + 1)/2 = 63$.

When data are in the form of a frequency distribution and a mean is wanted, the members of any class interval are assigned the midpoint value of that class. Essentially, this implies that the observations are assumed to be distributed uniformly in that class. This assumption is also made when determining the median.

To determine the median for a frequency distribution, refer to the grade data in frequency distribution form, Table 2-3-4. The appropriate computations are readily apparent by reference to Figure 3-6-1. Since the median is such that equal numbers of observations lie on either side and each observation is given the same area, the basic problem is to divide the histogram into two equal areas. We need 63 in each. Starting from the lower end, we observe 3 grades less than 50, $3 + 4 = 7$ less than 55,..., 61 less than 80, and $61 + 30 = 91$ less than 85. Clearly we require only $\frac{2}{30}$ of this last rectangle to partition it as needed (see Figure 3-6-2). The required median grade is then $80.5 + (2/30) \times 5 = 80.83$, since 80.5 is the lower boundary of the rectangle being partitioned and 5 is the class interval.

By our definition, there is always exactly one median and it is easy to find. Unlike the mean, it is not affected by extreme values and it can be found even for frequency distributions with open-ended classes provided it is not in such a class. In some cases, the median may be found without the necessity of quantitative measurement. Thus, if a group of steers in a show ring can be ranked by a judge, then the one in the middle of the array is the median. The median height of

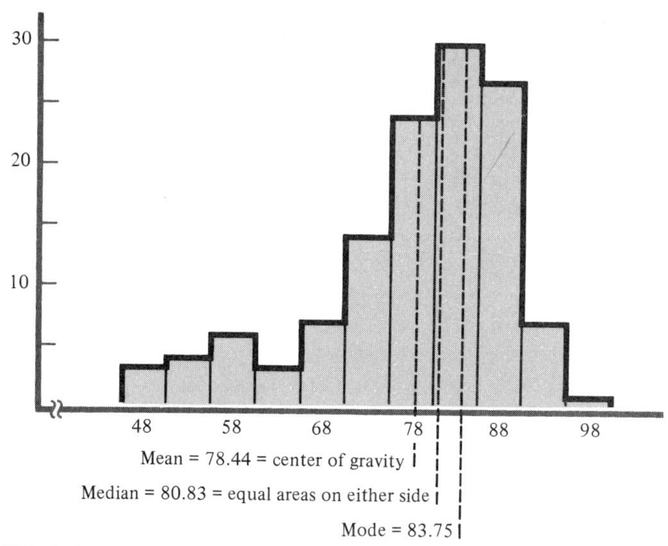

FIGURE 3-6-1 Location of the Mean, Median, and Mode for the Grade Data of Figure 2-3-3.

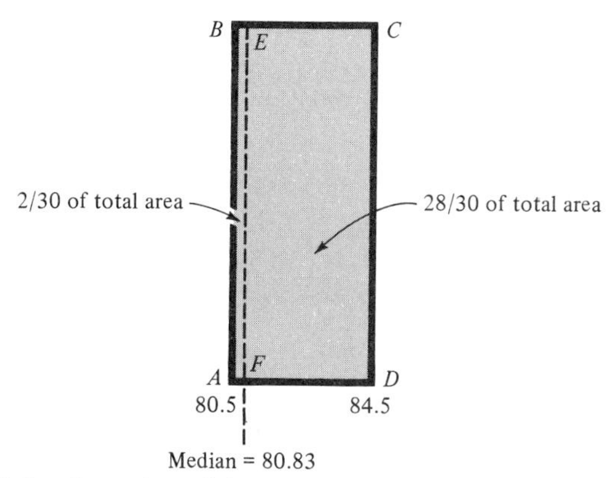

FIGURE 3-6-2 Location of the Median in Relation to Class Boundaries.

members of a basketball team could be found by a visual ranking followed by, at most, two measurements.

A disadvantage of the median that is not possessed by the mean is that the original observations must be available when pooling data. In other words, an overall median cannot be obtained satisfactorily from a set of medians, whereas an overall mean can be obtained as a weighted average of means. Medians vary more in repeated sampling than do means; that is, medians are less stable. This is a serious drawback when it is desired to draw an inference concerning a population.

EXERCISES

3-6-1 Find the mean and median of the following grades made by 25 students in an exam: 81, 85, 73, 56, 74, 62, 84, 86, 78, 93, 77, 64, 68, 77, 75, 70, 78, 86, 73, 69, 72, 86, 84, 83, 75.

3-6-2 The following are the lengths in millimeters of the lower molar of 16 members of the species *Hyopsodus paulus*, a fossil mammal of interest in the study of evolution: 4.87, 4.77, 4.93, 4.71, 4.68, 4.52, 4.77, 4.85, 4.80, 4.65, 4.45, 4.95, 4.76, 4.90, 4.72, 4.49. Find the mean and median. (Data from Olson and Miller, 1958.)

3-6-3 Find the mean and median heart rate of 13 cats after they received 10 milligrams of procaine: 170, 126, 105, 135, 186, 198, 140, 160, 138, 120, 150, 168, 123. (Data from Tanikawa, 1954.)

3-6-4 Find the mean and median cholesterol content expressed as milligrams per 100 milliliters in the blood of 11 diabetic adolescent girls both before and after receiving training in physical fitness. (Data from Larsson et al., 1962.)

Before	170	213	216	200	199	141	182	168	194	144	177
After	200	265	188	194	219	195	225	194	205	148	215

3-6-5 Find the mean and the median of the following measurements of percent copper in 10 bronze castings: 85.54, 85.72, 84.48, 84.98, 84.54, 84.72, 84.72, 86.12, 86.47, 84.98. (Data are from Wernimont, 1947. Reprinted by permission of the American Society for Quality Control, Inc.)

3-6-6 The following are a household's monthly gas bills for 1964, in dollars. Determine the mean monthly bill: 30.34, 28.18, 23.00, 17.60, 7.16, 5.12, 4.40, 5.58, 5.34, 11.46, 11.80, 28.50.

3-6-7 Below are given the commodity price indices for 1964 (U.S. Bureau of Labor statistics) based on 1957–59 as 100, for:

Meat, poultry, and fish: 98.3, 98.3, 97.2, 97.0, 96.6, 98.9, 99.2, 101.4, 100.6, 99.5, 99.0.

Dairy products: 105.0, 104.8, 104.5, 104.1, 103.9, 104.0, 104.3, 104.4, 104.6, 105.3, 105.3, 105.6.

Fruits and vegetables: 112.4, 113.7, 115.1, 115.7, 115.7, 120.2, 122.3, 117.3, 112.2, 111.7, 113.0, 114.5.

Find the mean index for each of the three commodities.

3-6-8 The average quiz grades obtained by members of four sections of a class were 84, 78, 81, and 80. The four sections had 36, 32, 37, and 30 students, respectively. Find the weighted average grade.

3-6-9 For the number of heads observed when tossing 5 coins 100 times, given in Table 2-3-2, compute the mean and median.

3-6-10 Determine the median age of licensed drivers in the United States using the data of Table 2-3-5.

3-6-11 Find the median age of passenger cars in the United States based upon the data of Exercise 2-4-5.

3-6-12 Find the mean and median without grouping and with grouping for the liver weights of the 80 hens of Exercise 2-3-4.

3-6-13 Find the mean and median with and without grouping for the ages of National League baseball pitchers given in Exercise 2-3-5.

3-6-14 Find the mean and median with and without grouping of the average June temperatures in Ohio given in Exercise 2-3-6.

3-6-15 Find the mean and median for the high-card point counts of the 200 bridge hands given in Exercise 2-3-7 with grouping.

3-6-16 Find the mean and median of the vertical load at ultimate strength for concrete corbels of Exercise 2-3-8.

3-7 QUANTILES AND MIDRANGE

A *quantile* or *fractile* is a measure of the location of a point such that a specified fraction of the data lies to its left. For example, the median is the quantile measuring the location of the point such that 50 percent of the data lie to its left. Thus, this quantile is a measure of central tendency in itself. Several quantiles may also be used simultaneously to provide measures of location.

Percentiles divide a ranked set of observations into hundredths, *deciles* into tenths, *quartiles* into quarters, and so on. For example, the 10th percentile, designated by P_{10}, is the point such that 10 percent of the observations are less than or equal to this value, whereas the 90th percentile, P_{90}, has 90 percent of the observations less than or equal to it.

Percentiles are regularly used with students' scores on standard aptitude tests, graduate record exams, and similar testing instruments. Percentiles quoted in conjunction with such tests are usually computed from the set of scores that were used to provide national or otherwise specified norms.

To illustrate the computation of the 10th percentile or first decile, we again refer to the grade data of Table 2-3-3 and, more particularly, to these data in frequency-distribution form, Table 2-3-6. Since $n = 126$, P_{10} is the 12.6th item from the left. Clearly there is no such observation. However, there are 7 less than 55.5 and 13 less than 60.5. Hence, P_{10} should be just short of 60.5, or, more particularly, (12.6 − 7)-sixths of the way into the 55.5 to 60.5 interval. In other words, $P_{10} = 55.5 + 5(12.6 - 7)/6 = 60.2$. The multiplier 5 is the length of the class interval, the denominator 6 is the number of observations in the interval being partitioned to provide the $12.6 - 7 = 5.6$ observations needed to complete the quota of 12.6 observations. Figure 3-6-2 for location of the median is a graphic illustration using a different percentile.

Quartiles are percentiles that often receive special attention. Thus $P_{50} = Q_2$, the second quartile, is the median, a measure of central tendency. Also, $P_{25} = Q_1$ and $P_{75} = Q_3$ are used to define the midrange, Equation 3-7-1, a measure that uses two observations. The sample midrange estimates the mean of a population and is another measure of central tendency.

$$\text{Midrange} = \frac{P_{25} + P_{75}}{2} = \frac{Q_1 + Q_3}{2} \tag{3-7-1}$$

Computation of P_{25}, P_{75}, and the midrange follows and should be clear on reference to Table 2-3-6. Refer also to Figure 3-7-1 and note that 50 percent of the

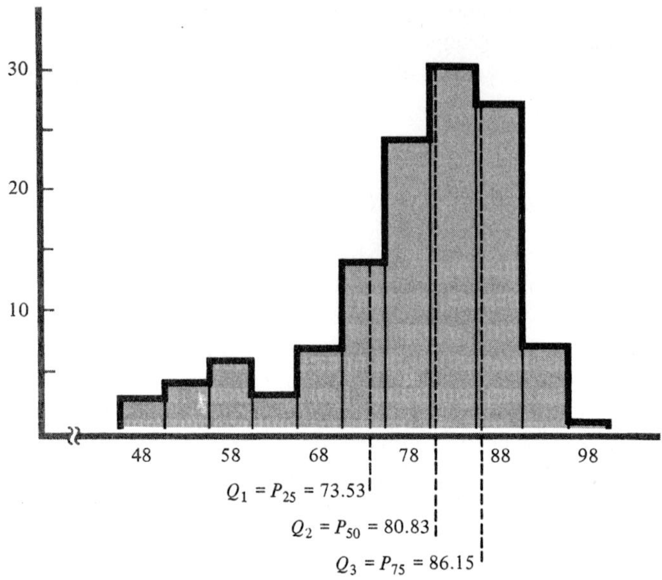

FIGURE 3-7-1 Location of the Quartiles for the Grade Data of Figure 2-4-3.

observations, and so 50 percent of the area of the histogram, must lie between these quartiles.

$$Q_1 = P_{25} = 70.5 + \frac{31.5 - 23}{14} 5 = 73.54$$

$$Q_3 = P_{75} = 85.5 + \frac{94.5 - 91}{27} 5 = 86.15$$

Such computations may be done from either end of the distribution. Thus,

$$Q_3 = P_{75} = 90.5 - \frac{31.5 - 8}{27} 5 = 86.15$$

Finally,

$$\text{Midrange} = \frac{73.54 + 86.15}{2} = 79.84$$

The midrange has an advantage over the median as an estimator of central tendency for small samples, up to $n = 5$, but the median is the better estimator for larger samples. The criterion used in drawing such conclusions is based on the extent to which such estimates spread out in repeated sampling.

The idea of using the midrange as a measure of central tendency raises the question, "What are the two best percentiles for such an estimate?" The answer is that they are known but their choice involves a number of considerations. A more comprehensive text, such as Dixon and Massey (1969, section 9-4), gives a more complete treatment.

The method for computing quantiles when the data are in the form of a frequency distribution has been shown. When this form has not been used, we find, as in the case of the median, that quantiles are a little harder to define verbally. Again, we resort to the "neither less than nor more than" terminology.

Quantiles are much used in education, commerce, and industry where large numbers of observations are involved. They also have interesting statistical applications with quite small samples where they can often provide easily computed and very satisfactory estimates of different parameters. The midrange as an estimate of the mean is one example.

EXERCISES

3-7-1 Find P_{25} and P_{75} of the frequency distribution prepared for the liver weight of hens of Exercise 2-3-4. Find the midrange and compare with the mean and median as computed in Exercise 3-6-12.

3-7-2 From the prepared frequency distribution of the ages of National League baseball pitchers given in Exercise 2-3-5, find P_{25}, P_{75}, and the midrange. Compare the last result with the mean and median computed in Exercise 3-6-13.

3-7-3 Find P_{25} and P_{75} of the distribution prepared from the temperature data of Exercise 2-3-6. Find the midrange and compare with the mean and median found in Exercise 3-6-14.

3-7-4 Use the prepared distribution of the high-card point count of the 200 bridge hands of Exercise 2-3-7 to determine P_{25}, P_{75}, P_{10}, and P_{90}. Find the midrange and the average of P_{10} and P_{90} and compare both with each other and the values of the mean and median as computed in Exercise 3-6-15.

3-7-5 Repeat Exercise 3-7-4 for the corbel data of Exercise 2-3-8 and the mean and median computed in Exercise 3-6-16.

3-8 THE MODE, GEOMETRIC MEAN, AND HARMONIC MEAN

These measures are more specialized than those previously discussed and are used when the nature of the data suggests that they are more realistic.

The *mode* is that observation which occurs most frequently; the *modal class* of a frequency distribution is that with the most observations. A mode does not always exist; if it does, it may not be unique. Thus,

2, 4, 6, 6, 6, 7, 7	has one mode, namely 6
2, 5, 5, 7, 8, 8, 9	has two modes, 5 and 8
2, 2, 3, 3, 5, 5, 6, 6	has no mode

The mode is not a satisfactory measure of location for use with ungrouped measurement data.

For a *modal class*, a particular value is not necessarily specified. This is particularly true with qualitative or categorical data. Thus in Table 2-3-1, the modal class is "whole life."

The *geometric mean* of a set of numbers $y_1, \ldots, y_n$ is defined by Equation 3-8-1.

$$G = \sqrt[n]{y_1 y_2 \cdots y_n} \tag{3-8-1}$$

This equation tells us that with n numbers, we are to multiply them and then take the nth root. For example, when $n = 2$, the geometric mean is the square root of their product; when $n = 3$, it is the cube root; and so on. One usually resorts to the use of logarithms for help in doing the necessary arithmetic.

The geometric mean is used with numbers that tend to increase geometrically rather than arithmetically; that is, each number is the same multiple of the preceding one. For example, some kinds of biological material may tend to double their size in a fixed length of time regardless of how large they are to begin with. Economic growth may act similarly. Under such circumstances, the results of using an arithmetic average of the actual increases in growth for a number of years to predict next year's growth or to interpolate missing values will be misleading.

To illustrate the use of the geometric mean and the need for alternatives to the arithmetic mean, suppose an individual has a salary of \$2,000 in 1945, \$4,000 in 1955, and \$18,000 in 1965, and that we are asked about the 10-year average rate of change in income.

The initial salary doubled in the first 10 years, an increase of \$2,000; $4\frac{1}{2}$ times that salary gave the final salary, a further increase of \$14,000. Pretty obviously, salary hasn't been increasing arithmetically; a geometric increase seems closer to reality, although it, too, offers a less than perfect explanation of the nature of the increase.

The arithmetic mean of the actual changes is $\$(2{,}000 + 14{,}000)/2 = \$8{,}000$. This suggests $\$(2{,}000 + 8{,}000) = \$10{,}000$ as an estimate of the 1955 salary and \$18,000 for 1965. (Note that the estimate for 1955 is also the arithmetic mean of the first and last salaries, as it should be.) This is quite badly off the true 1955 salary of \$4,000 because a fixed increase, to be added each 10-year period, simply does not explain what is happening to salaries. An average increase of \$8,000 per 10-year period is unrealistic for interpolating between 1945 and 1965 or for predicting beyond 1965.

The rates of change have been seen to be 2 and 4.5. Their arithmetic mean is $(2 + 4.5)/2 = 3.25$. Applying this average rate, we find estimates of the 1955 and 1965 salaries to be $\$2{,}000(3.25) = \$6{,}500$ and $\$6{,}500(3.25) = \$21{,}125$, respectively. Neither of these compares favorably with the actual values of \$4,000 and \$18,000. This arithmetic average of the rates of change is simply not very useful.

The geometric mean of the rates of change is $\sqrt{2 \times 4.5} = 3$. Applying this rate, we estimate the 1955 and 1965 salaries as $\$2{,}000 \times 3 = \$6{,}000$ and $\$6{,}000 \times 3 = \$18{,}000$, respectively. Our estimates are better than previous ones, although that for 1955 is still not very good.

Finally, the geometric mean of the salaries is

$$\$\sqrt[3]{2{,}000 \times 4{,}000 \times 18{,}000} = \$5{,}240$$

Clearly this mean compares most favorably with the $4,000 salary of 1955.

The real problem here is to decide how the change is taking place and then to use a method appropriate to the nature of the change. The salary increases are neither arithmetic nor geometric, and we must approach the problem empirically, using trial and error to find a mean that is close to the 1955 salary so that we can, in turn, devise a method for interpolation or for projecting salaries beyond 1965.

If the salaries were increasing by some roughly constant multiplication factor, then we would have a geometric increase; hence the geometric mean of all the salaries would be an appropriate mean, and the geometric mean of the rates would be a rate which could be used appropriately to interpolate or to predict.

The geometric mean is unique and uses all the observations. However, it cannot be used to average numbers which include a 0 or are of different signs.

The *harmonic mean* of a set of numbers $y_1, \ldots, y_n$ is the reciprocal of the arithmetic mean of the reciprocals of the numbers. It is defined by Equation 3-8-2.

$$H = \frac{1}{\sum (1/y_i)/n} \tag{3-8-2}$$

In practice, we compute $1/H = \sum (1/y_i)/n$ and invert.

The harmonic mean is used when observations are expressed inversely to our requirements or to their importance. For example, we may want to compare costs per unit of product but find them expressed as units of product per unit price; or we may wish to compute a travel time for a trip involving various distances at different speeds and so need to know time per mile rather than miles per hour. For such problems, a harmonic mean is generally applicable. Again, a desirable property of statistics that estimate parameters is that they should vary but little in repeated sampling. The implication is that small variation involves a large amount of information, an inverse relation sometimes calling for harmonic means.

As illustration, suppose you buy $1 worth of bananas at 5 cents each and $1 worth at 10 cents each. The average price paid was not $(5 + 10)/2 = 7\frac{1}{2}$ cents because more bananas were bought at 5 cents than at 10 cents. In other words, a weighted average is needed. Since 20 bananas were bought at 5 cents each and 10 at 10 cents each, the weighted average is $(20 \times 5 + 10 \times 10)/(20 + 10) = 6\frac{2}{3}$ cents. Note that the harmonic mean provides the same value

$$\frac{1}{H} = \tfrac{1}{2}(\tfrac{1}{5} + \tfrac{1}{10}) = \tfrac{15}{100} \qquad \text{and} \qquad H = \tfrac{100}{15} = 6\tfrac{2}{3} \text{ cents}$$

EXERCISES

3-8-1 Find the geometric mean of (a) 2 and 18, (b) 4, 8, and 16. If the increase continues geometrically, what is the next value to be expected in each case?

3-8-2 The net rate of interest earned on investment funds by United States life insurance companies in 1927, 1937, and 1947 was 5.05, 3.69, and 2.88 percent, respectively. Determine the average rate of change in interest rate during the two 10-year periods using the geometric mean. Determine the average change in interest rate using the arithmetic mean. Do you think the rates progress in an arithmetic or a geometric pattern?

3-8-3 Determine the harmonic mean of (a) 4, 6, and 8 and (b) 110 and 240.

3-8-4 A grocer bought from a wholesaler $330 worth of peaches at $1.10 per lug on Monday; $475 worth of peaches at $.95 per lug on Wednesday; and $690 worth of peaches at $1.15 per lug on Friday. Use the harmonic mean to determine the average purchase price per lug of peaches. Also, find the weighted average cost of a lug of peaches. Does this differ from the geometric mean? Why? What is the unweighted average cost of a lug of peaches? Of the three means, which do you recommend as most reasonable and why?

3-9 SELECTING A MEASURE OF CENTRAL TENDENCY

The arithmetic mean, median, and mode are different measures of the center of a set of values. Consequently, their definitions and limitations must always be clearly in mind. The choice of a measure should be made at the planning stage of an investigation with its purposes and needs in mind, and not after an examination of the data. Experience has shown that measures chosen at the planning stage and based on as much past experience as is available are more likely to lead to correct conclusions than are measures suggested by the particular set of data under consideration.

For categorical data, where a scale is not provided, the modal class will often be the obvious choice. This was so for the insurance data of Table 2-3-1 where no evident scale could be associated with the data.

In a study of salaried individuals where salary categories may be "to the nearest hundred dollars," we have a frequency distribution where there is an underlying scale. When a scale exists, the mean and median are measures of central tendency which are preferable to the mode. Salary distributions are often skewed toward large or extreme values, and these have an effect on the mean whereby its use could lead to misconceptions about the economic health of the individuals under study. On the other hand, extreme values have a lesser effect on the median, since, by definition, as many salaries are above this figure as are below. In general, the median is preferred to the mean for skewed data.

When a frequency distribution includes open-ended intervals, as for the age data of Table 2-3-5, the mean cannot be computed but there is no problem with the mode or the median.

For index numbers, much used in commerce, industry, and government, a time variable is usually involved, and this may suggest the sort of growth that makes the geometric mean the most appropriate measure of a center.

When several sets of data are to be combined, the algebraic properties of the mean give it a decided advantage in that a weighted average is easily computed. Neither the median nor the mode can be so handled.

If, instead of simply wishing to describe our data, we are primarily concerned with drawing inferences from a sample to a population, then we try to have a general concept of what the population to be sampled is like. Figure 2-4-1 suggests that data often fit into general categories or shapes—symmetric, bimodal, etc.—which can, in turn, be approximated and, consequently, described by mathematical formulas, generally called *models*. For any category the mathematical formula will leave certain constants unspecified, generally substituting symbols for them. For example, the mean may be one of the constants needed in a formula but it may simply be represented by the symbol μ. These constants are the parameters of the distribution, and the researcher's problem may be to estimate their numerical values for a particular population. This is a problem of inference. Now, with a model, a mathematical treatment of the problem of inference—presently that of choosing a measure of central tendency—is possible, and we can be concerned with various mathematical properties of the possible measures. Among these properties is stability—the degree of closeness one value has to others in repeated sampling. So, finally, we want to know which among the mean, median, or any other proposed measure of central tendency is the most stable. Obviously, the most stable one can be described as "best" in terms of stability. The arithmetic mean is superior to all other measures of central tendency in this regard. Measuring stability is the subject of the next chapter.

While it is desirable for estimates to be close to one another in repeated sampling, it is also desirable for them to be close to the population value they are estimating. When the mean of all possible values from repeated sampling is the population value being estimated, the formula that defines the method of computing the estimate is said to be *unbiased*. Otherwise, we have a *biased* estimator. This concept is discussed at some length in Chapter 13 and elsewhere.

Amenability to mathematical treatment and computational ease tend to go hand in hand. *Linear functions*, that is, weighted sums of observations, make for computational ease and for relatively simple mathematical treatment. The mean and median are such functions. For the mean, the weights are all $1/n$. No numbers are raised to powers, no products are computed, and no roots are taken, as for the geometric mean, which is a nonlinear function. For the median, the weights are mostly 0s since we use only the middle value with a weight of 1 or two middle values with weights of $\frac{1}{2}$.

Chapter 4

MEASURING VARIATION

4-1 INTRODUCTION

As in the case of locating a distribution, data summaries such as frequency distributions and graphical devices can help in judging the extent to which the observations in a population or sample vary. However, such judgments are not likely to have defining mathematical formulas and so will vary with the individual making the judgment. Something is needed which is not dependent on the person using the method, and which, like the arithmetic mean, can be dealt with algebraically.

In the discussion of the location of a distribution, it was observed that observations usually tend to cluster about a central value. We now need a numerical measure that describes how closely or loosely the observations so cluster.

This chapter discusses measures of variation, or *spread,* in common use. Some algebraic development of results is presented. However, it is the results that are important to the elaboration of the text, and the reader with limited time may choose to skip over the development.

4-2 MEASURING SPREAD

Measures of location are partial summaries of the information in a set of data but tell us nothing about the spread, scatter, dispersion, or variation among the individual observations. Before defining a measure to describe this phenomenon, let us examine some sets of data.

The three sets of data below have a common mean, namely 10. Their spacing is in general accord with their numerical values.

Data Set 1								8 8	9	10	11	12 12							
Data Set 2					5	6		8		10		12		14	15				
Data Set 3	1	2			5					10					15			18	19

Obviously these sets differ in the extent to which the numbers spread. We might use the ranges and say the spreads are $12 - 8 = 4$, $15 - 5 = 10$, and $19 - 1 = 18$, respectively. A precise bit of information is provided but it involves only two observations, the most extreme ones. This is so regardless of the sample size, and so the range must be relatively inefficient.

Consider another three sets of data with the same mean and the same ranges, 4, 10, and 18, as for the first three sets. These are again spaced in accord with their numerical values.

Data Set 4								8	9	10 10 10	11	12							
Data Set 5					5		7		9	10	11		13		15				
Data Set 6	1				5			8		10		12			15				19

If we now compare any set of observations with the earlier set with the same range, we see the observations here clustering about their mean where earlier they spread more into the tails. In defining variability, this type of information should be used since it provides an improvement over the range as a definition.

4-3 THE STANDARD DEVIATION AND THE VARIANCE

In searching for an alternative to the range for measuring variation, we would like a definition that uses all the data and can be treated algebraically. A first suggestion might be to use the distances of the observations from a central value such as the mean; for example, we might use the $(y_i - \bar{y})$'s. The use of these *deviations from the mean* to define a measure of variation is reasonable. Though some are positive and others negative, as numbers they leave an impression as to the extent to which the original data vary.

If we begin by considering the sum of the deviations $\sum (y_i - \bar{y})$, we find that this quantity is always 0. This is simply another statement of the fact that $\bar{y}$

corresponds directly to the center of gravity in mechanics as discussed in Section 3-4. This relationship may be shown as follows.

$$\begin{aligned}\sum_{i=1}^{n} (y_i - \bar{y}) &= y_1 - \bar{y} + y_2 - \bar{y} + \cdots + y_n - \bar{y} \\ &= y_1 + y_2 + \cdots + y_n - n\bar{y} \\ &= y_1 + \cdots + y_n - \cancel{n}\frac{y_1 + \cdots + y_n}{\cancel{n}} = 0\end{aligned}$$

Consequently, the sum of the deviations is no help in measuring variation.

This difficulty with deviations may be met by using absolute values, that is, by calling all deviations positive, or by squaring them. Both approaches are reasonable, but it turns out that squares result in measures with more desirable properties. Thus, the most used and, in many ways, best measures of variation, the standard deviation and its square, the variance, are based on squared deviations. The variance is always computed first and its square root provides the standard deviation. These measures utilize all of the observations and the way they are dispersed.

Suppose we have a sample of n values, $y_1, y_2, \ldots, y_n$, with mean $\bar{y}$. The *sample variance,* designated by s^2, or the *sample mean-squared deviation,* is defined by Equation 4-3-1, and the *sample standard deviation* or *root-mean-squared deviation* is defined by Equation 4-3-2.

$$s^2 = \frac{\sum (y_i - \bar{y})^2}{n - 1} \qquad (4\text{-}3\text{-}1)$$

$$s = \sqrt{s^2} \qquad (4\text{-}3\text{-}2)$$

Clearly s^2 is a measure of variation. It is 0 only if all observations are identical and so exhibit no variation. If the observations cluster closely about the mean, s^2 will be relatively small, whereas if they are widely scattered, it will be relatively large. To illustrate this, consider two data sets used in Section 4-2, set 1 having little variation and set 3 having considerable variation (Table 4-3-1).

The variances are $s^2 = \frac{18}{6} = 3$ and $s^2 = \frac{340}{6} = 56\frac{2}{3}$, respectively. Unfortunately, these values are not very helpful in any obvious way since they are no longer in the original unit of measurement.

Using Equation 4-3-2, we find the standard deviations to be $s = \sqrt{3} = 1.73$ and $s = \sqrt{56\frac{2}{3}} = 7.53$. These values are in the same units as the original observations, and, consequently, interpretable as measures of spread more readily than are variances.

Referring again to Equation 4-3-1, observe that since any sum of squares will tend to increase with increase in sample size, it seems proper to use an average when defining spread. This should keep the resulting value to the same order of magnitude, regardless of the size of the sample, and thus justify the use of the

TABLE 4-3-1

	Data Set 1			Data Set 3	
y_i	$y_i - \bar{y}$	$(y_i - \bar{y})^2$	y_i	$y_i - \bar{y}$	$(y_i - \bar{y})^2$
8	−2	4	1	−9	81
8	−2	4	2	−8	64
9	−1	1	5	−5	25
10	0	0	10	0	0
11		1	15	5	25
12	2	4	18	8	64
12	2	4	19	9	81
Sum: 70	0	18	70	0	340
s^2:		18/6 = 3			340/6 = $56\frac{2}{3}$

adjective "standard." The most obvious average of the squared deviations would use n as a divisor, as is sometimes done. However, in repeated sampling, the s^2's computed with divisor n tend to underestimate the corresponding population value. To avoid this difficulty, one might suggest that we use a small divisor and so get somewhat larger values as estimates of the population variance. It can be shown that when $n - 1$ is the divisor, the population variance is neither underestimated nor overestimated. This argument about underestimation and overestimation does not apply in the case of the standard deviation and s, as defined in Equation 4-3-2, underestimates its population counterpart, on the average. However, it is now common practice to use $n - 1$ in defining the variance and standard deviation, as in Equations 4-3-1 and 4-3-2. Any potential problems arising from the use of s as defined have been taken care of by other means. When n is large, the difference between using n or $n - 1$ as divisor is small and can be of little importance if s^2 and s are to be used for purely descriptive purposes.

The divisor $n - 1$, called the *degrees of freedom* and abbreviated *df*, can be justified on other grounds. Thus, suppose we wish to write out a set of n deviations from the mean. The first $n - 1$ deviations are no problem and we can be as free as we wish in our choice, but the last one must be constructed so that all sum to 0; otherwise, we do not have deviations from the mean. Now since we are free to choose only $n - 1$ deviations, we may argue that this should determine the divisor in any definition using all the deviations. Note that this argument also justifies the term degrees of freedom.

Another argument which is essentially the same calls for us to observe Equation 4-3-1 as n increases. Thus, for $n = 1$, we do have a sample but it provides no information about the variability of observations. For $n = 2$, the two deviations from the mean must be equal in value though opposite in sign since

$\sum (y_i - \bar{y}) = 0$. There is, then, really only one deviation. The argument continues algebraically to show that for any n, there are really only $n - 1$ quantities which can be said to truly differ. Hence, $n - 1$ is a reasonable divisor.

Finally, the term root-mean-square deviation, often abbreviated *rms*, can now be seen to be quite meaningful.

The computation of deviations as required for the numerator of Equation 4-3-1 can be a lengthy procedure, with rounding errors almost certainly introduced. Let us consider Equation 4-3-3, which is simply the numerator of Equation 4-3-1. This is a meaningful *definition* formula for a sum of squares of deviations, but we require a more convenient computing formula.

$$\mathrm{SS} = \sum (y_i - \bar{y})^2 \tag{4-3-3}$$

The quantity SS, the sum of the squares of the deviations from the mean, is usually called simply the *sum of squares* unless there is a possibility of confusion. To shorten the work and improve the accuracy of computing SS, we use Equation 4-3-4, which yields the same results as Equation 4-3-3.

$$\mathrm{SS} = \sum y_i^2 - \frac{(\sum y_i)^2}{n} \tag{4-3-4}$$

No deviation need be computed, rounding due to division by n is done only once and then at the end of the computations, and the most elementary desk computer finds $\sum y_i$ and $\sum y_i^2$ simultaneously.

This computing formula is derived from the definition formula as follows, making use of the definition of $\bar{y}$ twice after the fourth equals sign.

$$\begin{aligned}
\sum (y_i - \bar{y})^2 &= (y_1 - \bar{y})^2 + (y_2 - \bar{y})^2 + \cdots + (y_n - \bar{y})^2 \\
&= y_1^2 - 2y_1\bar{y} + \bar{y}^2 + y_2^2 - 2y_2\bar{y} + \bar{y}^2 + \cdots + y_n^2 - 2y_n\bar{y} + \bar{y}^2 \\
&= y_1^2 + y_2^2 + \cdots + y_n^2 - 2\bar{y}(y_1 + y_2 + \cdots + y_n) + n\bar{y}^2 \\
&= \sum y_i^2 - 2\frac{\sum y_i}{n}\sum y_i + n\frac{(\sum y_i)^2}{n^2} \\
&= \sum y_i^2 - \frac{(\sum y_i)^2}{n}
\end{aligned}$$

Using the computing formula directly on the y_i's, we find

$$s^2 = \frac{718 - (70)^2/7}{6} = \frac{18}{6} = 3$$

and

$$s^2 = \frac{1{,}040 - (70)^2/7}{6} = \frac{340}{6} = 56\tfrac{2}{3}$$

as before.

The term $\sum y_i^2$ is called the *unadjusted* or *uncorrected* sum of squares since it is a sum of squares about the origin. It is thus distinguished from SS, which is called the *adjusted* or *corrected* sum of squares, the adjectives being used only when there would otherwise be ambiguity.

The quantity $(\sum y_i)^2/n$ is a sum squared and divided by n and called the *correction term* or *correction factor*, although it is not a factor of anything. It is also called the correction or adjustment for the mean, an appropriate designation in that it adjusts $\sum y^2$, a sum of squares about the origin, to be a sum of squares about $\bar{y}$.

While the standard deviation serves a descriptive purpose by measuring spread, it may still seem to have little other practical use. In fact, it does help us make decisions as to whether or not observed variation should be attributed to the fortunes of sampling or to real differences in sampled populations. For example, a doctor might have two randomly selected groups of individuals that will be observed throughout the coming year for the number and severity of the colds they have. One group will be provided with large doses of vitamin C, and the other will be given no vitamin C. At the end of the year, there are bound to be differences between the groups in both number and severity of colds even if vitamin C is neither good nor bad as a cold treatment. Here the standard deviation can help us decide whether or not the observed difference is a real one, that is, attributable to vitamin C, or simply a random sampling difference.

Let us, then, consider additional aspects of the standard deviation as we try to gain more understanding of its usefulness, particularly in terms of interpreting the nature of a distribution.

Suppose we add 100 to each of the numbers in data set 1 (Table 4-3-1). Has the variability changed? Obviously not, since the spacings among the numbers are still the same, and so our measure, the standard deviation, should not and does not change. More formally, we have the following theorem.

Theorem 4-3-1

Addition of the same constant to each observation in a set increases the mean by that constant but does not change the standard deviation of the observations.

To show this, suppose we have a set of observations $y_1, \ldots, y_n$ with mean $\bar{y}$ and variance s_y^2. Let $y_i' = y_i + c$, where c is a constant, for all i. (Read y′ as "y-prime.") Then

$$\begin{aligned}\sum y_i' &= y_1 + c + y_2 + c + \cdots + y_n + c \\ &= \sum y_i + nc\end{aligned}$$

Hence,

$$\bar{y}' = \bar{y} + c$$

In other words, if all our observations are increased by a constant c, the mean is also increased by c.

Next we use Equation 4-3-1 and compute the variance of the y_i''s.

$$s_{y'}^2 = \frac{\sum (y_i' - \bar{y}')^2}{n-1}$$
$$= \frac{\sum [(y_i + c) - (\bar{y} + c)]^2}{n-1}$$
$$= \frac{\sum (y_i - \bar{y})^2}{n-1}$$
$$= s_y^2$$

Consequently, $s_{y'} = s_y$, and the standard deviation has not been changed.

Now suppose we multiply each observation in data set 1 by 10. The spacings among the numbers have changed; our measure of variation should also change. The range is now from 80 to 120, just 10 times the original value. Conveniently, the same effect should be shown by the standard deviation, and this is the case.

Theorem 4-3-2

Multiplication of each observation in a set by the same constant k gives a new set of observations with mean and standard deviation k times the corresponding values for the original set.

To show this, suppose we have observations $y_1, \cdots, y_n$ with mean $\bar{y}$ and variance s_y^2. Let $y_i' = ky_i$. Then

$$\sum y_i' = ky_1 + \cdots + ky_n$$
$$= k(y_1 + \cdots + y_n)$$

and

$$\bar{y}, = k\bar{y}$$

Now we use Equation 4-3-1 to compute $s_{y'}^2$.

$$s_{y'}^2 = \frac{\sum (y_i' - \bar{y}')^2}{n-1}$$
$$= \frac{\sum (ky_i - k\bar{y})^2}{n-1}$$
$$= \frac{\sum k[(y_i - \bar{y})]^2}{n-1}$$
$$= \frac{k^2 \sum (y_i - \bar{y})^2}{n-1}$$
$$= k^2 s_y^2$$

And, finally

$$s_{y'} = ks_y$$

We often change the location and spread of a set of observations by quoting percentages. The preceding theorems show that we may change or *transform* observations so that they have any desired mean and standard deviation. Thus if we give an exam and the mean grade is 59 and the standard deviation is 13, but feel we would like the grades to have a mean of 70 and a standard deviation of 10, the change is easily made. For example, we might have several sections of a particular course with the students assigned at random to sections so that no real differences in average performance or variation are to be expected. Now we might consider that observed differences are attributable to the examining and grading procedures of the different instructors. Under these circumstances, we could consider it desirable to use a scheme that gave all sets of grades the same mean and standard deviation.

Suppose we have a set of observations $y_1, \ldots, y_n$, with mean $\bar{y}$ and standard deviation s_y. The first step in transforming them to a set with prescribed mean and standard deviation is to standardize them.

A *standardized observation* is one from a set with zero mean and unit standard deviation, that is, a standard deviation of 1.

To standardize observations, we first subtract the mean $\bar{y}$ from each of the original values. That the mean of these deviations is 0 follows from Theorem 4-3-1 using $c = -\bar{y}$. We next divide each deviation by s_y so that these final values, say y''s, have a standard deviation of 1. This follows from Theorem 4-3-2, with $k = 1/s_y$. Equation 4-3-5 describes the transformation algebraically.

$$y_i' = \frac{y_i - \bar{y}}{s_y} \tag{4-3-5}$$

Note that since standardized observations have unit standard deviation, they are no longer dependent on the original unit of measurement. In fact, no matter what the original unit of measurement, we arrive at the same set of standardized values.

It is now a simple matter to change these standardized observations to observations with any desired mean and standard deviation. For example, *standardized scores*, sometimes called *z-scores*, have mean 500 and standard deviation 100. For the new values to have the right standard deviation, we need only multiply y_i' by 100; next we add 500 and they have the correct mean. Equation 4-3-6 described this transformation.

$$\begin{aligned} z_i &= 100y_i' + 500 \\ &= \frac{y_i - \bar{y}}{s_y} 100 + 500 \end{aligned} \tag{4-3-6}$$

More generally, if we wish to transform y's to new values, say x's, with mean $\bar{x}$ and standard deviation $s_{\bar{x}}$, Equation 4-3-7 is appropriate.

$$x_i = \frac{y_i - \bar{y}}{s_y} s_x + \bar{x} \tag{4-3-7}$$

For example, if we wished the observations 8, 8, 9, 10, 11, 12, and 12 with mean 10 and standard deviation $\sqrt{3}$ to be transformed so as to have mean 70 and standard deviation 10, the appropriate transformation would be

$$x_i = \frac{y_i - 10}{\sqrt{3}} 10 + 70$$

The transformed values would be 58.5, 58.5, 64.2, 70.0, 75.8, 81.5, and 81.5.

Perhaps a better idea of the usefulness of the standard deviation for describing the nature of a distribution can be seen from the use of *Chebyshev's theorem.* Chebyshev (1821–1894) was a Russian mathematician whose very general theorem is as follows.

Chebyshev's Theorem

For any set of data and positive number k, the proportion of the number of observations lying within k standard deviations of the mean is certain to be at least $1 - 1/k^2$.

For $k = 1$, $1 - 1/k^2 = 0$ and the theorem is not very informative. For $k = 2$, $1 - 1/k^2 = \frac{3}{4}$ so that at least $\frac{3}{4}$ or 75 percent of any set of observations must lie within 2 standard deviations of the mean, a span of 4 standard deviations. For $k = 3$, $1 - 1/k^2 = 1 - \frac{1}{9}$, so that at least $\frac{8}{9}$ or 89 percent of any set of observations must lie within 3 standard deviations of the mean, a span of 6 standard deviations.

For the preceding transformed data $\bar{y} \pm 2s_y = 70 \pm 2 \times 10 = (50, 90)$. According to the theorem, at least 75 percent of the data should lie in this interval, whereas 100 percent are actually so observed.

For the set of grades given originally in Table 2-3-3, the mean is 78.45 and the standard deviation is 10.86. Let us compute intervals of 2 and 3 standard deviations about the mean. These are $78.45 \pm 2(10.86) = (56.73, 100.17)$ and $78.45 \pm 3(10.86) = (45.37, 111.13)$. Of course, upper limits above 100 have no real meaning, since 100 is the top grade. By counting, we find 118 observations, 94 percent of them, in the interval 56.73 to 100 where we have been guaranteed to find at least 75 percent; and all observations, 100 percent of them, are in the interval 45.37 to 100 where we have been assured that there would be at least 89 percent.

Many naturally occurring distributions follow approximately the so-called normal distribution. In other words, measurements made on many naturally occurring characteristics such as heights, weights, test scores, etc., often have a nearly symmetrical, bell-shaped distribution reasonably well described by the normal distribution discussed in Chapter 10. If we are reasonably confident that a particular set of observations follows this specific distribution, then we can make stronger statements about percentages of observations in intervals. A stronger statement would specify a smaller percentage than that given by the very general Chebyshev theorem which applies to any and all sets of observations. About 66 percent of the observations from a normal distribution lie within 1 standard deviation measured from the mean, about 95 percent lie within 2, and about 99 percent within 3 standard deviations of the mean. Notice that these are statements about the normal distribution, that is, about a population, and not about a sample. Consequently, they apply only approximately to samples and are best when associated with large ones.

For the 126 grades, 95 grades or 75 percent, 118 or 94 percent, and 100 percent lie in intervals of 1, 2, and 3 standard deviations, respectively, on either side of the mean. The corresponding population percentages are 67, 95, and 99 percent. In this case, we might suspect that the true distribution was somewhat more *peaked* than the normal, since 75 percent rather than 67 percent lie within 1 standard deviation of the mean. We must, of course, allow some disparity due to random sampling. On the other hand, we have also observed that the distribution is not symmetrical but somewhat skewed to the left as might be expected where there is a sample mean of 78.45 and a possible range of 0 to 100.

It should now be apparent that the standard deviation can be a meaningful descriptive number when one tries to summarize the information in a set of numerical values. It gives an idea as to how observations are spread about the mean, and whether the distribution tends to be heaped up or flat, particularly when related to the familiar normal distribution.

If we do not wish to talk about a sample at hand but abstractly about a future random sample, then we define the variance by Equation 4-3-8, using capital Y's.

$$s^2 = \frac{\sum (Y_i - \bar{Y})^2}{n - 1} \tag{4-3-8}$$

The distinction is that Equation 4-3-1 provides us with a constant for a particular sample, whereas if we wish to talk about the sample variance as still another random variable, we write Equation 4-3-8.

To define a population variance and standard deviation, first consider a finite population of N observations $y_1, y_2, \ldots, y_N$. The number of spots on the upturned face of a die is such a population, consisting of $y_1 = 1$, $y_2 = 2, \ldots,$ $y_6 = 6$. The last subscript tells us that $N = 6$. Finite populations may be sampled with replacement or without replacement.

A sample of n observations is obtained by rolling a die n times with all six values being available at each roll. This is *sampling with replacement* because a value once obtained is returned to the population and made available again for the next sampling. The sample may be of any size and the sampled population becomes essentially infinite even though it consists of only a finite number of distinct values.

When sampling with replacement is involved, we define the population variance and the standard deviation by Equations 4-3-9 and 4-3-10, respectively.

$$\sigma^2 = \frac{\sum (y_i - \mu)^2}{N} \tag{4-3-9}$$

$$\sigma = \sqrt{\sigma^2} \tag{4-3-10}$$

The symbols σ and σ^2, read as sigma and sigma square(d), use the lowercase Greek letter sigma. They represent parameters, that is, population constants.

In the case of the tossed die, the population variance and standard deviation may be computed as follows:

$$\sigma^2 = \frac{(1 - 3.5)^2 + (2 - 3.5)^2 + \cdots + (6 - 3.5)^2}{6} = \frac{35}{12}$$

$$\sigma = \frac{35}{\sqrt{12}}$$

For *sampling without replacement*, an observation once drawn is not replaced to be sampled again. Thus, if we are sampling markets to measure the number of pounds of apples sold in a specified time period, we want any particular market to be in the sample only once. For this sampling procedure, we generally use the symbols S^2 and S for the population variance and standard deviation and define them by Equations 4-3-11 and 4-3-12.

$$S^2 = \frac{\sum (y_i - \mu)^2}{N - 1} \tag{4-3-11}$$

$$S = \sqrt{S^2} \tag{4-3-12}$$

The use of two definitions for the population variance is not as arbitrary as it seems. This can be demonstrated by tossing a die to obtain a sample of size $n = 2$. If you compute all possible sample s^2's together with the frequencies with which they occur and find their average, it will be $\sigma^2 = \frac{35}{12}$. This is the case of sampling with replacement, and we are led to state that s^2 is a "good" estimate of σ^2. However, if you write the numbers $1, 2, \cdots, 6$ on six discs, draw two simultaneously so as to be sampling without replacement, and find the average of all possible s^2's, this average is $S^2 = 3\frac{1}{2}$. Now we are led to say that s^2 is a "good" estimate of S^2. Thus, each definition of population variance seems appropriate when related to the type of random sampling being used, that is, according to whether it is with replacement or without replacement. In other words, we have used only one definition for s^2 but whether it is a good estimate of σ^2 or of S^2 depends on how the random sampling was done.

To define the variance of a population with a continuous variable so that the computation can be performed, mathematical techniques other than simple arithmetic are needed.

EXERCISES

4-3-1 Determine the variance and standard deviation for the grades of students as given in Exercise 3-6-1. Compute the intervals $\bar{y} \pm s$, $\bar{y} \pm 2s$, and $\bar{y} \pm 3s$. Count the numbers of observations in each interval and compute corresponding proportions or percentages. Compare these values with those given (a) by Chebyshev's theorem for any set of data and (b) for a normal population.

Convert these grades to scores with mean 70 and standard deviation 10.

4-3-2 Find the variance and standard deviation of the data in Exercise 3-6-2. Use both the definition and computing formulas but subtract 4 from each observation since all lie between 4 and 5. The results should differ only because of rounding errors, if any. (You may also ignore the decimal point, that is, multiply each observation by 100, while doing the computing and then divide the standard deviation by 100 to get a value in millimeters. This part would be an application of Theorem 4-3-2.)

4-3-3 Compute the variance and standard deviation for the heart-rate data of Exercise 3-6-3, and/or any of the sets of data given in Exercises 3-6-4 (two sets) to 3-6-7 (three sets).

4-3-4 The data in Table 2-3-2 are in the form of a frequency distribution. Write definition and computing formulas for the variance and standard deviation when data is so presented. (See Equation 3-4-4 for some guidance.) To compute s^2, what must be the divisor of SS? Use your formulas to compute the variance and standard deviation of these data.

4-3-5 One may compute the mean (Equation 3-4-4), variance, and standard deviation for grouped data such as in Table 2-3-4 by assigning the class midpoint as the value of all observations in that interval. The formulas developed in Exercise 4-3-4 would be used. Compute the variance and standard deviation for the data in Table 2-3-4 and compare these values with those computed in this section for the same data but without the use of grouping.

4-3-6 Compute the variance and standard deviation for the frequency distribution prepared (a) in Exercise 3-6-12 for the liver-weight data of Exercise 2-3-4 and/or (b) in Exercise 3-6-13 for the baseball data of Exercise 2-3-5 and/or (c) in Exercise 3-6-14 for the meteorological data of Exercise 2-3-6 and/or (d) in Exercise 3-6-15 for the bridge-hand data of Exercise 2-3-7 and/or (e) in Exercise 3-6-16 for the concrete-strength data of Exercise 2-3-8.

4-3-7 Compute the intervals $\bar{y} \pm s$, $\bar{y} \pm 2s$, and $\bar{y} \pm 3s$ from the results of your computations in Exercise 4-3-6. Refer now to the original data and, for each set of data, find the number of observations in each of the appropriate intervals. Convert these to percentages and compare with corresponding values given (a) by Chebyshev's theorem for any set of data and (b) for the normal distribution.

4-3-8 Show algebraically that the correction term $(\sum y_i)^2/n$ may also be written in the less desirable form of $n\bar{y}^2$.

4-3-9 Show that the variance s^2 may also be computed by the formula

$$s^2 = \frac{n \sum y_i^2 - (\sum y_i)^2}{n(n - 1)}$$

The exercises that follow are more particularly for those seeking, at this time, a bit more insight into the distinction between populations and samples, and between sampling with and without replacement.

4-3-10 Suppose we toss a die twice, recording the number of spots on the upturned face for both the first die and the second. How many so-ordered pairs of values are possible? What are they? Compute s^2 for each pair.

You now have a population of s^2's generated by a sampling-with-replacement scheme. In answering the following question, you may use the s^2's as computed or put them in the form of a frequency distribution.

Find the average of the s^2's. This is the mean of a population of s^2's. An appropriate symbol would be μ_{s^2}. Compare the result with σ^2 for the population: $y_1 = 1$, $y_2 = 2, \ldots, y_6 = 6$ when sampling is with replacement.

4-3-11 Six chips are available with the numbers 1, 2, . . . , 6 marked on them. Write out all possible samples for $n = 2$ values drawn simultaneously. Keep an accounting as in Exercise 4-3-10, but remember that the first chip is not available for a second drawing in the same sample. Compute all s^2's and find their average. Compare with S^2 for sampling without replacement. You have used the same population as in Exercise 4-3-10, but this time the sampling scheme has been without replacement.

4-4 THE RANGE

If a set of numbers is designated by $y_1, y_2, \ldots, y_n$, then the same set ordered by increasing magnitude, that is, their array, is written as $y_{(1)}, y_{(2)}, \ldots, y_{(n)}$. The spread of a set of observations may be defined conveniently, but roughly, by the range. This uses only two of them.

The *range* of a set of observations is defined by Equation 4-4-1.

$$\text{Range} = y_{(n)} - y_{(1)} \tag{4-4-1}$$

If we are thinking of random samples to be obtained and, consequently, of random variables rather than a sample at hand, then the $Y_{(i)}$'s are called *order statistics* and the *range*, as a random variable, is defined by Equation 4-4-2.

$$\text{Range} = Y_{(n)} - Y_{(1)} \tag{4-4-2}$$

The range is, then, the difference between the maximum and minimum observations.

Data sets 1, 2, and 3 in Section 4-2 have ranges $y_{(7)} - y_{(1)}$ of $12 - 8 = 4$, $15 - 5 = 10$, and $19 - 1 = 18$, respectively. Data sets 4, 5, and 6 have the same ranges as the first three but exhibit different variability. Thus, we see that the range utilizes only a portion of the sample information, that in the maximum and minimum values, and none of the information concerned with variation among the observations within this interval. For the same reason, it is an unstable measure of variation except when the sample is very small. This means that in repeated sampling from the same population and for samples of the same size, the range will show more variability than many other possible measures of spread.

The range is dependent on the number of observations in the sample. Samples of large size are more likely to include the less frequent observations from a population, usually the smaller and the larger ones. Consequently, one expects the range to increase with increasing sample size. This makes it undesirable as a measure for comparing variation in samples of differing size; here the standard deviation is more suitable. However, tables of factors are available which, when multiplied by the range, yield values comparable to the standard deviation; these factors are appropriate when samples are from the normal distribution.

The range is a measure of spread that is easy to understand and to compute. It is not generally the most desirable measure of variation but is used for very small samples where it may not vary from sample to sample a great deal more than does the standard deviation. Statistical quality control where the sample size is often five or fewer makes use of it for this reason as well as for computational convenience.

EXERCISES

4-4-1 Observe the range for the sets of data given in Exercises 3-6-1 to 3-6-4 (two sets), and 3-6-5 to 3-6-7 (three sets). Compare these ranges with the computed standard deviations by preparing a table giving sample sizes arranged from smallest to largest, with corresponding standard deviations, ranges, and ratios of range to standard deviation. Does it seem as though there might be a relationship between sample size and range or sample size and range/standard deviation?

The exercises that follow involve algebraic manipulations. They are not necessary to an understanding of the text but may be of interest to the algebraically inclined student.

4-4-2 Show algebraically that the range is not changed by the addition of a constant to each observation.

4-4-3 Show algebraically that when each observation in a set is multiplied by the same constant k the range of the new values is k times the original range.

4-4-4 Show that for $n = 2$, $s^2 = \sum_1^2 (y_i - \bar{y})/1 = (y_1 - y_2)^2/2$. This shows that there is a one-to-one relationship between the variance (or standard deviation) and the range and that, consequently, one is no more variable than the other in repeated sampling for $n = 2$.

4-5 OTHER MEASURES OF VARIATION

In Section 4-3 we observed that the sum of the deviations from the mean is 0, making it useless as a measure of variation. However, squaring the deviations eliminated the problem and led to the variance and standard deviation.

An alternative to squaring is to ignore the signs of the deviations so that all are considered to be positive regardless of whether y_i or $\bar{y}$ is larger. The instruction to ignore the sign is indicated by two vertical bars about the deviation; write $|y_i - \bar{y}|$. This suggests the definition of the average or mean deviation or mean absolute deviation given in Equation 4-5-1.

$$\text{Mean (absolute) deviation} = \frac{1}{n}\sum |y_i - \bar{y}| \qquad (4\text{-}5\text{-}1)$$

To make this value comparable to s when the observations are from a normal distribution, it must be multiplied by $\sqrt{\pi/2} \approx 1.253$, where $\approx$ means "approximately equals."

This measure of spread is easily associated with the real world but has not been much used. Recently, there has been more interest in its possible practical applications. Values from different samples cannot be pooled without reverting to the original data.

The range is one of many similar quantities which are used as measures of spread. We have already observed that data sets 1 and 4, 2 and 5, 3 and 6 have the same ranges, by pairs, though the observations in the first set of each pair clearly exhibit more spread. An obvious alternative would be to use $y_{(n-1)} - y_{(2)}$, here $y_{(6)} - y_{(2)}$, moving in one observation from each end. Now the two samples in each pair yield different values. One might go on to consider the general problem of finding the "best two" observations for measuring variability; for example, see Dixon and Massey (1969, section 9-4).

For larger samples, the interquartile range is sometimes used as a measure of dispersion. Quartiles were defined in Section 3-7. The interquartile range is given by Equation 4-5-2.

Interquartile range $= P_{75} - P_{25} = Q_3 - Q_1$ (4-5-2)

Half this value, $(Q_3 - Q_1)/2$, called the *semi-interquartile range* or *quartile deviation*, is also used as a measure of dispersion. These measures are less affected by extreme values than is the range, and so they are more stable in repeated sampling.

More generally, percentiles may be used to measure dispersion. Thus, if P_{10} is the value such that 10 percent of the observations are smaller, and similarly for other subscripts, than the 10 to 90 percentile range is defined by Equation 4-5-3.

10–90 percentile range $= P_{90} - P_{10}$ (4-5-3)

The semi 10 to 90 percentile range is half this value, $(P_{90} - P_{10})/2$.

Percentile statistics, including the interquartile range, are primarily descriptive measures. With an appropriate multiplier, they become comparable to the standard deviation when the observations are from the normal distribution.

Statistics constructed from order statistics are not too easy to handle algebraically, but much mathematical theory using them has been developed; for example, more than two order statistics can be used to measure variability, and they can be used to measure location.

There are several measures of variation that are expressed in *relative units*. These may be used when all observations are naturally greater than zero. Most important is the coefficient of variation, or coefficient of variability (CV) defined by Equation 4-5-4.

$$\textit{Coefficient of variation} = \frac{100s}{\bar{y}} \quad \text{percent} \tag{4-5-4}$$

This is the sample standard deviation expressed as a percentage of the sample mean. For the grade data of Table 2-3-3, we have $\bar{y} = 78.45$ and $s = 10.86$. Hence,

$$\text{CV} = \frac{10.86}{78.45} 100 = 13.8 \text{ percent}$$

This coefficient is often used as a measure of precision. An experimenter usually has a good idea of the relative variability to be expected in a given type of experiment and thus can compare the results of different experiments in the same research area. Also, the coefficient of variability enables one to compare the variability of different sets of data based on different units of measurement, say pounds versus grams. The relative variabilities of sets of measurements which differ greatly in their means can be compared by the coefficient of variability. Thus, plant breeders are concerned, for various reasons, with some lines of corn that produce much corn and others that produce but little. High-yielding lines tend to show more variation than low-yielding lines. A comparison of relative variabilities is often valuable to the breeders.

A second measure of relative variability is the *coefficient of quartile variation* V_q, defined by Equation 4-5-5.

$$V_q = \frac{P_{75} - P_{25}}{P_{75} + P_{25}} 100 = \frac{Q_3 - Q_1}{Q_3 + Q_1} 100 \tag{4-5-5}$$

For the grade data of Table 2-3-3, using the relevant computations of Section 3-7, we have

$$V_q = \frac{86.15 - 73.53}{86.15 + 73.53} 100 = 7.9 \text{ percent}$$

This is a convenient substitute for the coefficient of variation, especially when there are many observations. For small numbers of observations, we might prefer the ratio of the midrange to the range, Equation 4-5-6.

$$\text{Midrange over range} = \frac{\frac{1}{2}[Y_{(n)} + Y_{(1)}]}{Y_{(n)} - Y_{(1)}} \tag{4-5-6}$$

EXERCISES

4-5-1 Determine the interquartile range, the semi-interquartile range, the coefficient of quartile variation, and the midrange/range for the data of Exercises 2-3-4 to 2-3-8, using the relevant computations of Exercises 3-7-1 to 3-7-5, respectively.

4-5-2 For the data of Exercises 3-7-4 and 3-7-5, using the computations completed there, determine the 10 to 90 percentile and semipercentile ranges.

4-5-3 Compute the coefficient of variation and the midrange/range for the relatively small samples of Exercises 3-6-1 to 3-6-7.

Chapter 5
PROBABILITY

5-1 INTRODUCTION

Statisticians are certainly interested in the compilation and summarization of data. However, they are especially concerned with sampling and inferential processes.

Reasoning from sample data to appropriately defined collections of which the data are a part, that is to populations, is inductive inference. The resulting inferences have an associated element of uncertainty or chance that goes back to the sampling procedure. In addition, the validity of any inductive inference can be neither proved nor disproved in the deductive manner of mathematical theorems.

If the element of chance in the sampling process is subject to recognized laws of probability, then probability theory can help with the inferences. In particular, under various hypothesized situations we can compute probabilities for sets of events of interest to us.

In this chapter, we consider some elementary probability calculations of use in going beyond descriptive to inferential statistics.

5-2 ELEMENTARY PROBABILITY

We all have an acquaintance with certain probability notions. Thus, we use expressions like "probable" or "unlikely" to discuss some aspect of tomorrow's weather; or we talk about the odds associated with the outcome of an approaching election or sporting event. An important feature of such expressions is that they apply to future occurrences or events. More particularly, we feel that we recognize a situation, know the possible outcomes that may follow, and have some feeling as to which outcomes were most frequent and which least frequent when the situation arose in the past. This feeling is reflected in the adjective or adverb we choose to qualify our expression.

Sometimes we replace historical frequency by what we regard as logic. Thus, suppose we toss a newly minted coin. We argue that there are only two possible outcomes and that they are equally likely when chance governs the outcome. Most of us are prepared to say that the probability that the coin will fall heads is .5; we give the same value for the probability that the coin will fall tails. We may write $P(\mathrm{H}) = .5 = P(\mathrm{T})$. We are saying we believe that, in the long run, or over very many such tosses or trials, the number of heads observed will not differ appreciably from the number of tails; about half the trials will result in heads, half in tails.

If our coin has two heads, then we say that the probability it will fall heads is certainty; we write $P(\mathrm{H}) = 1$. That the coin will fall tails is impossible; we write $P(\mathrm{T}) = 0$. The use of probability values of 0 and 1 in these situations is a matter of convention.

Many real situations involve more than two outcomes. Thus, we can toss a die and observe 1, 2, 3, 4, 5, or 6 spots on the upturned face; or we can toss 10 coins and observe 0 heads and 10 tails, 1 head and 9 tails, . . . , or 10 heads and 0 tails, recording simply 0, 1, 2, . . . , or 10 heads as the outcome. These outcomes are distinct in that when one occurs, no other one can.

A set of possible outcomes is said to consist of *mutually exclusive* outcomes if, when one occurs in a trial, no other one can at that trial.

A set of outcomes is said to be *exhaustive* when no other possibilities exist.

When a trial has n possible, equally likely, and mutually exclusive outcomes, and if n_1 of these are considered favorable, that is, falling in a certain category, then the probability of a favorable outcome is n_1/n. In other words, we compute probabilities as in Equation 5-2-1.

$$P(\text{favorable outcome}) = \frac{\text{number of favorable outcomes}}{\text{number of possible outcomes}} \qquad (5\text{-}2\text{-}1)$$

The definition is considered to be the classical approach to probability computations. It is a reasoned definition based on a deductive process. The probabilities are available prior to the conduct of an experiment, and so are called *a priori* probabilities. The definition has limitations, as when logic does not provide for equally likely outcomes or when there are an infinite number of possible outcomes. The following situation provides an instance where a reasoned set of probabilities cannot be assumed.

Suppose we have a bent, damaged, or unevenly weighted coin and are uncertain about the probability that it will fall heads. Thus we do not feel free to argue that the two possible outcomes are equally likely; neither probability appears to be $\frac{1}{2}$. We must *estimate* the probabilities by an experiment that uses the coin.

To estimate the probability of our damaged coin falling heads, we describe an experiment in which we toss a coin, making sure it spins rapidly before falling.

The principal requirement of any *random experiment* is that we can visualize it as being repeated time and time again under identical circumstances, and that it will result in one of a number of possible outcomes, but we cannot, in advance, predict the outcome with certainty. This is where chance enters.

For our bent-coin experiment, we then observe and record the result or outcome, namely, the side of the coin which appears uppermost. Perhaps we conduct this experiment 100 times, recording 43 heads and 57 tails. We now estimate $P(\mathrm{H})$ to be exactly .43 and $P(\mathrm{T})$ to be exactly .57, the two observed relative frequencies. At the same time, we know that if we try another 100 tosses, it is unlikely that we will get 43 heads and 57 tails again.

Let us toss the damaged coin another 100 times and reestimate $P(\mathrm{H})$ by totaling the number of heads in the two sets of tosses and dividing by 200. We still have only an estimate of the true $P(\mathrm{H})$. Nevertheless, we are convinced that the longer the experiment continues, the more reliable is our relative frequency as an estimate. It becomes more and more stable and, if we could continue the experiment indefinitely, our estimate of $P(\mathrm{H})$ would converge to some "true" value, presumably not .5, whose existence we assume and will call the probability of a head and write as $P(\mathrm{H})$.

In summary, if we toss any coin once and take the relative-frequency estimate of $P(\mathrm{H})$, we will necessarily get either 0 or 1. If we toss it 10 times, our estimate will almost always be other than 0 or 1. In a long run of tosses, the relative frequency will come closer and closer to the true value. If you conduct such an experiment, you are led to conclude that *convergence* of the relative frequency is a reality when the experiment is such that chance operates. Thus, the notion of probability as the *limit* of a relative frequency covers many situations that do not lend themselves to the classical approach.

We will keep in mind both approaches to probability as we develop the topic.

Recently a considerable amount has been said and written by a new school of thought on *personal* or *subjective* probabilities. This school interprets a probability as a measure of one's personal belief concerning a future outcome. Such probabilities would seem most appropriate to single, nonrepeatable events where there is little direct evidence to let us compute or otherwise estimate a probability. Such would be the case if we wanted a probability for the success of a new television program in a particular time slot.

5-3 RANDOM EXPERIMENTS AND SAMPLE SPACES

Statisticians use probability theory to approach practical problems. For example, at one time in the previous section we assumed a coin with $P(\mathrm{H}) = .5$. If we further assume that a coin has no memory, then one toss is no different from another. These minimum assumptions provide an idealized representation or *mathematical model* of how a tossed coin behaves. Using them, we can compute the probability that four tosses will result in exactly 0, 1, 2, 3, or 4 heads. The practical problem could be one in gambling. Alternatively, we might decide that the idealized representation was appropriate to the study of available data on the sex distribution in four-children families. The problem here is concerned with the inheritance of sex within families.

It is clear that we require a formal, manageable probability theory. We begin this with a discussion of random experiments and sample spaces.

For a random experiment, it is necessary to know how it is to be conducted and what is to be observed. We shall temporarily limit ourselves to random experiments with a finite number of outcomes. Thus, in advance of the experiment, we must be able to provide an exhaustive list of mutually exclusive outcomes.

For example, our random experiment may be a single toss of a coin. The coin is to be tossed spinning at least 1 foot into the air and is to fall on a smooth, hard, level surface. The observation is to be made on the nature of the visible face of the coin when it comes to rest. The possible outcomes are a head and a tail. If a head shows, then a tail cannot and the outcomes are mutually exclusive. There are only two faces and we do not allow the coin to remain on edge, if, indeed, such is possible, so our list or *set* is exhaustive. We shall use H and T to symbolize the outcomes head and tail.

If the random experiment calls for tossing a penny and a quarter, the list or set of outcomes can be presented symbolically as HH, HT, TH, TT, where the first position refers to the penny, the second to the quarter. Again we have a mutually exclusive and exhaustive set of outcomes.

A *set* is simply a collection of things.

No further definition is needed. For the present, we will deal with finite sets of outcomes or of symbols used to designate outcomes.

The individuals comprising the set are *elements* or *sample points*.

In our penny-quarter experiment, we have the sample points HH, HT, TH, TT. Collectively, they constitute our set or *sample space*, designated by S. The following representation is often used here.

$$S = \{\text{HH, HT, TH, TT}\}$$

A *subset* consists of some of the elements or sample points. It is also called an *event*.

One subset of the set S consists of those elements that include at least one head. Call this the event A.

$$A = \{\text{HH, HT, TH}\}$$

The symbols $\in$ and $\notin$ are used to designate "is an element of" and "is not an element of"; the stroke / symbolizes negation. Thus, HH $\in S$ and HH $\in A$ while TT $\in S$ but TT $\notin A$.

A *sample space* is then a set of elements or sample points used to list the possible outcomes of an experiment. For each outcome, there is exactly one element or sample point.

For the two tossed coins, we may write $S = \{\text{HH, HT, TH, TT}\}$ for the sample space or may display it in any convenient representation; for example, see Figure 5-3-1.

In the penny-quarter experiment, it is possible to visualize more than one sort of outcome or element and, consequently, more than one sample space. For example, we might consider an outcome to be simply the number of heads so that there are only three outcomes, namely, 2 heads, 1 head, or 0 heads. Here, a

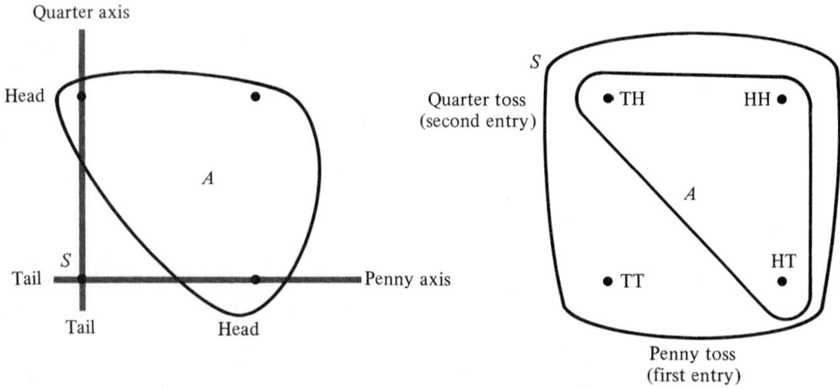

FIGURE 5-3-1 Sample Points and Sample Space for a Two-Coin Experiment.

sample space is $S_1 = \{(2, 0), (1, 1), (0, 2)\}$, the first entry being the number of heads and the second the number of tails, or, simply, $S_H = \{2, 1, 0\}$. Again, we have in S_1 or S_H a set of elements such that each outcome corresponds to exactly one sample point. Thus, a definition which specifies what an outcome is, for example, number of heads, determines what events are to be sample points in a sample space.

It is often convenient to begin with a sample space for which an outcome is an *indecomposable* result; that is, we can conceive of no way to decompose it into more elementary outcomes. The ordered pair in the penny-quarter experiment, where the sample space is S, is an indecomposable outcome. Any other conceivable outcome will be an event consisting of a subset of sample points from such a sample space.

EXERCISES

5-3-1 A random experiment consists of tossing a single die and observing the upper face. What are the possible outcomes of the experiment? Write a set of elements such that each outcome corresponds to exactly one element. What name is applied to this set of elements? What sample points make up the event consisting of an odd number on the upper face? The event consisting of any number divisible by 2? By 3? Can two different events have a point or points in common? Can they have no points in common?

5-3-2 A tetrahedron is like a three-sided pyramid, so, including its base, it has four sides. Number these 1, 2, 3, and 4. A random experiment consists of rolling a tetrahedron twice and recording the ordered pair of numbers hidden on the two rolls. How many possible outcomes are there? Exhibit a sample space for the experiment. Do the sample points have equal probabilities of occurring? What sample points are in the event A described as consisting of two even numbers? The sample point (3, 3) has an even sum. Is $(3, 3) \in A$?

5-3-3 In two tosses of a tetrahedron, we may be interested in the outcome as measured by the sum of the numbers on the two hidden faces. Exhibit a sample space for the experiment. How many elements are in your set? Are all sample points equally likely? Is an odd sum as likely as an even sum? What sample points from Exercise 5-3-2 go into each sample point in this exercise?

5-3-4 A housewife selects three packages of dry yeast from a shelf. A package may be defective or nondefective, as she will learn from experience. Exhibit a sample space S of indecomposable outcomes. How many sample points does it contain? If the outcome of the experiment is simply the number of defective packages, what is an appropriate sample space S_1? How many sample points from S go into each of the events constituting an outcome in S_1?

5-3-5 A study is to be made of families with exactly four living children. The purpose of the study is concerned with the distribution of the sexes within families. Exhibit an appropriate sample space of indecomposable results. How many sample points are required for the list of outcomes to be exhaustive? How many sample points go into the event "at least one boy"? How many sample points go into the event "more than two girls"? Are the two just-described events mutually exclusive? How many sample points are in the event "last child was a girl"?

5-4 PROBABILITIES

To each sample point of a finite sample space, we next assign a *probability*. This will be a real number or weight, subject to certain conditions. The assignment is by means of a set of relations or a *measure* or *probability function*. For example, we write

$$P(\mathrm{H}) = 1/2 \qquad P(\mathrm{T}) = 1/2$$

as the probability function for a fair coin. The elements H and T have been assigned weights or probabilities of 1/2.

For a fair die, the elements 1, 2, . . . , 6 are each assigned a probability of 1/6, and we have

$$P(1) = 1/6,\ P(2) = 1/6, \ldots, P(6) = 1/6$$

The probability function pairs exactly one weight or probability with each point of a sample space. We then have a *function* defined on the set constituting our sample space, the assigned values being those of the function.

The conditions to which our probabilities are subject are called *postulates* or *axioms*. If E_i is a sample point in S, if the set of E_i's is exhaustive, and if we denote the probability of E_i by $P(E_i)$, then we write the two postulates:

1 $P(E_i) \geq 0$
That is, a probability must be *nonnegative*.
2 $\sum P(E_i) = 1$
That is, the sum of the probabilities of all the sample points in S is unity.

We now define the probability of an event.

The *probability P of an event A*, a subset of sample points, is the probability that the outcome of a random experiment will be some element of the event A. This probability is the sum of the probabilities of the sample points in A.

Suppose we toss a die and observe the outcome as the number of spots on the uppermost face. Then

$$S = \{1, 2, 3, 4, 5, 6\}$$

For a fair die, the probability function may be written as

$$P(E_i) = 1/6 \qquad \text{for } E_i \in S$$

In other words, the weight or probability to be associated with each sample point in S is 1/6. Observe that $P(E_i) \geq 0$ and $\sum P(E_i) = 1$, as required by the postulates.

To find the probability that the sample point is an element of the event of an even number, we need

$$P(\text{even}) = P(2) + P(4) + P(6) = 1/6 + 1/6 + 1/6 = 1/2$$

To find the probability that the sample point is an element of the event of being greater than 4, we need

$$P(5) + P(6) = 1/6 + 1/6 = 1/3$$

Let us consider one more example by supposing that a random experiment consists of tossing a red and a blue tetrahedron, each with sides numbered 1, 2, 3, and 4, and observing the numbers on the hidden faces. An outcome or sample point is to be the resulting ordered pair of integers. A sample space is exhibited in Figure 5-4-1.

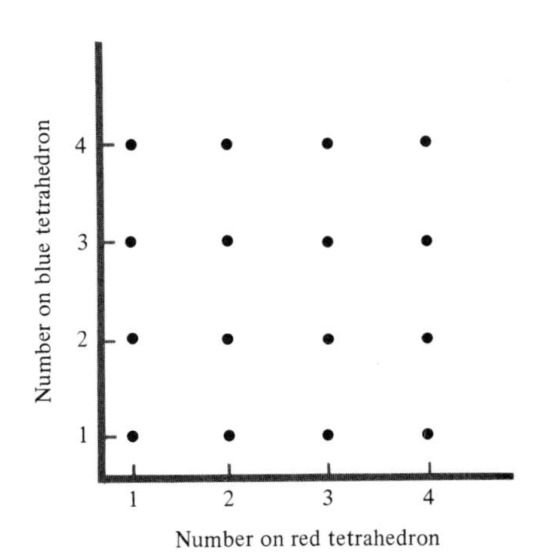

FIGURE 5-4-1 Sample Points and Sample Space for a Two-Tetrahedron Experiment.

A reasonable probability function calls for $P(i, j) = 1/16$, where (i, j) is the ordered pair of integers, an indecomposable result, describing a point in the sample space.

Probabilities for many different events can now be computed. Thus,

$$\begin{aligned} P(i, j = \text{even, even}) &= P(2, 2) + P(2, 4) + P(4, 2) + P(4, 4) = 4(1/16) = 1/4 \\ P(i = j) &= P(1, 1) + P(2, 2) + P(3, 3) + P(4, 4) = 4(1/16) = 1/4 \\ P(i \text{ or } j < 2) &= P(1, 1) + P(1, 2) + P(1, 3) + P(1, 4) + P(2, 1) \\ &\quad + P(3, 1) + P(4, 1) = 7(1/16) = 7/16 \end{aligned}$$

For the same random experiment, the outcome might have been specified as the sum of the integers on the hidden faces of the two tetrahedrons. Notice how easy it is to construct a sample space and probability function from the one already at hand. We find:

$$
\begin{aligned}
P(i + j = 2) &= P(1, 1) = 1/16 \\
P(i + j = 3) &= P(1, 2) + P(2, 1) = 2/16 \\
P(i + j = 4) &= P(1, 3) + P(2, 2) + P(3, 1) = 3/16 \\
P(i + j = 5) &= P(1, 4) + P(2, 3) + P(3, 2) + P(4, 1) = 4/16 \\
P(i + j = 6) &= P(2, 4) + P(3, 3) + P(4, 2) = 3/16 \\
P(i + j = 7) &= P(3, 4) + P(4, 3) = 2/16 \\
P(i + j = 8) &= P(4, 4) = 1/16
\end{aligned}
$$

Thus, the set of elements which lists the possible outcomes, that is the sample space, consists of the integers 2 through 8. This may be written conveniently as 2[1]8, which is read as "two, by steps of one, to eight." The probability function may be written using one formula for $i + j = h$, say, when h = 2[1]5, and a second formula when h = 5[1]8. Both formulas are valid for $h = 5$. They are

$$P(h) = \frac{h - 1}{16} \qquad \mathrm{h} = 2[1]5$$

$$P(h) = \frac{9 - h}{16} \qquad \mathrm{h} = 5[1]8$$

For the event that the sum is at least 6, the probability is computed as

$$P(6) + P(7) + P(8) = 3/16 + 2/16 + 1/16 = 6/16$$

EXERCISES

5-4-1 A coin and a die are tossed. The outcome of this random experiment is recorded as the ordered pair consisting of the letter H or T, according to whether the coin falls a head or a tail, and one of the integers 1[1]6, depending on the number of spots on the upturned face of the die. Exhibit a sample space for the experiment. Assuming the coin and die to be fair, what is the reasonable probability function for the sample space?

Using your own probability function, compute probabilities for the following events:

(a) (H, even number).
(b) (T, number greater than 3).
(c) A tail is observed.
(d) The observed number is greater than 3.

5-4-2 Consider families in which there are three children. Exhibit a sample space in which the outcome is a triple of letters representing the sex, that is, boy or girl, in the order of birth of the children. Assume each point in the sample space has equal probability.

Compute probabilities for each of the following events:

(a) The first child is a boy.
(b) There are two girls.

(c) There is at least one boy.
(d) The middle child is a girl.
(e) The middle child is the only girl.
(f) The oldest and youngest children are boys.
(g) The oldest and youngest children are the same sex.

5-4-3 A poll was taken to obtain information as to whether or not the available voters of a community preferred an elected or appointed school board. The results are tabled as follows:

	Married With:				Single		
	Children		No Children				
Voter Prefers	Man	Woman	Man	Woman	Man	Woman	Total
Elect	11	21	8	18	7	11	76
Appoint	3	4	2	2	2	2	15
No opinion	2	3	8	15	3	8	39
Totals	16	28	18	35	12	21	130

Our sample space consists of the set of 18 points constituting the cells of the table. A probability function assigns to any cell a probability equal to the ratio of the number of voters in that cell divided by the total number of voters, namely, 130. If a voter is to be selected at random from among those polled, what is the probability that this voter will be:

(a) A married man with no children who favors an elected school board
(b) A single woman who holds no opinion as to how the school board should be formed
(c) A single person who prefers an appointed school board
(d) A married woman with no children who holds an opinion as to how the school board should be formed
(e) A married person with no opinion as to how the school board should be formed
(f) A person who prefers an appointed school board
(g) A married man
(h) A single person
(i) A person who holds an opinion as to how a school board should be formed

5-5 UNIONS AND INTERSECTIONS

A sample space S is a set of elements corresponding to the mutually exclusive outcomes in the exhaustive list of outcomes of a random experiment. The probability of an event E, a subset of S, has been defined as the sum of the probabilities assigned to the elements in this subset. However, the elements in a subset E_1 need not be entirely distinct from those in an E_2, so that events are not necessarily mutually exclusive or *disjoint*. Consequently, there is no reason to expect that the addition of probabilities of events will always give meaningful answers. We now pursue the matter of probabilities of events.

Suppose we have a sample space consisting of the names of 155 master's candidates and the single language called for by each plan of study. Thus, one sample point might be (M. G. Rogers, Spanish). The 155 sample points provide the following four events: 35 sample points in E_1, where the student takes French, 30 in E_2 for German, 10 in E_3 for Russian, and 80 in E_4 for Spanish. Clearly these events are disjoint.

By definition, the event described as E_1 *or* E_2 includes the sample points in E_1 and those in E_2. Thus, it includes students in either French *or* German. This event is also called the *union* of E_1 and E_2, written as $E_1 \cup E_2$. Since E_1 and E_2 are disjoint, we have the Equation 5-5-1.

$$P(E_1 \cup E_2) = P(E_1) + P(E_2) \qquad \text{for } E_1 \text{ and } E_2 \text{ disjoint} \qquad (5\text{-}5\text{-}1)$$

More generally, for any number of disjoint events $E_1, \ldots, E_k$, with union $E_1 \cup E_2 \cup \cdots \cup E_k$ or $\bigcup_{i=1}^{k} E_i$, we have Equation 5-5-2.

$$\begin{aligned} P(E_1 \cup \cdots \cup E_k) &= P\Big(\bigcup_{i=1}^{k} E_i\Big) \\ &= P(E_1) + \cdots + P(E_k) \\ &= \sum_{i=1}^{k} P(E_i) \qquad \text{for disjoint } E_i\text{'s} \end{aligned} \qquad (5\text{-}5\text{-}2)$$

Recall the summation notation introduced in Section 3-3.

When E_1 and E_2 are not disjoint, we must be careful to include in the union those points which are also in both E_1 and E_2. At the same time, a count of the points in the union must include these points only once. The equations just given for computing probabilities for unions do not hold for nondisjoint events.

If $E_1, \ldots, E_k$ are disjoint and exhaustive, that is, if $E_1 \cup \cdots \cup E_k = S$, then the k events are called a *partition* of S. In this case, the following is true.

$$P\Big(\bigcup_{i=1}^{k} E_i\Big) = \sum_{i=1}^{k} P(E_i) = P(S) = 1$$

For any event E there is always the complementary event consisting of all the points in the sample space not in E. This *complement* is designated by $\bar{E}$, read "E-bar." Thus $\{E, \bar{E}\}$ is a partition of S and $P(E) + P(\bar{E}) = 1$. Consequently, we have Equations 5-5-3.

$$P(\bar{E}) = 1 - P(E) \qquad \text{and} \qquad P(E) = 1 - P(\bar{E}) \qquad (5\text{-}5\text{-}3)$$

This relationship is often useful in computing probabilities. Thus, in tossing two dice, the probability of the two faces being alike is 6/36. Hence,

$$P(\text{faces differ}) = 1 - 6/36 = 30/36 = 5/6$$

It has not been necessary to look at the 30 sample points in the complement. While this is not difficult for this problem, there are many situations where it is clearly advantageous to compute first the probability of the complement of an event under discussion rather than to compute directly the probability of the event.

It is often helpful to use *Venn diagrams* to visualize unions and complements. First draw a rectangle, Figure 5-5-1, or a freeform figure, to indicate a sample space or a set of points intended for general discussion. This is called the *universal set*. Within this, we draw a circle or other figure to represent an event. See Figures 5-5-2 and 5-5-3.

FIGURE 5-5-1 Venn Diagram Illustrating a Universal Set.

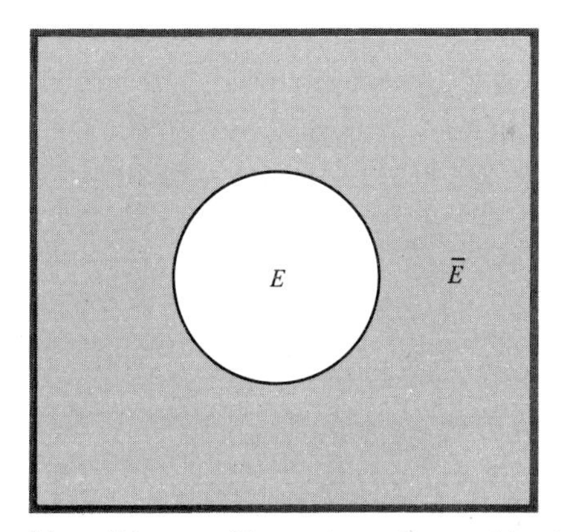

FIGURE 5-5-2 Venn Diagram Illustrating a Set and Its Complement.

Note that Figures 5-5-2 to 5-5-4 illustrate partitions. Also from Figure 5-5-3, observe that $\bar{E}_1$ contains E_2 and E_3, whereas $\overline{E_1 \cup E_2 \cup E_3}$ excludes E_1, E_2, and E_3. It follows then that $\overline{E_1 \cup E_2 \cup E_3} \neq \bar{E}_1 \cup \bar{E}_2 \cup \bar{E}_3$. In words, the complement of the union, say E_4, does not equal the union of the complements in this case. In fact, since $\bar{E}_1$ includes E_2 and E_3, and $\bar{E}_2$ includes E_1 and E_3, it follows that the union of the complements, $\bigcup_{i=1}^{3} \bar{E}_i$, is the universal set in Figure 5-5-3.

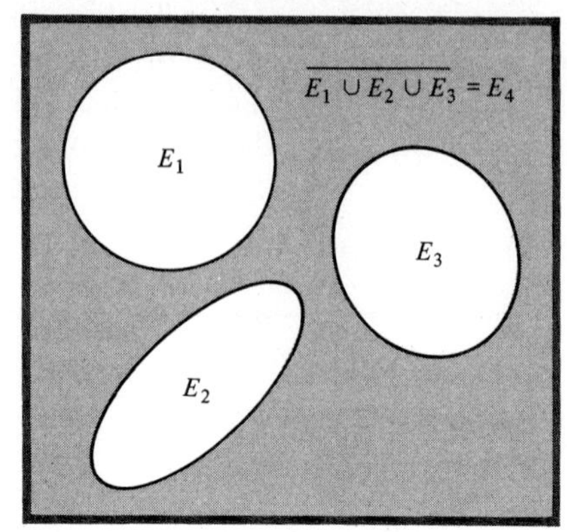

FIGURE 5-5-3 Venn Diagram Illustrating Disjoint Sets and the Complement of Their Union.

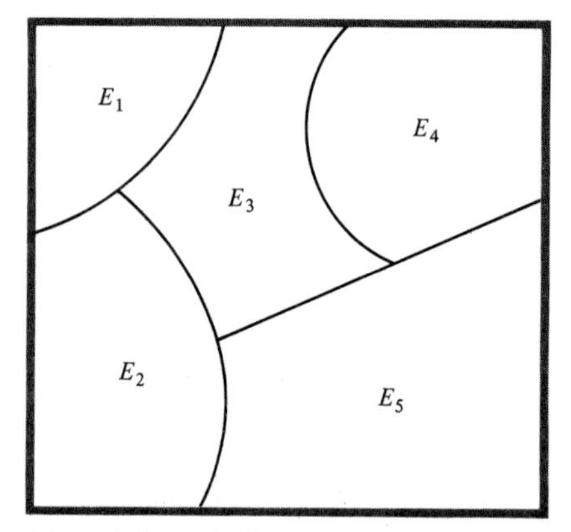

FIGURE 5-5-4 Venn Diagram Illustrating a Partition.

In Figure 5-5-3, the space might be intended to represent the 155 students with E_1 including those taking French, E_2 those in German, E_3 those in Russian, and $\overline{E_1 \cup E_2 \cup E_3} = E_4$, say, for those in Spanish. Observe that those not taking French include those taking German, Russian, and Spanish; also, those not taking German include those taking French, Russian, and Spanish. Finally, those not taking French, or not taking German, or not taking Russian must include everyone involved.

Two events need not consist of disjoint sets but may have points in common. See Figures 5-5-5 to 5-5-8. For example, if a random experiment calls for drawing a card from a deck of 52, then the sample space may be represented by a 4 by 13 lattice or two-way listing, where the first dimension is related to the suit and the second to the value of the card. If E_1 is the event of a black card, it consists of

26 points; if E_2 is the event of an honor, i.e., an A, K, Q, or J, then it consists of 16 points. Clearly, E_1 and E_2 have points in common. These points are called the intersection of E_1 and E_2.

The *intersection* of two sets, E_1 and E_2, consists of those points common to both and is written as $E_1 \cap E_2$.

Where unions call for all points in one *or* another set, and hence also points in both, intersections call for only points simultaneously in one *and* the other.

For the suit by value lattice and E_1, the event of a black card, and E_2, the event of an honor, $E_1 \cap E_2$ consists of cards which are black *and* honors. Clearly, there are exactly eight, whereas $E_1 \cup E_2$ has 34 points. This can be seen by direct count or as $26 + 16 - 8$, where -8 is to take care of the points in $E_1 \cap E_2$, points counted twice in $26 + 16$.

If two sets have no points in common, we write $E_1 \cap E_2 = 0$ and say that the intersection is the *empty set*, also written as $\emptyset$. For any two sets of a partition, $E_i \cap E_j = 0$ because the intersection of any two disjoint or mutually exclusive sets is the empty set.

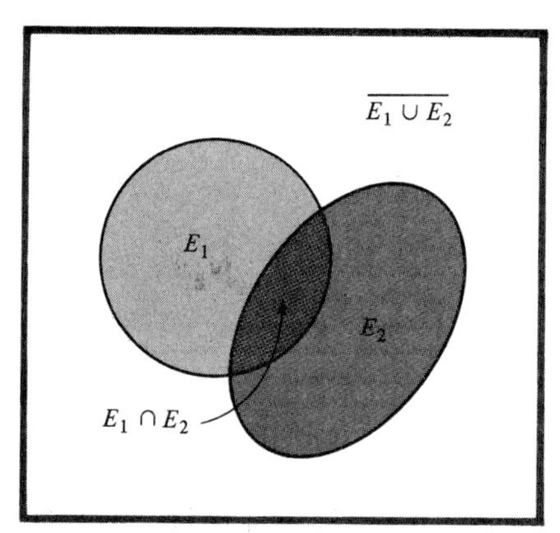

FIGURE 5-5-5 Venn Diagram Illustrating $E_1 \cap E_2$.

Since the probability of an event is the sum of the weights or probabilities assigned to the sample points that constitute the event, it is clear from Figure 5-5-5 that when $E_1 \cap E_2 \neq 0$, then addition of the probabilities of E_1 and E_2 does not provide $P(E_1 \cup E_2)$ because the probabilities associated with the points in $E_1 \cap E_2$ would be included twice. Thus, if E_1 is drawing a black card in a single draw, then $P(E_1) = 26/52$; if E_2 is drawing an honor, then $P(E_2) = 16/52$; but for the event consisting of drawing either a black card or an honor, $P(E_1 \cup E_2) = 34/52 \neq (26 + 16)/52$ because $E_1 \cup E_2$ contains only 34 points. However, we could add the probabilities of the points in E_1 and E_2 and then subtract the probabilities of those that appeared twice because they were in $E_1 \cap E_2$. For the card problem, the points in $E_1 \cap E_2$ are 8 in number, namely, the 8 black honors, so

$$\begin{aligned} P(\text{black or honor}) &= P(\text{black}) + P(\text{honor}) - P(\text{black honor}) \\ &= \frac{26 + 16 - 8}{52} \\ &= \frac{34}{52} \end{aligned}$$

More generally, we have Equation 5-5-4, the *addition theorem* of probability.

$$P(E_1 \cup E_2) = P(E_1) + P(E_2) - P(E_1 \cap E_2) \tag{5-5-4}$$

Note that Equation 5-5-4 includes Equation 5-5-1 as the special case for which $E_1 \cap E_2$ is empty.

Figures 5-5-6 to 5-5-8 are other illustrations concerning unions and intersections.

From the discussions of unions and intersections, it is now apparent that the "or" of unions is an "and/or," since it is intended to include points which are in both sets, say E_1 and E_2, or simply in one of them. For intersections "and" means that the points involved must be in both sets.

Equation 5-5-4 may be generalized to any number of sets; Equation 5-5-5 is the case for three.

$$\begin{aligned} P(E_1 \cup E_2 \cup E_3) = P(E_1) + P(E_2) + P(E_3) - P(E_1 \cap E_2) - P(E_1 \cap E_3) \\ - P(E_2 \cap E_3) + P(E_1 \cap E_2 \cap E_3) \end{aligned} \tag{5-5-5}$$

EXERCISES

5-5-1 Of 190 college students, 35 take only French, 30 take only German, 10 take only Russian, 80 take only Spanish, 20 take both French and Spanish, and 15 take both French and German.

(a) Propose a sample space that adequately accounts for each individual student and the student's language program.

(b) Draw a Venn diagram where the events consist of the information relative to language but not to the individual. Let F, G, R, and S be the events containing the sample points corresponding to all students taking French, German, Russian, and Spanish, respectively.

(c) Do F, G, R, and S constitute a partition of the sample space?

(d) How many sample points are in each of the following sets:

(1) R	(2) G	(3) S	(4) F
(5) $\bar{R}$	(6) $\bar{G}$	(7) $\bar{S}$	(8) $\bar{F}$
(9) $R \cup G$	(10) $\overline{R \cup G}$	(11) $R \cap G$	(12) $\overline{R \cap G}$
(13) $F \cup S$	(14) $\overline{F \cup S}$	(15) $F \cap S$	(16) $\overline{F \cap S}$
(17) $F \cup G$	(18) $\overline{F \cup G}$	(19) $F \cap G$	(20) $\overline{F \cap G}$
(21) $F \cup G \cup S$	(22) $\overline{F \cup G \cup S}$		

(23) What is another name for $\overline{F \cup G \cup S}$?

(24) What is another name for $F \cap G \cap S$?

(25) What is another name for $\overline{F \cap G \cap S}$?

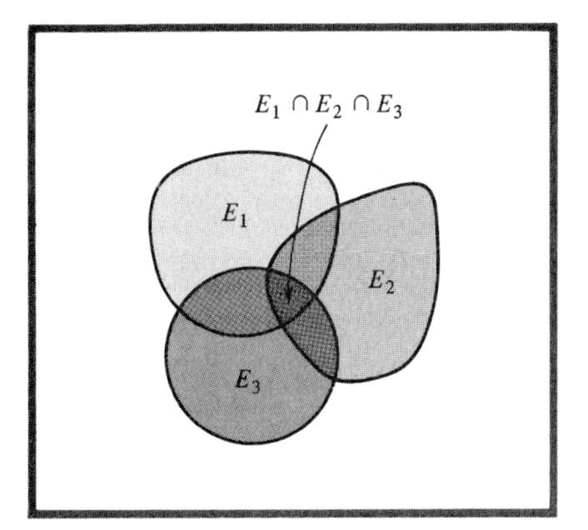

FIGURE 5-5-6 Venn Diagram Illustrating $E_1 \cap E_2 \cap E_3$.

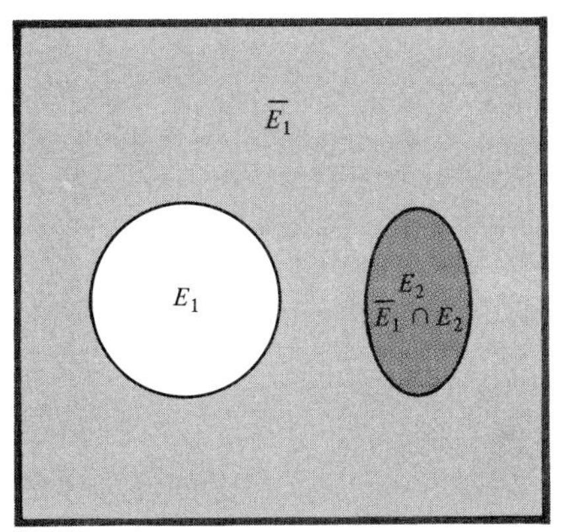

FIGURE 5-5-7 Venn Diagram Illustrating $\bar{E}_1 \cap E_2 = E_2$ When $E_1 \cap E_2 = 0$.

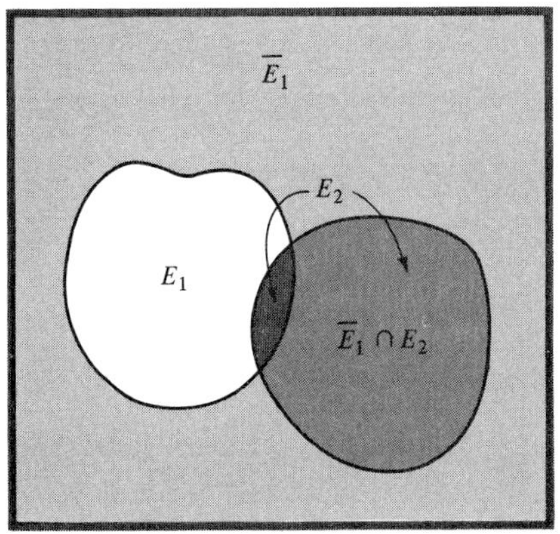

FIGURE 5-5-8 Venn Diagram Illustrating $\bar{E}_1 \cap E_2$ When $E_1 \cap E_2 \neq 0$.

(e) If we are to consider probabilities associated with possible language programs for a student randomly selected from the 190 language students, what probability function should be associated with the sample points accounting for each student and that student's program?
(f) What are the probabilities of the events F, G, R, S, $F \cup G$, $F \cup R$, $F \cup S$, $G \cup R$, $G \cup S$, $R \cup S$, $G \cup R \cup S$, $F \cap G$, $F \cap R$, $F \cap S$, $G \cap R$, $G \cap S$, $R \cap S$, $G \cap R \cap S$?
(g) Propose a partition of the sample space which provides the essential information of the first paragraph. Do not introduce any new letters.

5-5-2 Two dice, one red and one green, are tossed. The outcome is recorded as the ordered pair of numbers appearing on the uppermost faces of the red and green dice, respectively.

Let E be the event that the sum of the two numbers is even. Let T be the event that at least one of the numbers is a 3.

(a) Describe each of the following events. How many sample points are included in each?
(1) $E \cup T$ (2) $E \cap T$ (3) $E \cup \bar{T}$ (4) $\bar{E} \cap T$
(b) What are the probabilities that the outcome of a single toss will be included in each of the events symbolized above? Use your own symbolization of the events to show how the probabilities are computed.
(c) Among the sets $E \cup T$, $E \cap T$, $E \cup \bar{T}$, $\bar{E} \cap T$, is there a partition of the sample space?

5-5-3 Suppose that 260 people who have recovered from one heart attack agree to participate in a program intended to evaluate a preventive treatment proposed for such cases. The individuals are divided into three groups: a treated group, receiving an anti-coagulant; a placebo group, receiving a pill of no known value; and an untreated group, receiving no medication. The results might be as in the accompanying table. Let the

Group	No Subsequent Attack	Subsequent Attack: Survived	Subsequent Attack: Died	Totals
Treated	58	15	17	90
Placebo	58	14	18	90
Untreated	54	11	15	80
Totals	170	40	50	260

following letters represent the events described:

A: no subsequent attack and, therefore, a survivor
B: survived a subsequent attack
C: died following a subsequent attack
D: member of the treated group
E: member of the placebo group
F: member of the untreated group

(a) Use the above set notation and that for union, intersection, and complementation to symbolize the following descriptions:
(1) All survivors
(2) All survivors in untreated group
(3) The untreated patients who did not survive

(4) Those who did not survive (in terms of A and B)
(5) Those still alive at the end of the experiment who were not in the treated group
(6) Survivors of a subsequent attack who were in the placebo group
(7) Patients in the treated group who had a subsequent heart attack
(b) What are the probabilities that an individual randomly selected from a listing of the 260 participants will be a member of each of the sets described above? Use your own symbolization of the sets to indicate how the probabilities are computed.
(c) Among the events $A, \ldots, F$, is there a partition of the sample space?

Chapter 6

INDEPENDENT AND DEPENDENT EVENTS

6-1 INTRODUCTION

If we toss two distinguishable coins, then each toss is called a *trial.* The possible outcomes of the two-trial experiment may be represented by the sample space

$$S = \{HH, HT, TH, TT\}$$

If the two coins of our experiment are fair ones, then one outcome is as likely as another and, since the outcomes are mutually exclusive and the set is exhaustive, then the probability of any one is 1/4.

Nothing need be said as to which coin is to be tossed first or whether they may be tossed simultaneously. The order of tossing is immaterial, since the outcome for one coin can have no effect on the outcome for the other. Such trials and the resulting events are said to be independent.

If, however, we know that one of the coins has fallen heads, then our sample space is restricted to $S_1 = \{HH, HT, TH\}$, since TT is ruled out as a possible sample point, and the probability of each event is now 1/3.

Thus, if we ask for the probability that the outcome of such an experiment is two heads, then the answer is that the probability is 1/4 if we know nothing about the true outcome but it is 1/3 if we are aware that there is at least one head. In the latter case, the choice of the sample space has depended on our knowledge of the partial outcome and our computing of the probability has been conditioned by this knowledge. The events of exactly two heads and of at least one head are said to be dependent events.

This chapter deals with independent and dependent events and the computation of their probabilities.

6-2 INDEPENDENT EVENTS

Suppose an experiment is to consist of rolling a die and drawing a card from a deck of 52 cards. A joint event involving the die and the deck is defined as favorable if the outcome of the experiment results in a 1 or a 2 on the die and an honor card from the deck. We wish to know the probability of a favorable outcome.

There are two ways of observing the event of a 1 or a 2, namely, to roll a 1 or to roll a 2. There are 16 honors in the deck and so 16 ways in which an honor may be drawn. Since the rolling of the die has no effect on the drawing of a card, each of the die outcomes may be associated with each of the card outcomes. Thus, there are 2×16 equally likely favorable outcomes constituting the joint event.

Since there are 6 faces on the die and 52 possible cards, there are 6×52 equally likely outcomes for the experiment. Now, applying Equation 5-2-1, we find that the probability of a favorable outcome of the experiment, that is, of a 1 or a 2 on the die *and* an honor card, is as follows:

$$\begin{aligned} P(1 \text{ or } 2, \text{honor}) &= \frac{2 \times 16}{6 \times 52} \\ &= \frac{2}{6}\frac{16}{52} \\ &= P(1 \text{ or } 2)P(\text{honor}) \end{aligned}$$

More generally, if we are concerned with an event A that can occur successfully in m_1 out of a possible m ways and an event B that can occur successfully in n_1 out of a possible n ways, and if the occurrence of A or $\bar{A}$ has no effect on the occurrence of B or $\bar{B}$, then Table 6-2-1 shows the number of ways that the various possible joint events involving A, $\bar{A}$, B, and $\bar{B}$ can occur. Note that there are $m_1 n_1$ ways in which the event $A \cap B$ may occur, that is, for A and B to occur simultaneously, since any one of the m_1 A-type outcomes can occur with any one of the n_1 B-type outcomes because the occurrence of one has no effect on the occurrence of the other. We reason similarly for the other three cells of the table.

TABLE 6-2-1 / Frequency of Possible Joint Events

	Event		
Event	B	$\bar{B}$	Total
A	$m_1 n_1$	$m_1(n - n_1)$	$m_1 n$
$\bar{A}$	$(m - m_1)n_1$	$(m - m_1)(n - n_1)$	$(m - m_1)n$
Total	mn_1	$m(n - n_1)$	mn

Now, if we are interested in the event that A and B occur simultaneously, that is, in $A \cap B$, we simply apply Equation 5-2-1. A little additional effort leads to Equation 6-2-1.

$$P(A \cap B) = \frac{m_1 n_1}{mn} = \frac{m_1}{m}\frac{n_1}{n} = P(A)P(B) \qquad (6\text{-}2\text{-}1)$$

Bear in mind that Equation 6-2-1 has been shown to be valid only when the separate events A and B that make up the joint event $A \cap B$ cannot influence one another. Such separate events are said to be independent.

Mutually exclusive and independent events are different concepts and not to be confused. Mutually exclusive events are events such that the occurrence of one rules out the possibility of any of the others occurring. Thus, a coin can fall a head or a tail, but it cannot fall both. In other words, $H \cap T = 0$. Clearly, mutually exclusive events are very much dependent and we are never concerned with the simultaneous occurrence of two or more of them, although we might well be concerned with unions of such events.

EXERCISES

6-2-1 Consider families consisting of exactly two children. Assume that the probability of a child being born a boy is 1/2, regardless of whether the child is the first or second. A family is to be drawn at random from a very large set of two-children families.
(a) What is the probability that the older child will be found to be a girl? Is this event independent of the event that specifies the sex of the first child?
(b) What is the probability that both children will be girls?
(c) What is the probability that both children will be the same sex?
6-2-2 Sketch a representation of a sample space for an experiment consisting of drawing 2 cards from a set of 13 consisting of diamonds only. Let each point represent an ordered pair when sampling is without replacement.
(a) What is the probability of drawing one ace? One deuce? No aces at all? No deuces at all?

(b) What is the probability of drawing an ace and a deuce in any order?
(c) What is the probability that the pair of cards includes neither an ace nor a deuce? That the pair includes either an ace or a deuce but not both?
(d) What is the probability that the pair will include a face card?
(e) What is the probability that the pair will include a card other than a face card?
6-2-3 Consider families with exactly three children and assume that the probability of a child being born a male is 1/2, regardless of position within the family.
(a) Present a reasonable sample space to handle problems relative to sex distribution within the family.

A family is drawn at random from a very large set of such families.
(b) What is the probability that the children will all be of the same sex?
(c) What is the probability that the children will not all be of the same sex?
(d) What is the probability that there will be at least one girl in the family?
6-2-4 Two children play a game where ties are not possible. They play until one or the other has won three games; that child is considered the winner. This is essentially what happens in the baseball World Series.
(a) Present a sample space that shows who "won" and which of the individual games made up the winning.
(b) If the children are sufficiently evenly matched that the outcome of each and every game can be said to be determined by chance, independently of the result of previous games, what probabilities will you assign to each of the sample points?
(c) What is the probability that there will be a winner after exactly three games? Exactly four games? Exactly five games?
(d) What is the probability that the winner will win by winning three games consecutively?
6-2-5 Four dice are tossed with the observed outcome being the ordered sequence of the number of spots appearing on each of the four uppermost faces of the dice.
(a) How many possible outcomes are there?

Consider the event consisting of the sum of the number of spots on the four faces involved in the experiment.
(b) In how many ways can a 4 be observed? A 5? A 6? A 7? An 8?
(c) In how many ways can the resulting number be no larger than 8?
(d) What is the probability that, in random sampling, the mean number of spots will be no more than 2?
(e) Do you think this is a small probability?
Much of this problem was suggested in Section 3-4, shortly after Equation 3-4-3.

6-3 DEPENDENT EVENTS AND CONDITIONAL PROBABILITIES

A deck of cards may be considered to consist of black honors, black nonhonors, red honors, and red nonhonors. The numbers in each category may be found by direct count and placed in a two-way table such as Table 6-3-1.

Let us suppose that a card has been selected at random and is known to be black. Under these circumstances, what is the probability that it is an honor? This required probability is written as $P(H \mid B)$, which is read as the probability of H occurring given that B has already occurred. In other words, | is read "given." Such a probability is called a *conditional probability*.

Here, we are clearly being told that we are to restrict ourselves to the top

TABLE 6-3-1 / Cards Categorized as: Black, B; Red, $\bar{B}$; Honor, H, and Nonhonor, $\bar{H}$

Color	H	$\bar{H}$	Totals
B	8	18	26
$\bar{B}$	8	18	26
Totals	16	36	52

line of Table 6-3-1. Consequently, the appropriate probability $P(H \mid B)$ is 8/26, by Equation 5-2-1. In terms of the probabilities associated with the table, we see that we can just as well write $P(H \mid B) = P(H \cap B)/P(B)$.

More generally, for any two events, we may write the following definition of conditional probability.

If A and B are two events, then the *conditional probability* of A, given that B has occurred, is given by Equation 6-3-1 provided that $P(B) \neq 0$.

$$P(A \mid B) = \frac{P(A \cap B)}{P(B)} \qquad (6\text{-}3\text{-}1)$$

Let us apply Equation 6-3-1 to the problem of Section 6-1 where we know one coin has fallen heads and ask for the conditional probability that both are heads, knowing that at least one is a head. The events are that both are heads, that is, A = HH, and that at least one is a head, that is, B = HH or HT or TH. The intersection of A and B is $A \cap B$ = HH, since HH is the only sample point in common. We are asking for the conditional probability $P(A \mid B)$, and use Equation 6-3-1.

$$P(A \mid B) = \frac{P(A \cap B)}{P(B)} = \frac{1/4}{3/4} = \frac{1}{3}$$

As a further result, we may write Equations 6-3-2, the second one as a consequence of the symmetry of the system.

$$\begin{aligned} P(A \cap B) &= P(B)P(A \mid B) \\ &= P(A)P(B \mid A) \end{aligned} \qquad (6\text{-}3\text{-}2)$$

This is called the *multiplication theorem* of probability.

Let us apply Equation 6-3-2 by considering a card problem. We will draw two cards at random without replacement; we are interested in the probability that both will be aces.

At the first draw, $P(\text{ace}) = 4/52$. If an ace is, in fact, drawn then, at the second draw, there are only three remaining aces in the 51 cards. We write

P(second card is an ace | one ace drawn) = 3/51. Now, to find the probability of two aces in two draws, we apply Equation 6-3-2.

$$P(\text{two aces}) = \frac{4}{52}\frac{3}{51}$$

This is certainly an easier approach than trying to visualize a sample space for the problem.

Although Equation 6-3-2 involves a conditional probability, we may relate this equation to Equation 6-2-1 and see that the computation of the probability of a joint event consisting of two independent events as called for by Equation 6-2-1 is simply the special case that exists when $P(B) = P(B \mid A)$ and $P(A) = P(A \mid B)$. Independence of two events may, then, be stated as a definition.

Two events A and B are said to be *independent* if and only if Equations 6-3-3 hold. (If one holds, the other must.)

$$\begin{aligned} P(A \mid B) &= P(A) \\ P(B \mid A) &= P(B) \end{aligned} \qquad (6\text{-}3\text{-}3)$$

Now, if we apply these equations to the example in Table 6-3-1, with the notation used there, we find

$$P(B) = \frac{26}{52} = \frac{1}{2} \quad \text{and} \quad P(B \mid H) = \frac{8}{16} = \frac{1}{2}$$

and

$$P(H) = \frac{16}{52} = \frac{4}{13} \quad \text{and} \quad P(H \mid B) = \frac{8}{26} = \frac{4}{13}$$

The events B and H are seen to be independent.

On the other hand, for the coin problem of this section, we have

$$P(A) = \frac{1}{2}\frac{1}{2} = \frac{1}{4} \quad \text{and} \quad P(A \mid B) = \frac{1}{3}$$

and

$$P(B) = \frac{3}{4} \quad \text{and} \quad P(B \mid A) = 1$$

In the case of $P(B \mid A)$, notice that B is the event of at least one head, whereas A is that of exactly two; B certainly happens under this condition.

The multiplication theorem of probability can be generalized to any number of events. For three events, say A, B, and C, Equation 6-3-4 holds.

$$P(A \cap B \cap C) = P(A)P(B \mid A)P(C \mid A \cap B) \qquad (6\text{-}3\text{-}4)$$

EXERCISES

6-3-1 We now return to the two-children family of Exercise 6-2-1 and ask the following questions.

(a) Suppose we know that the first child is a boy. What is the probability that the second is a girl? Is the event that describes the sex of the second child independent of the event describing the sex of the first?
(b) Suppose we know that one child is a boy. What is the probability that there is at least one girl in the family? Is the event "at least one girl" independent of the event "one child is a boy"?
(c) Suppose we know that one child is a girl. What is the probability that there are two girls? Are these two events independent?
(d) Suppose we know that one child is a boy. What is the probability that he has a sister? Is the event of a "sister" independent of the event "one child is a boy"?

6-3-2 In Exercise 6-2-2, we considered certain card problems for which a suitable sample space was requested. Additional questions follow.
(a) What is the probability that the pair will consist of an ace and a deuce, given that it does include an ace? (Observe a suitably restricted sample space and also use Equation 6-3-1, first carefully defining A and B.) Is the event "an ace and a deuce" independent of the event "does include an ace"?
(b) If it is known that one card is a face card, what is the probability that the pair has only one face card? Are these two events independent?

6-3-3 Exercise 6-2-3 dealt with three-children families. The following additional questions are also concerned with such families.
(a) If it is known that there is a boy in the family, what is the probability that he is the only boy? Is this a case of independent events?
(b) If it is known that the oldest child is a boy, what is the probability that the other children are both girls? Is this a case of independent events?
(c) If it is known that there are at least two girls in the family, what is the probability that the other child is a girl? Is this a case of independent events?

6-3-4 The following questions are asked about the game in Exercise 6-2-4 where two children played until one or the other won three games.
(a) If it takes five games to produce a winner, what is the probability that the winner will win the last three? Carefully define the events A, B, $A \cap B$ called for in Equation 6-3-2 as they apply to this question. Are events A and B independent?
(b) If the winner wins by winning three games consecutively, what is the probability that it took five games? Carefully define the events A, B, $A \cap B$ called for in Equation 6-3-1 as they apply to this question. Are events A and B independent?

6-3-5 A student who has just received a bachelor's degree decides to continue in graduate school. She applies for a teaching assistantship (TA) at two universities. The probability that she will be offered a TA at the first of these is 1/2, at the second is 1/3, and at both is 1/4.
(a) What is the probability that the student will be offered at least one TA?
(b) Through the grapevine, the student learns that she is being offered a TA but not the name of the university. What is the probability that she will receive two offers?

6-3-6 As a consequence of a census of available voters, the accompanying table, showing the probability distribution by sex and by preference of type of school board, is made available.
(a) Let M symbolize the event of obtaining a male in a single random selection. What is standard notation for the event of selecting a woman?
(b) Let E and A symbolize the events of obtaining an "elect" and an "appoint" voter in a single random selection. What is standard notation for the "don't care's"?

	Elect	Appoint	Don't Care	Total
Male	.28	.08	.04	.40
Female	.30	.18	.12	.60
Total	.58	.26	.16	1.00

(c) Using the symbols of (a) and (b), provide appropriate symbols for all cells in the table.
(d) Is the probability of obtaining a male in a single random selection independent of obtaining an individual who favors an elected school board?
(e) What is the probability that a randomly selected individual will favor an elected school board if we are told the person is a male?

6-3-7 Suppose a vaccination procedure calls for a first shot of vaccine to which the probability of a reaction is .6. If there is a detectable reaction to this shot, no further vaccine is given. However, if there is no detectable reaction, then a shot of a second vaccine is given. For those receiving this vaccine, the probability of a reaction is .9.
(a) What is the probability that a randomly selected individual will react to neither vaccine?
(b) What is the probability that a randomly selected individual will react only to the second vaccine?

6-3-8 A manufacturing plant has 1,000 items ready to ship. Of these, 10 are defective because of faulty parts, 35 because of errors in assembly, and 3 for both causes.
(a) If an item is selected at random, what is the probability that it will be defective for one reason or another?
(b) If an item selected at random turns out to be defective, what is the probability that this is due to a faulty part? Due to faulty assembly?
(c) Should the last two probabilities sum to unity? Explain.

Chapter 7

PERMUTATIONS AND COMBINATIONS

7-1 INTRODUCTION

The classical approach to probability is based on the ratio of the number of favorable outcomes to the total number of possible outcomes.

At worst, this requires a list and a count of all possible outcomes and a tagging and counting of those described as favorable. Fortunately we rarely have to prepare such a list since we are able to count by means of permutations and combinations.

This chapter is about permutations and combinations and their use in counting problems and the computation of related probabilities.

7-2 PERMUTATIONS

The ordering Mary, Jane, and Susan is a permutation of a particular set of three girls' names; ace of hearts and 7 of spades is a permutation of two cards; the letters ABC may be permuted as ABC, ACB, BAC, BCA, CAB, or CBA.

A *permutation* is an ordering of things.

Many problems in probability and statistics involve permutations. Sometimes all the elements in the permutation will be different; at other times, this will not be so. A permutation may involve all the elements available or only a subset. In general, it is required to know how many permutations are possible.

Permuting all when all are different. Suppose three people find three theater seats together. In how many ways may they occupy the seats?

Any one of the three may take the first seat. Next, any one of the two remaining may sit in the second. Finally, the last person must sit in the third seat. In all, there are $3 \times 2 \times 1$ seating arrangements.

A tree diagram is sometimes useful in showing the principle for finding the number of permutations. For the preceding example, let the people be A, B, and C; the appropriate *tree diagram* or *dendrogram* is shown in Figure 7-2-1.

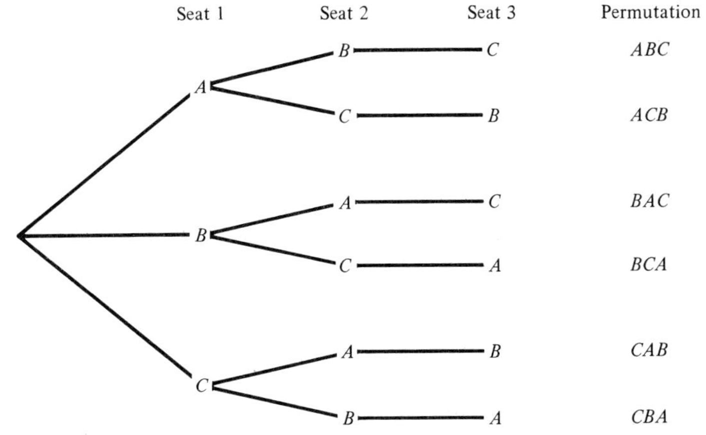

FIGURE 7-2-1 Tree Showing Three People in Three Seats.

We begin with seat 1 and show the three possible occupants. If A sits in seat 1, then A is not available for the other seats. More particularly, seat 2 may now be occupied by B or C. If B sits in seat 2, then C must sit in seat 3. The first permutation is seen to be ABC. If C sits in seat 2, then B must sit in seat 3. The second permutation is, then, ACB. We have exhausted this branch and must go back to seat 1 with B.

Notice how easy it is to proceed with this method for showing all options by eliminating the objects in a sequential procedure. The chance of an error depends simply upon how carefully we proceed.

More generally, if there are n ways of carrying out a first operation, $n - 1$

ways for a second, . . . , until finally there is a single way for the final operation, it follows that the complete set of operations can be carried out in $n(n - 1) \cdots 1$ ways. The resulting number is called a *factorial* and is written $n!$ as in Equation 7-2-1.

$$n! = n(n - 1) \cdots 1 \qquad \text{for } n \text{ a positive integer} \tag{7-2-1}$$

Selecting and permuting some when all are different. Suppose we draw five cards from a deck of 52, looking at each card as it is drawn, so that a hand is an ordered quintuple. How many hands are possible?

The first card may be any one of 52, the second any one of $52 - (2 - 1) = 51, \ldots$, and the fifth and last any one of $52 - (5 - 1) = 48$. In all, there are $52 \times 51 \times 50 \times 49 \times 48$ hands. This may be written somewhat more conveniently in factorial notation as follows:

$$(52 \times 51 \times 50 \times 49 \times 48) \times \frac{47!}{47!} = \frac{52!}{47!}$$

We have simply multiplied the number of hands by a 1 written as $47!/47!$ so as to make the resulting numerator a $52!$.

For simplicity, we may write ${}_nP_r$ or P_r^n as the number of permutations of n things taken r at a time. This gives us Equation 7-2-2.

$${}_nP_r = \frac{n!}{(n - r)!} \tag{7-2-2}$$

Notice that if all 52 cards are involved in a permutation, there are $52!$ possibilities. Hence ${}_nP_n = 52!$. But if we rely on Equation 7-2-2, then ${}_nP_n = n!/(n - n)! = n!/0!$. To be consistent with our notation, we must define $0!$ as unity; that is, $0! = 1$.

The general principle here is one of multiplication. If an operation can be performed in n_1 ways and any one of these is to be followed by a second operation that can be performed in n_2 ways, then together the complete operation can be performed in $n_1 \times n_2$ ways. The idea is readily extended to cover any number of operations.

Permuting all when not all are different. Suppose we consider families of two boys and three girls. How many different boy-girl permutations are possible if a permutation is determined by the birth dates of the children? For illustration, the permutation BBGGG says the two boys are the oldest children.

If the five letters were different, there would be $5!$ permutations. But two letters are B and the number $5!$ treats them as though they were a B_1 and a B_2, distinct from one another, to be permuted. Clearly, in the case of the B's alone, B_1B_2GGG and B_2B_1GGG have been counted as two permutations. Similarly, all other pairs of positions that could be occupied by the two B's have contributed two permutations. Thus, on the basis of the two boys alone, we have counted twice as many permutations as are actually distinguishable if the subscripts are absent. Also, the three G's have been permuted as though they were G_1, G_2, and G_3. Every time they occupied a set of three positions, they supplied $3! = 6$ permutations. Hence, the true number of permutations in which we do not distinguish between boys or among girls is $5!/(3!\,2!) = 10$. The permutations are BBGGG, BGBGG, BGGBG, BGGGB, GBBGG, GBGBG, GBGGB, GGBBG, GGBGB, and GGGBB.

Somewhat more generally, we may consider that we have n cells to fill with two kinds of objects, r of type B and $n - r$ of type G. We need to select r cells for the type B objects, and G's will occupy the remaining cells. There are $n(n - 1) \cdots (n - r + 1)$ ways of selecting the r cells in sequence for the B's, but any one set of r cells has been counted many times since permutations were used. In other words, a set consisted of r cells and the order in which they were chosen. However, the B's are not distinguishable so that a particular set of r cells should be counted only once with the order of their choice being ignored. Looking at one such set, we see that the B's in it can be permuted in $r!$ ways. The same is true for any other such set. Hence, the number of ways to select and fill the r cells should be

$$\frac{n(n - 1) \cdots n - r + 1}{r!} = \frac{n!}{(n - r)!\,r!}$$

The G's simply occupy the remaining positions, so our computation is complete.

The general permutation formula for two kinds of objects, all being involved in the permutation, is given by Equation 7-2-3. We write $\binom{n}{r}$ and read it as "the number of permutations of n objects where r are of one kind and $n - r$ are of another" or "the number of ways of selecting r objects from a set of n objects."

$$\binom{n}{r} = \binom{n}{r,\, n - r} = \frac{n!}{r!\,(n - r)!} \tag{7-2-3}$$

The first notational form is probably found more frequently than the second.

More generally, if n_1 objects are alike and of one kind, n_2 are alike and of another, . . . , and n_k are alike and of still another kind, and if $n_1 + \cdots + n_k = n$, then the number of permutations of them all together is given by Equation 7-2-4.

$$\binom{n}{n_1, \ldots, n_k} = \frac{n!}{n_1!\,n_2! \cdots n_k!} \tag{7-2-4}$$

Note that if $n_1 = \cdots = n_k = 1$, we are permuting all when all are different, and Equation 7-2-4 gives $n!$ as the answer, as it should.

EXERCISES

7-2-1 Consider the first hand drawn under "permuting some when all are different." Since it is an ordered quintuple, the same cards differently ordered have been counted as other hands. How many times has this same set of cards been counted in the total number of hands?

7-2-2 I have a choice of eight reasonable routes to Chicago. How many ways can I make the round trip and not return by the same route? How many ways can I make the round trip if I'm willing to return by the same route?
7-2-3 Three boys go to a restaurant where seven different toppings are available for a basic pizza. In how many ways may they order separate pizzas so each has a different topping? If the boys do not restrict themselves to different toppings, how many ways may they order? In the latter case, how many of the possible orders are identical? How many of these orders will consist of exactly two pizzas with the same topping and one different? (*Hint:* First select the boy to receive the "different" topping.)
7-2-4 Four people go to a theater and find nine empty seats. In how many ways may they seat themselves?
7-2-5 A coin was tossed 20 times with 13 tosses showing heads and 7 showing tails. How many possible permutations of 13 heads and 7 tails are there?
7-2-6 Consider families of six children consisting of four boys and two girls. How many permutations are possible if we consider a family to be ordered on the basis of birth dates? How many such families will have a boy as the oldest child? How many will start and end with a boy?
7-2-7 An experimenter has eight cultivars (varieties) of corn on which he wishes to use five fertilizer treatments, applying each in three different ways. If a treatment consists of a cultivar-fertilizer combination, how many different treatments will he have? If he wants to use each treatment on four separate experimental plots, how many plots will he need for the experiment?
7-2-8 An experiment on frozen pork chops uses 7 methods of wrapping, 3 storage temperatures, and 2 types of storage, with measurements being made on 11 characteristics. If 6 pork chops are to be used for each treatment and all measurements are to be made on each pork chop, how many measurements will be made?
7-2-9 An experimenter has 13 treatments to compare and 13 experimental units to which she will apply them. In how many ways can she assign the treatments to the units? (This is called a randomization.) If she has 4 other sets of 13 experimental units, in how many ways may the whole experiment be randomized? (A set of units is always to be kept intact.)

7-3 COMBINATIONS

Combinations differ from permutations in that the order of making a *selection* is not a consideration. Thus there is only one way of selecting all things, say n, when all are different, whereas there are $n!$ permutations.

Suppose we plan a trip and can select three paperbacks from a set of 30 for reading. There are $30 \times 29 \times 28$ ways of selecting them if different orderings of the same three are to be called different selections. In fact, each set of three different books has been counted $3!$ times under these circumstances. Consequently, the number of sets of three different books is $30 \times 29 \times 28 / 3! = 30! / 27!\, 3!$.

In general, the number of combinations of n objects selected r at a time is designated by ${}_nC_r$, C_r^n, or $\binom{n}{r}$ and computed by Equation 7-3-1.

$$ {}_nC_r = \binom{n}{r} = \frac{n!}{r!\,(n-r)!} \qquad (7\text{-}3\text{-}1) $$

This problem was discussed in Section 7-2 under "permuting all when not all are different." Equation 7-3-1 is Equation 7-2-3 with additional notation. There, we had to permute two or more kinds of objects using Equations 7-2-3 and 7-2-4 when permutations within a kind had no meaning. Thus, combinations are permutations on which certain restrictions have been imposed.

Permutations and combinations are much used in statistics. For example, when an experiment is being planned for the comparison of a number of treatments, the treatments are assigned to the experimental units in a random manner. Randomness will allow us to use probability theory in drawing inferences. At the same time, no treatment will be favored, consciously or unconsciously. It becomes necessary to know how many permutations are possible. In sample survey work, randomization is a device used to draw samples which will be representative, on the average. The order of selection may or may not be important. Again, we need to know the number of possible samples. Permutations and combinations provide us with answers to these and many other questions arising in statistics.

EXERCISES

7-3-1 There are 10 members of a swimming team that are available to participate in a 200-yard freestyle relay, a 4-person event. How many different teams are possible if the order of their swimming is not considered? How many different teams are possible if the event is a medley relay and the team requires a swimmer for each of backstroke, breaststroke, butterfly, and freestyle? Assume each of the 10 can swim any stroke required.

7-3-2 A coin is tossed eight times and three heads and five tails are observed. How many sequences consist of three heads and five tails? How many possible head-tail sequences are there? What is the probability of exactly three heads and five tails if tossing is random?

7-3-3 A student decides he can afford three escape novels and two of science fiction. He has 25 of the former to choose from and 12 of the latter. How many choices of different books are possible?

7-3-4 A bridge hand consists of 13 cards. How many different hands are possible from a standard 52-card deck? How many of these hands will consist solely of red cards? How many will have exactly four red cards? What is the probability of an all red hand if the cards have been randomly dealt? Exactly four red?

7-3-5 If a fair coin is tossed five times, what is the probability of getting five heads? What is the probability of getting exactly two heads?

7-3-6 A buyer will accept a lot of 100 transistor radios if 5 are selected and not more than one fails to operate. Suppose that 8 of the 100 will not operate. How many samples of different radios are possible? How many samples have no defectives? How many have exactly one defective? What is the probability that the buyer will accept the lot?

7-3-7 An experimenter has 10 pairs of mice available for an experiment. Mice within a pair are expected to respond alike if treated alike; the experimenter plans to treat a randomly selected mouse from each pair and then observe whether the response to a certain stimulus is greater in the treated or untreated mouse of each pair. How many possible randomizations are there? The experimenter plans to say the treatment is effective if the treated mouse in 8, 9, or 10 pairs gives the greater response. What is the probability of saying the treatment is effective if, in fact, it is not?

7-3-8 A gambler hopes he has an unfair die and sets out to estimate the probabilities associated with each of the six faces. He intends to begin by tossing the die $n = 100$ times and record the frequency with which each of 1, 2, . . . , 6 occurs, say $n_1, \ldots, n_6$, where $\sum n_i = n$. The values turn out to be 15, 21, 24, 16, 11, and 13, respectively.

In how many ways can this event occur? Different outcomes depend upon the order in which the six faces appear. Does the notation ${}_nC_{n_1, n_2, \ldots, n_k}$ seem appropriate for this result?

7-3-9 In Section 3-4, a die was tossed four times and the mean of the observed results computed to be 4. In how many ways can this result be obtained? The same set of numbers observed in a different order is considered to be a different way.

Chapter 8

RANDOM VARIABLES AND PROBABILITY DISTRIBUTIONS

8-1 INTRODUCTION

In Chapters 5 and 6, the concept of a random experiment led to the notion of a sample space. The idea of a function, subject to certain postulates, which assigned weights called probabilities to the points of the sample space, was next introduced. We then had a probability function which allowed us to compute probabilities for events.

The choice of values for probabilities was made with reference to the real world. Thus we said that for a single toss of a fair coin $P(\text{H}) = 1/2 = P(\text{T})$, and that for a single roll of a fair die $P(1) = \cdots = P(6) = 1/6$. Basically, we have stated a scientific theory about the way a tossed coin or die behaves to explain observations in an elemental situation and have done so in mathematical language. We have proposed a *mathematical* or *probabilistic model*.

Scientific theories or models are our way of depicting and explaining how observations come about. Such theories are simplified statements containing essential features and make for easier comprehension and communication. In statistics, we use a mathematical approach since we quantify our observations. Our mathematical model must still fit the facts if it is to have any usefulness in leading us to take effective action. In other words, a mathematical model is not for impressing the uninitiated but is intended to be a useful tool and must be modified or discarded as real data indicate such a need.

The set of possible outcomes for a random experiment is not always finite in number. For example, if we are counting insects to determine the effectiveness of one or more insecticides, we count by integers but may not necessarily see an endpoint, say M, such that we can say, "There will be no counts greater than M." Here, the most convenient solution may be to say only that we can put the possible counts in a one-to-one relation with the integers. This makes them a *denumerable infinity*. Thus, we have a *discrete variable* that allows for a denumerable infinity of outcomes. As another example, consider the problem of listing the possible outcomes of an experiment calling for measurements of height, weight, or age. Clearly, such variables do not proceed through a range by discrete steps but do so continuously throughout an interval. We are, however, limited to intervals dependent upon our measuring devices. Thus, we measure and record to the nearest centimeter, kilogram, or month. Here, we have a *continuous variable.*

Our notions of sample space, probability function, and model must be developed to cover these situations which are clearly realistic. This chapter deals with random variables and probability distributions for both discrete and continuous variables.

8-2 RANDOM VARIABLES AND PROBABILITY FUNCTIONS FOR DISCRETE VARIABLES

A random experiment, one which can be visualized as indefinitely repeatable under identical conditions and where chance is involved in the outcome, although this outcome is not predictable, terminates in an outcome which is placed in correspondence with a point in a sample space. Thus, for a two-coin experiment, the outcomes are two tails, a tail and a head, a head and a tail, or two heads. The sample space can be represented as (TT, TH, HT, HH), so that the correspondence is clear. Now suppose that a unique or single value of a real-valued variable, that is, a variable which takes only real numbers as possible values, has been assigned to each sample point. For the two-coin problem, we might summarize the data by using the number of heads and so assign the values (0, 1, 1, 2), respectively, to the points in the sample space.

Any function or association that assigns a unique, real value to each sample point is called a chance or *random variable.* The assigned values are the values of the random variable.

Random variables are symbolized by capital letters, most often Y, and their values by lowercase letters. The outcome of a random experiment determines a point in the sample space, called the *domain* of the random variable, and the function transforms each sample point to one of a set of real numbers, that is, to a point on the y axis. This set of real numbers is called the *range* of the random variable.

If we let Y represent the number of heads in two trials, then Y is clearly a variable and is a random variable because its value is determined by the outcome of a random experiment. It transforms HH from S to 2 on the y axis, HT from S to 1, TH from S to 1, and TT from S to 0 on the y axis. The sample space with its four points is the domain of Y, while the values 0, 1, and 2 are its range. Also, we might say, "Let Y be the random variable measuring red-blood-cell count" under specified experimental conditions; or we might say, "Suppose the random variable Y, the sum of the number of spots in the roll of two dice, takes the value $y = 11$."

Many other examples of random variables come to mind: the number or weight of fruit harvested from a peach tree; the peak rate of runoff from a watershed in a calendar year; and so on.

Some variables are based on *qualitative characteristics*, like eye color. Now, an individual falls into one or another category of the characteristic. For example, a person is blue-eyed or not. Here we might have a random variable that assigns a 1 to an individual with blue eyes and a 0 otherwise. Thus, a random sample provides a number of 1s and 0s as values of this discrete random variable, and the sum of the sample values is a count of the number of blue-eyed people in the sample.

Other characteristics are *quantitative*, like the number of insects captured in an insect trap in a 24-hour period or the weight of a person or the number of gallons it takes to fill your gas tank when you stop at a service station. Here, the idea of a number as the value of the random variable is implicit. Such a variable may be either discrete or continuous.

For both quantitative and qualitative characteristics, we eventually quantify the outcome of the random experiment in that we assign a real value to each sample point. This is required by our definition of a random variable.

The association of probabilities with the various values of a discrete random variable is done by reference to the probabilities in the sample space and through a system of relationships or a function called a probability set function or, simply, a probability function.

A discrete *probability function* is a set of ordered pairs of values of a random variable and the corresponding probabilities.

Let the discrete random variable Y assume the values $y_1, \ldots, y_n$; then the system of relations can be written as in Equation 8-2-1.

$$P(Y = y_i) = p_i \tag{8-2-1}$$

This is read as "The probability (P) that the random variable (Y) takes ($=$) the value y-sub-i (y_i) is ($=$) p-sub-i (p_i)." The symbols in parentheses are from Equation

8-2-1 and are preceded by their interpretation. The set of ordered pairs (y_i, p_i) constitutes a probability function with numerical values to be provided for the y_i's and p_i's.

Illustration 8-2-1

Suppose we have a red and a blue tetrahedron with faces numbered 1, 2, 3, and 4, and the random experiment consists of tossing them and recording the ordered pair of numbers on the hidden faces. The sample space can be represented by the 16 ordered pairs (1, 1), (1, 2), . . . , (4, 4), where the first entry refers to the red tetrahedron. The sample space may also be represented as in Figure 8-2-1.

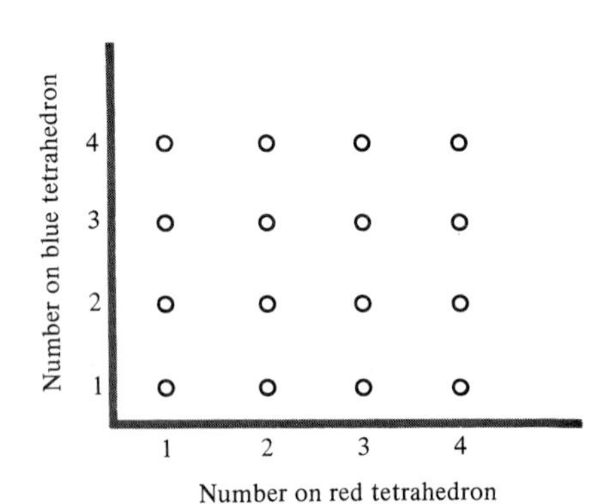

FIGURE 8-2-1 Sample Space for Two Tetrahedra.

Let the first entry of any ordered pair be designated by the letter i, the second by j. A sample point is now represented by (i, j). Let Y be the random variable defined by $Y = i + j$ that assigns that unique real value to the sample point (i, j). Thus for (1, 1), $y = 1 + 1 = 2$; for (2, 3), $y = 2 + 3 = 5$; and so on.

Reasonable probability values for the sample points are assigned by Equation 8-2-2.

$$P(i, j) = 1/16 \qquad i, j = 1[1]4 \qquad (8\text{-}2\text{-}2)$$

The latter bit of notation in this equation states that both i and j go from 1, by steps of 1, to 4.

Finally, the probabilities assigned by Equation 8-2-2 provide probabilities for each value of Y since each Y specifies an event in the sample space, and the set of events is seen to be a partition as defined in Section 5-5. Thus, the event for which $y = 2$ consists of the single point (1, 1); for $y = 3$, the sample points are (1, 2) and (2, 1); for $y = 4$, they are (1, 3), (2, 2), and (3, 1); for $y = 5$, they are

(1, 4), (2, 3), (3, 2), and (4, 1); for $y = 6$, they are (2, 4), (3, 3), and (4, 2); for $y = 7$, they are (3, 4) and (4, 3); and for $y = 8$, the single point is (4, 4). The events are seen to be disjoint and exhaustive, which assures us that we have a partition. The values of the random variable Y and their probabilities are presented in Equation 8-2-3, Table 8-2-1, and Figure 8-2-2.

$$P(Y = y) = \begin{cases} \dfrac{y-1}{16} & \text{for } y = 2, 3, 4, 5 \\[2ex] \dfrac{8-(y-1)}{16} & \text{for } y = 5, 6, 7, 8 \end{cases} \tag{8-2-3}$$

Any one of Equation 8-2-3, Table 8-2-1, or Figure 8-2-2 will serve as probability function. Notice that either part of Equation 8-2-3 gives the correct probability for $y = 5$.

TABLE 8-2-1 / Sums for Two Tetrahedron Faces with Their Probabilities

Sums, y	2	3	4	5	6	7	8
Probabilities, $P(Y = y)$ or $f(y)$	1/16	2/16	3/16	4/16	3/16	2/16	1/16

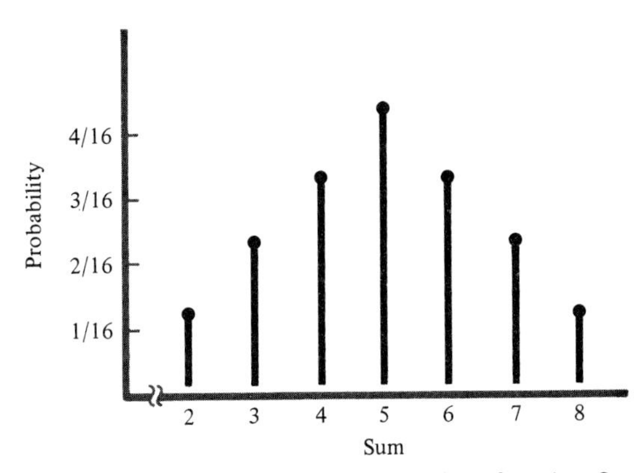

FIGURE 8-2-2 Graph of Probability Function for the Sum of Two Tetrahedron Faces.

EXERCISES

8-2-1 One tetrahedron with faces numbered 1, 2, 3, and 4 is tossed twice, and the recorded outcome consists of the ordered pair of numbers on the hidden faces at the first and second tosses.

(a) Let the random variable Y take on the value 0 if the sum of the numbers in the ordered pair is even and 1 if odd. What is the probability function for this variable?

(b) Let the random variable Y take on the value 2 if both numbers in the ordered pair are even, 1 if exactly one is even, and 0 if neither is even. What is the probability function for this variable?
(c) Let the random variable Y be the number of divisors in the sum of the two faces. (An integer is a divisor if it is an integral multiple of the number; thus, the divisors of 6 are 1, 2, 3, and 6.) What is the probability function for this variable?
8-2-2 A history test includes a list of names, namely, Castro, Kissinger, Sadat, and Wilson, together with a list of geographical areas, namely, America, Cuba, Egypt, and England. The problem is to match the names and places correctly. Suppose the student guesses the answers.
(a) Present a sample space for the possible outcomes of the experiment. What probability is assigned to each point in the sample space?
(b) Let the random variable Y be the number of correct answers for each outcome. Prepare a probability function for Y.
8-2-3 A random experiment consists of tossing a fair coin four times and recording the ordered sequence of heads and tails observed.
(a) Present a sample space for the outcome of the experiment. What probability is associated with each outcome?
(b) Let the random variable Y be the number of heads observed at each outcome. Prepare a probability function for Y.
8-2-4 Four discs numbered 1, 2, 3, and 4, but otherwise identical, are placed in a container. Two are then drawn simultaneously, or without replacement, as a sample. The outcome consists of the pair ordered as observed.
(a) Present a sample space for the outcome of the experiment. What probability should be assigned to each point in the sample space?
(b) Let the random variable Y_1 be the sum of the numbers on the two discs. Present a probability function for Y_1.
(c) Let the random variable Y_2 be 0 if the sum is even and 1 if odd. What is the probability function for Y_2?
(d) Let the random variable Y_3 be 2 if both numbers are even, 1 if exactly one is even, and 0 if neither is even. What is the probability function for Y_3?
(e) Let the random variable Y_4 be the number of divisors in the sum of the two numbers on the two discs. What is the probability function for Y_4?

Note: This is a case of sampling without replacement. Compare this with Exercise 8-2-1, a case of sampling with replacement.

8-2-5 In a lot of 12 items, 3 are defective; a random sample of 4 items is drawn. Let Y be the number of defective items in the sample. Prepare a table of the probability distribution of Y.
8-2-6 Of six tennis balls in a bag, two are known to be from last year and soft. The balls are drawn one at a time from the bag and observed until both soft ones are removed.
(a) Present a sample space for the experiment. What probabilities are to be assigned to each sample point?
(b) Let the random variable Y be the number of trials required to get the two soft balls out of the bag. What is the probability distribution of Y?

8-3 THE RECTANGULAR OR UNIFORM DISTRIBUTION

As pointed out in Section 8-1, the set of possible outcomes of a random experiment is not always finite in number. In particular, an experiment may terminate in an outcome that requires a measurement such as a weight or the life of a continuously burning light bulb. Such an outcome exists on a continuous scale rather than on one that proceeds by discrete steps. The sample space or domain over which the random variable is to be defined must have an infinite number of sample points. Clearly, we have a continuous random variable that requires a range consisting of an interval or union of intervals of real numbers. This range may, of course, serve as the sample space itself. Of necessity, however, we must measure to the nearest value of some particular and practical unit such as a gram or an hour.

It is obvious that a probability function for a continuous random variable cannot be a set of relations which assign finite amounts of probability to each sample point; for no matter how small each assigned probability may be, it still takes only a finite number of these to sum to unity. Our notion of a probability function must be revised.

Suppose we consider an experiment where a balanced spinner or pointer, such as in Figure 8-3-1, is set in motion and allowed to stop without interference on the circumference of a circle. Each point on the circle has equal probability of being the stopping point. However, there are infinitely many such points so that we are unable to write a nonzero value for the probability. What mathematical representation will describe the probabilities associated with the system?

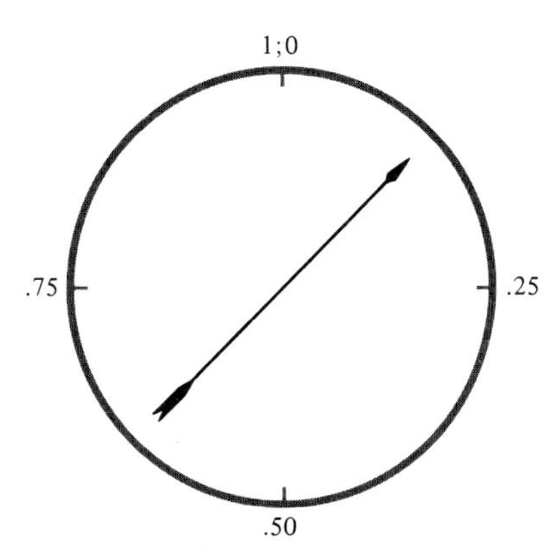

FIGURE 8-3-1 A Balanced Spinner.

First, let us declare the circumference of the circle to be of unit length. Now, the continuous random variable that locates the stopping point of the spinner may take on any real value in the interval between 0 and 1, the unit interval, so that the sample space can be described as the set of real numbers on the unit interval. We write this as $S = \{Y:0 \leq y \leq 1\}$ and read it as, "The sample space S is such that the random variable Y may assume any value y in the unit interval, including the endpoints 0 and 1." Recall that $\leq$ is read as "less than or equal to."

Since all possible stopping points must have equal probability, the probability behavior may be represented graphically by drawing a line of constant height above the unit interval as in Figure 8-3-2. The horizontal line tells us that one y-value is as likely as another.

An event has already been defined as a subset of the sample space. When the variable is a continuous one, it becomes convenient to define an event as an interval. Thus the man on the midway has pegs equally spaced at small intervals around his wheel, the pointer stops in an interval, and the outcome is this interval rather than a point.

The wheel operator wants us to believe that all intervals have equal frequency or probability when the pointer is spun and allowed to come to rest. In Figure 8-3-2, areas above equal intervals are equal, so it becomes logical to relate areas and probabilities. This is essentially what was proposed for bar charts and histograms, namely, that the truest representation of the data was given when areas were made proportional to frequencies.

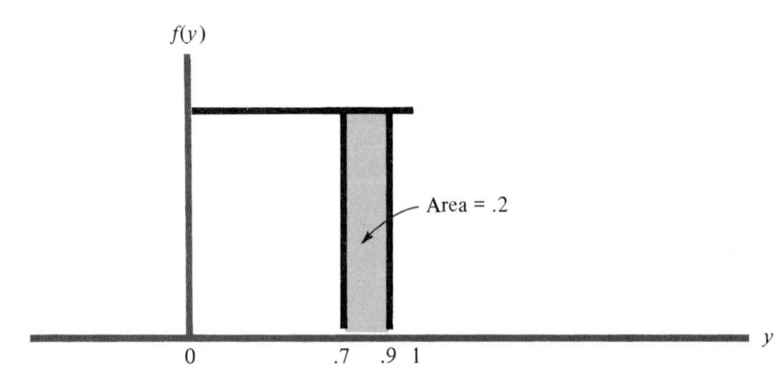

FIGURE 8-3-2 (Probability) Density Function for the Uniform Distribution on the Unit Interval.

Since we have already required probability functions to be such that the sum of the probabilities allotted to the parts of a partition of the sample space be unity, it is also logical to assign unit area to the part of the plane of Figure 8-3-2 above the unit interval and below the line that suggests the nature of the probability behavior. Thus, the line must have unit height. In turn, the event that calls for the pointer to stop between .7 and .9 has an associated area of .2 units and thus a probability of .2. More generally, the area above any interval on the y axis and under the line intended to describe the probability behavior of the system must equal the probability that the random variable will take a value in that interval. Note that if we had assigned 10 units as the length of the circumference, the height of the line would be necessarily $\frac{1}{10}$ unit.

The mathematical representation of the probability behavior in this problem is given by the equation of the line in Figure 8-3-2, namely, Equation 8-3-1.

$$f(y) = \begin{cases} 1 & 0 \le y \le 1, \text{ the unit interval} \\ 0 & \text{elsewhere, i.e., for values of } y \text{ outside the unit interval} \end{cases} \tag{8-3-1}$$

More generally, we have the interval (a, b) on the real line and Equation 8-3-2.

$$f(y) = \begin{cases} \dfrac{1}{b - a} & a \leq y \leq b \\ 0 & \text{elsewhere, i.e., for values of } y \text{ outside the interval } (a, b) \end{cases} \tag{8-3-2}$$

The original illustration in this section began with a circle of unit circumference where the points 0 and 1 coincide, rather than with a unit interval where 0 and 1 are endpoints, and this appears to raise a point for argument. However, for a continuous random variable, the probability associated with any point is 0, as we have seen. Consequently, it does not matter whether we say "less than or equal to," written $\leq$, or simply "less than," written $<$, when considering the endpoints of an interval.

A function such that areas, under the corresponding curve and above intervals of the real line, are equal to probabilities that a random variable will take on a value within the interval is called a *probability density function, pdf,* or, simply, a *density function.* Clearly it must be non-negative and the total area must equal unity.

The function in Equation 8-3-2 is called the *uniform* or *rectangular* (probability) density function; Equation 8-3-1 is a particular case.

The parameters, numbers which summarize information, of the uniform distribution are usually thought of as the endpoints of the interval on which $f(y) > 0$.

For any two intervals E_1 and E_2 on the real line and any density function, Equation 8-3-3 is true.

$$P(E_1 \cup E_2) = P(E_1) + P(E_2) - P(E_1 \cap E_2) \tag{8-3-3}$$

When E_1 and E_2 are disjoint, that is, $E_1 \cap E_2 = 0$, then $P(E_1 \cap E_2) = 0$.

A probability density function gives rise to a function which provides probabilities that the random variable will be less than a specified value, that is, $P(Y \leq y)$, the probability that the random variable Y takes a value less than or equal to the value y. This is the *cumulative probability distribution function*, the *cumulative distribution function* or *cdf*, the *probability distribution function*, or, simply, the *distribution function.*

For the uniform density function, we have the (probability) distribution function of Figure 8-3-3 and Equation 8-3-4.

$$P(Y \leq y) = F(y) = \begin{cases} 0 & y < a \\ \dfrac{y - a}{b - a} & a \leq y \leq b \\ 1 & b < y \end{cases} \tag{8-3-4}$$

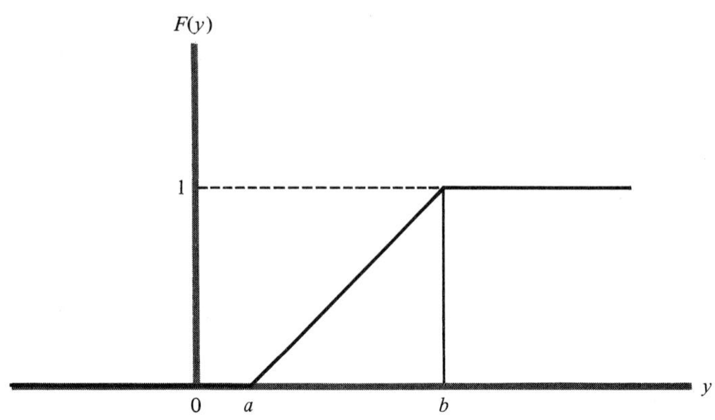

FIGURE 8-3-3 (Cumulative) Distribution Function for the Uniform Density Function on the Interval (a, b).

Note that the slope of this function measures the rate at which probability is being added as acceptable values of the random variable increase. Thus equal increases of the random variable in the interval (a, b), for which $f(y) > 0$, result in equal additions to the accumulating probability.

Since $P(Y \leq y) = F(y)$, we have the general result given by Equation 8-3-5 for finding the probability that a random value of Y will lie in the interval (y_1, y_2).

$$P(y_1 \leq Y \leq y_2) = F(y_2) - F(y_1) \tag{8-3-5}$$

For the uniform density function on the unit interval, the probability distribution is given by Equation 8-3-6.

$$P(Y \leq y) = F(y) = \begin{cases} 0 & y < 0 \\ y & 0 \leq y \leq 1 \\ 1 & 1 < y \end{cases} \tag{8-3-6}$$

The uniform density function has direct application to some practical problems. For example, in a genetics study, one might assume that X-ray-induced breaks are uniformly distributed along a chromosome, or that textile fibers under test break along their lengths according to a uniform distribution. The results could have important implications in both areas. Here again we are proposing a mathematical model to describe a real-world situation. If it does not fit the existing facts, it must be revised or discarded.

The uniform distribution is also important in theoretical statistics because it is involved in the so-called *probability transformation* and the associated theorem which assures us that proofs involving properties of this simple distribution on the unit interval are applicable to the same properties of continuous distributions in general, including distributions which are not suitable for use in their initial form.

EXERCISES

8-3-1 A random variable Y has a uniform distribution over the interval $0 \leq y \leq 10$.
(a) What would you guess to be the mean of the random variable?
(b) What is the probability that a random value of Y will be in the interval $4 \leq y \leq 9$?
(c) The interval $0 \leq y \leq 1$ or $9 \leq y \leq 10$?
(d) The interval $4 \leq y \leq 7$ or $6 \leq y \leq 8$?
(e) The interval $3 \leq y \leq 6$ and $8 \leq y \leq 9$?
(f) The interval $2 \leq y \leq 7$ and $5 \leq y \leq 8$?
(g) If a random value of Y is obtained, what is the probability that $y^2 \leq 1$?
(h) $y^2 \geq 9$?
(i) $4 \leq y^2 \leq 64$?
(j) $y^2 \geq 49$ or $y^2 \leq 9$?
(k) $y^2 \geq 49$ and $y^2 \leq 81$?

8-3-2 An employer goes to a certain restaurant for lunch, arriving between 12:30 and 1:45 P.M., with all times being equally likely. At least one of his employees is within sight of the restaurant at all times as he returns to work during the period between 12:45 and 1:15 P.M. What is the probability that an employee will see the employer on his way to eat?

8-3-3 A schoolgirl arrives at home each afternoon between 3:45 and 4:00 P.M. The mailman delivers the mail to the house between 3:00 and 4:30 P.M., with all times being equally likely. What is the probability that the mail will be delivered during the time within which the girl regularly arrives? Has it been necessary to assume a distribution for the girl's arrival time?

8-3-4 A circus parade is expected to pass the post office at some time between 1:00 and 3:00 P.M., all times being equally likely. A certain child will be at the same post office between 2:00 and 3:30 P.M., all times being equally likely. What is the probability that both the circus and the child will be at the post office sometime between 2:00 and 3:00 P.M., though not necessarily simultaneously? What assumption did you make in computing the probability?

Chapter 9

THE BINOMIAL AND OTHER DISCRETE DISTRIBUTIONS

9-1 INTRODUCTION

For many experiments there are only two possible outcomes. Thus a coin falls a head or a tail, a plant may be flowering or nonflowering, an opinion on a political issue may be for or against, a machined part may be defective or nondefective, and so on. Such experiments provide the smallest sample space to which a random experiment can apply.

A sequence of experiments will often be such that the outcome of one has no effect on the outcome of another; these trials are said to be *independent*. In addition, the probabilities for two-outcome experiments are frequently constant from trial to trial. A coin has no memory, so that the outcomes of successive trials are independent and the probabilities are constant. Or if opinion sampling is truly random and we sample in a sufficiently small time interval, then the trials may well be independent and the probabilities constant.

Independent trials with only two possible outcomes and constant probabilities are called *binomial* or, sometimes, *Bernoulli* trials, after J. Bernoulli (1654–1705), who was a pioneer in the field of probability.

This chapter is about the very important binomial distribution and a few other discrete distributions.

9-2 THE BINOMIAL DISTRIBUTION

An appropriate sample space for a single random Bernoulli trial consists of two points corresponding to outcomes E and not-E, the latter often being written as $\not{E}$ or as $\bar{E}$, the complement of E. The letters S and U for successful and unsuccessful, or S and F for success and failure, are also much used.

Let us assign a value of 1 to the point E in our sample space and a value of 0 to the point $\not{E}$. We now have a random variable that takes on the values 0 and 1; call it W.

To remain general, suppose we let the probability that W takes on the value 1 be p. Hence, we write $P(E) = P(W = 1) = p$, and, consequently, $P(\not{E}) = P(W = 0) = 1 - p$. The value $1 - p$ is often represented by q so that $p + q = 1$. We now have the probability function for the random variable W. It may be written in a single expression as Equation 9-2-1.

$$P(W = w) = p^w(1 - p)^{1-w} \qquad w = 0, 1 \tag{9-2-1}$$

Thus, if we substitute $w = 1$, we have $P(W = 1) = p^1(1 - p)^0 = p$ since $(1 - p)^0 = 1$; for $w = 0$, we have $P(W = 0) = p^0(1 - p)^1 = 1 - p$.

Most likely, many observations are available. Thus, we toss a coin many times, observing the sequence of heads and tails; or we examine many machined parts, observing which are defective and which nondefective. For n independent Bernoulli trials, we record an outcome as a sequence of n letters, say H's and T's, or as an ordered set of n numbers, called an *n-tuple*, each entry being a 1 or 0 in this case, the order being that in which the results of the random trials were observed. For example, if we toss a coin five times, we might record the result as THTHH or as the sample point or n-tuple (01011) where the random variable assigns a 1 to an H and a 0 to a T.

In general, if W_i is the random variable associated with the ith binomial trial, then $(W_1, \ldots, W_n)$ represents the random outcome of the full experiment. Our observation is a sequence of 1s and 0s since each W_i can assign only one of two possible outcomes, a 0 or a 1, at a trial. Also, there must be exactly 2^n sample points in the sample space. For example, see Figure 5-3-1 for $n = 2$.

To present such data more comprehensibly, we may add the numbers in the sequence to get the number of times the event E occurred, for example, the number of heads in the previous sequence of $n = 5$ coin tosses was 3. Consider the random variable $Y = \sum W_i$, that is, the number of 1s or the number of times E occurred in n binomial trials. In other words, Y assigns a unique, real value to each of the 2^n sample points, and, consequently, an observed value is determined as soon as we know the outcome of a random experiment. In the case of two tossed coins, Y assigns 0 to (0, 0), 1 to (0, 1), 1 to (1, 0), and 2 to (1, 1). For n trials, Y takes one of the values 0[1]n. We now need the probability distribution of the random variable Y.

When $Y = n$, the event E was observed at every trial, that is, $W_i = 1$, for all i. The observed n-tuple was $(1, \ldots, 1)$. Now $P(W_i = 1) = p$ for all i since the probability remains constant from trial to trial. Since the trials are also independent, the generalization of the multiplication theorem for probabilities, Equation 6-3-2, applies with the additional Equations 6-3-3 to give Equation 9-2-2.

$$P(Y = n) = pp \cdots p = p^n \tag{9-2-2}$$

When $Y = n - 1$, only once was the event $\not{E}$ observed. Suppose this occurred at the first trial. Then the outcome of the experiment is described by the n-tuple or sample point $E_1 = (0, 1, \ldots, 1)$, and its probability is given by Equation 9-2-3.

$$\begin{aligned} P(E_1) &= P(0, 1, \ldots, 1) \\ &= P(W_1 = 0)P(W_2 = 1) \cdots P(W_n = 1) \\ &= (1 - p)p \cdots p \\ &= (1 - p)p^{n-1} \end{aligned} \tag{9-2-3}$$

If the event $\not{E}$ had occurred at the second trial, then the outcome of the experiment is described by the n-tuple $E_2 = (1, 0, 1, \ldots, 1)$. Its probability is given by Equation 9-2-4.

$$\begin{aligned} P(E_2) &= P(1, 0, 1, \ldots, 1) \\ &= P(W_1 = 1)P(W_2 = 0)P(W_3 = 1) \cdots P(W_n = 1) \\ &= p(1 - p)p \cdots p \\ &= (1 - p)p^{n-1} \end{aligned} \tag{9-2-4}$$

Continuing, we see that the event $\not{E}$ could occur at any one of the n random trials, leading to a 0 at the corresponding position in the n-tuple and 1s elsewhere. Clearly, there must be n n-tuples or sample points, say $E_1, E_2, \ldots, E_n$, all of which have the same probability, that are assigned a value of $n - 1$ by the random variable Y. Hence, we have Equation 9-2-5.

$$\begin{aligned} P(Y = n - 1) &= P(E_1 \cup \cdots \cup E_n) \\ &= \sum P(E_i) \\ &= n(1 - p)p^{n-1} \end{aligned} \tag{9-2-5}$$

When $Y = n - 2$, the event $\not{E}$ is observed exactly twice. The experimental outcome described by the n-tuple $(0, 0, 1, \ldots, 1)$, say E_1', is one point in the sample space to which the value $Y = n - 2$ is assigned. Its probability is given by Equation 9-2-6.

$$\begin{aligned} P(E_1') &= P(0, 0, 1, \ldots, 1) \\ &= P(W_1 = 0)P(W_2 = 0)P(W_3 = 1) \cdots P(W_n = 1) \\ &= (1 - p)(1 - p)p \cdots p \\ &= (1 - p)^2 p^{n-2} \end{aligned} \tag{9-2-6}$$

The two 0s which occur when $Y = n - 2$ may occupy any two positions in the n-tuple. Equation 7-2-3 or 7-3-1 tells us that there are $\binom{n}{2}$ different choices of two spaces from an available n spaces. The remaining positions in any n-tuple are to be occupied by ones. Consequently there are $\binom{n}{2}$ different n-tuples or sample points, call them $E'_1, \ldots, E'_{\binom{n}{2}}$, with exactly two 0s and, hence, $\binom{n}{2}$ sample points to which the random variable Y assigns the value $n - 2$. The probability that $Y = n - 2$, then, is given by Equation 9-2-7.

$$\begin{aligned} P(Y = n - 2) &= P(E'_1 \cup \cdots \cup E'_{\binom{n}{2}}) \\ &= \sum P(E'_i) \\ &= \binom{n}{2} (1 - p)^2 p^{n-2} \end{aligned} \tag{9-2-7}$$

In general, $Y = n - r$ is the experimental outcome in which $\not{E}$ was observed r times and E was observed $n - r$ times. This may be represented by a sample point with r 0s and $n - r$ 1s. Any one of these has a probability of $(1 - p)^r p^{n-r}$, and there are $\binom{n}{r}$ such points. The probability that $Y = n - r$ is, then, given by Equation 9-2-8.

$$P(Y = n - r) = \binom{n}{r} (1 - p)^r p^{n-r} \tag{9-2-8}$$

Since we tend to use lowercase letters to represent values of a random variable, we may write Equation 9-2-8 as Equation 9-2-9.

$$P(Y = y) = \binom{n}{y} p^y (1 - p)^{n-y} \qquad y = 0, 1, \ldots, n \tag{9-2-9}$$

Equation 9-2-9 is called the binomial distribution. It is a probability (density) function. Recall that $\binom{n}{0}$ and $\binom{n}{n}$ give no problem, since 0! has been defined as equal to 1. This function is an appropriate probability model in many situations.

We have a *binomial distribution* when:

1 A single trial results in one of two possible outcomes, say E and $\not{E}$.
2 The probability, say p, that the outcome E occurs remains constant in repeated trials.
3 Repeated trials are independent.
4 A fixed number, say n, of trials are to be conducted.

The probability density of the random variable that is the number of times E occurs, say Y, is given by Equation 9-2-9. This is usually called the binomial distribution.

Some examples of the application of the binomial distribution follow.

Illustration 9-2-1

Four plants, constituting a random sample, are classified according to the presence or absence of a certain characteristic expected to be found with probability 13/16. What is the probability that the characteristic will be present in, at most, one of the plants?

The required probability is that of the two events: (a) exactly one showing presence and (b) none showing presence. Thus, the probability is given by two terms from the binomial distribution and is

$$\binom{4}{1}\frac{13}{16}\left(\frac{3}{16}\right)^3 + \left(\frac{3}{16}\right)^4 \approx .02$$

Here, we might have considered a genetics problem and come to the conclusion that the characteristic in question occurred with a probability of 13/16. We next obtained a random sample of four plants and found only one with the characteristic. This seems like a small number considering the value 13/16 is nearly 1, and raises the question whether or not our sample is unusual. Although we have only five possible experimental outcomes here, 0, 1, 2, 3, or 4 with the characteristic, we realize that, more generally, when n is large and many outcomes are possible, any one must have a low probability. Consequently, it is not appropriate to draw a conclusion by looking at the probability for a single outcome, so we rephrase our question to ask for the probability of the observed event or a more extreme one. This is the probability we have computed.

Now we note that if we do have a binomial distribution with $p = 13/16$, and find that out of a randomly selected sample of four plants at most one of the plants exhibits the characteristic, then we have observed an event that is likely to occur only about 2 times out of 100 on the average. Most of us would think that we had observed an unusual event or that the theory that led to the probability model was not valid. In the latter case, we would almost certainly question the validity of the hypothesis that $p = 13/16$. On the other hand, if we were unsure about the sampling scheme, we might want to review the assumption about the independence of the trials.

Illustration 9-2-2

On moving into a new neighborhood, I find that the family next door consists of four boys and four girls. Assuming that at each birth the probability of having a boy is 1/2 and that determination of the sex of the child at successive births involves independent events, what was the probability that the next-door family would consist of four boys and four girls if it was to consist of eight children?

For a binomial model, the probability must be as follows.

$$\binom{8}{4}\left(\frac{1}{2}\right)^4\left(\frac{1}{2}\right)^4 = \binom{8}{4}\frac{1}{2^8} \approx .27$$

This says that for families of 8, the observed distribution of sexes is likely to occur about 3 times out of 10 (27/100), on the average. The result does not seem to be unusual.

In a case like this, there are only nine possible kinds of families, from eight boys and no girls to no boys and eight girls. As was pointed out previously, where there are many possible values of a random variable, clearly no one value can have a very high probability associated with it, so that every value is an unusual one. In such cases, we often pool a number of outcomes—for example, ones that we categorize as extreme, as when a family consists of eight girls—to give an event and then talk about probabilities.

Alternatively, we might look at the probabilities associated with other possible families. Since $(1/2)^8$ will be a part of each probability, we need to look only at the coefficients. A quick check shows that $\binom{8}{4} = 70$ is the largest.

In many problems, one might well raise the question as to which event has the greatest probability. In general, if $p(E$ in a single trial$) = p$, if there are to be n independent trials, and if Y is the random variable that counts the number of times E occurs, then the probability is maximum when the count is the integer in the interval $[p(n + 1) - 1, p(n + 1)]$. For a family of eight, as in Illustration 9-2-2, this interval is $[\frac{1}{2}(9) - 1, \frac{1}{2}(9)] = (3\frac{1}{2}, 4\frac{1}{2})$, and hence $y = 4$ gives the term with maximum probability.

Also, observe that Equation 9-2-9 is the general term in the expansion of $[p + (1 - p)]^n$, given in Equation 9-2-10.

$$[p + (1 - p)]^n = p^n + \binom{n}{n-1} p^{n-1}(1 - p) + \binom{n}{n-2} p^{n-2}(1 - p)^2 + \cdots + \binom{n}{r} p^r(1 - p)^{n-r} + \cdots + (1 - p)^n \qquad (9\text{-}2\text{-}10)$$

This binomial expansion can be generated by direct multiplication. You may already have seen Equation 9-2-10 in an algebra course as the *binomial theorem* for the more general expansion of $(a + b)^n$.

Finally, the binomial distribution is of so much practical use that many related tables have been produced. Among these are tables prepared by the National Bureau of Standards (1950). These give probabilities for single terms and cumulative probabilities for sequences beginning with terms at the end of the distribution and over a wide range of values of n and p.

EXERCISES

9-2-1 A fair die is rolled three times. How many points are there in the sample space? If Y is the random variable representing the number of 6s observed, what is the probability distribution of Y? If four people try the experiment, what is the probability that they will not all roll three 6s?

9-2-2 A fair coin is tossed 10 times. Let Y be the random variable representing the number of heads. Present the probability distribution of Y. What is the probability that the 10 tosses will be either all heads or all tails? At most one head? At most one tail? At most two heads? That the H:T ratio be no less extreme than 8:2 either way?
9-2-3 A test has 10 multiple-choice questions, and each question has 4 options. If a student takes the test using only guesswork, what is the distribution of the number of questions correctly answered? What is the probability that five or more are answered correctly?
9-2-4 For Illustration 9-2-1, what number of plants having the characteristic involved has the greatest probability of occurring? What are the probabilities associated with its neighbors?
9-2-5 In the course of a large experiment, suppose that an investigator makes 20 independent tests of hypotheses. She does this with the understanding that each time she tests a hypothesis, the probability of her falsely declaring it to be unsupported by the data is .10.

If all 20 hypotheses are, in fact, true, what is the probability that she will falsely declare no more than two to be unsupported by the data?
9-2-6 A recent very large survey reported that 80 percent of a certain population had at least three-quarters of their hospital bills paid by insurance. Suppose that a doctor whose hospital patients may be considered to be a random sample from this population presently has eight hospitalized. What is the probability that all will have at least three-quarters of their hospital bills paid by insurance? That none will be covered this extensively? That no more than one will have to pay better than one-quarter of the hospital bills because of lack of insurance coverage?
9-2-7 A toy store finds that about one in five of its model kits is returned because of stated defects. In one day, six kits are sold. What is the probability that fewer than half of this day's sales will be returned?

The questions that follow are more involved or algebraic in nature.
9-2-8 Suppose a trial consists of tossing two fair coins and observing the number of heads X. What is the distribution of X? What is the probability that a random trial will produce no heads? At least one head?

If five of the two coin trials are made, let Y be the number of trials in which no heads were observed. What is the probability distribution of the number of trials producing no heads? What is the probability that at no time will a head be observed? How does this compare with the probability that no heads will be observed when a single coin is tossed 10 times? What is the probability that each trial will result in one head and one tail? Where is this probability in terms of the probability distribution of Y?
9-2-9 In an inheritance problem, a geneticist is not sure whether the observed have versus have-not pattern should exhibit a ratio representative of a true 1:1 or a true 3:1 ratio. He decides to base his conclusion on an experiment in which he will make 20 observations.

Suppose other evidence suggests that the 1:1 ratio is more likely true than is the 3:1 ratio. On this basis, he decides to reject the 1:1 ratio in favor of the 3:1 ratio only if the haves show a marked numerical superiority. He will reject the 1:1 ratio only if the probability, computed under the assumption that the 1:1 ratio is true, of the number observed in the have class, or any larger number, has a value less than .1. He completes the experiment and finds exactly 15 haves. What is the probability of 15 or more haves in a random sample of 20? Which ratio does he conclude to be the correct one?

If, on the other hand, supporting evidence had suggested that 3:1 was the more likely ratio, then he might well have made a decision against 3:1 only if the have-nots had a considerable numerical superiority. Suppose the rule is to reject 3:1 in favor of 1:1 only if the probability, computed using the 3:1 ratio, of as many or more have-nots as observed, is less than .1. Suppose 17 have-nots are found. What is the value of the probability he is to compute? What ratio does he conclude to be the correct one?

Again, suppose there is no reason to favor one ratio rather than the other. The investigator decides to accept the 1:1 ratio if there are 0, 1, 2, . . . , or 12 haves and to accept the 3:1 ratio if there are 13, 14, . . . , or 20 haves.

What is the probability of accepting 3:1 if 1:1 is true? What is the probability of accepting 1:1 if 3:1 is true?

This question is concerned with a minimax solution to a decision problem.

9-2-10 Show that $\binom{n}{r} = \binom{n}{n-r}$.

9-2-11 Consider the binomial expansion. When $n = 1$, we have $p + q$ with coefficients 1 and 1, respectively.

By direct multiplication, find the expansion when $n = 2$ and observe the coefficients. Multiply this expansion by $p + q$ and so have the binomial expansion for $n = 3$. Again, observe the coefficients. Repeat the procedure at least one more time.

Prepare a table of coefficients as follows, allowing one line for each value of n. Begin at the left margin and write down the coefficients for $n = 1$. Directly below these, write those for $n = 2$; a third column must, necessarily, be introduced. Below these, write those for $n = 3$, introducing still another column. Repeat for $n = 4$.

Note that any two consecutive numbers on a line add to the number, on the line below, directly under the second number in the sum. This provides a way of generating new lines of coefficients. The resulting triangular array is called *Pascal's triangle*.

9-2-12 In a binomial trial, what dimension is the space (point, line, flat surface, volume) required to represent the sample space? What dimension is required if the sample space is to represent the joint result of two trials? Three trials? What do you do if there are n trials and n is not specified?

9-3 THE HYPERGEOMETRIC DISTRIBUTION

Sampling a finite population with replacement, when there were only two possible outcomes at a trial, was one sampling scheme for which the binomial distribution provided a realistic model. However, it is often *necessary* to sample without replacement, as is the case for many card games, and in other situations, such a scheme seems at least desirable. For example, if our problem is to sample a shipment of dried yeast packages for defectives in order to judge its acceptability, the obvious sampling scheme is one without replacement.

Suppose then that we have a population of size N in which N_1 items are of a particular kind, perhaps nondefectives, and $N - N_1$ are of another, perhaps defectives. We are to draw a sample of n items and observe them as n_1 of the first kind and $n - n_1$ of the second.

In a single trial, the probability of getting an item of the first kind is N_1/N, but, at a second trial, the probability of this event will depend on the outcome of the first. Relating this fact to the binomial distribution, we find that the trials are not independent and that probabilities do not remain constant from trial to trial. The binomial model is not suitable here.

We have seen, Equations 7-2-3 and 7-3-1, that n distinct items can be drawn from a population of size N in $\binom{N}{n}$ ways. This is without regard to the order of their being selected. Our sample space has, then, $\binom{N}{n}$ points, all of which are equally likely and, as a consequence, each has probability $1/\binom{N}{n}$.

Let Y, as a random variable, be the number of items of the first kind to be found in a sample of size n. Then the random variable Y assigns a real value to each and every point in the sample space, its domain. Note that if Y takes the value n_1, then the sample will also contain $n - n_1$ items of the second kind.

Consider the sample probability when $Y = n_1$. There are $\binom{N_1}{n_1}$ ways of drawing exactly n_1 items of the first kind for the sample and $\binom{N - N_1}{n - n_1}$ ways to draw the second kind. Consequently, a sample of size n that includes exactly n_1 items of the first kind may be selected in $\binom{N_1}{n_1}\binom{N - N_1}{n - n_1}$ ways. Hence the probability associated with this sample is given by Equation 9-3-1.

$$P(Y = n_1) = \frac{\binom{N_1}{n_1}\binom{N - N_1}{n - n_1}}{\binom{N}{n}} \tag{9-3-1}$$

This probability (density) function is often called the hypergeometric distribution.

We have a *hypergeometric distribution* when:

1 A single trial results in one of two possible outcomes, say E and $\not{E}$.
2 The probability, say p, that the outcome E occurs varies in repeated trials.
3 Repeated trials are dependent.
4 A fixed number, say n, of trials are to be conducted.

The probability density of the random variable that is the number of times E occurs, say Y, is given by Equation 9-3-1. This is usually called the hypergeometric distribution.

It can be seen that construction of tables for the hypergeometric distribution is not as simple as for the binomial distribution and must depend upon four values: N, N_1, n, and n_1. Lieberman and Owen (1961) have prepared a reasonably extensive table of this distribution.

For large and even small population sizes and not too large sample sizes, the binomial distribution may serve as a reasonable approximation to the hypergeometric distribution. When sampling is from a small finite population and/or the sample consists of more than 5 percent of the population, the finite population correction, given later in Section 13-6, should be used in conjunction with binomial-distribution procedures.

Illustration 9-3-1

An opaque bag contains eight Red Delicious and two Yellow Delicious apples. Suppose you are allowed to reach in the bag and select two apples at random. What is the probability that they will both be Red Delicious? One of each? Both Yellow Delicious?

For this problem, $N_1 = 8$, $N - N_1 = 2$, and $N = 10$. For the first part, $n = 2$, $n_1 = 2$, and $n - n_1 = 0$. Hence,

$$P(\text{both red}) = \frac{\binom{8}{2}\binom{2}{0}}{\binom{10}{2}} = \frac{\frac{8(7)}{2} \times 1}{\frac{10(9)}{2}} = \frac{28}{45}$$

For one of each, $n_1 = 1$ and $n - n_1 = 1$. Hence,

$$P(\text{one red, one yellow}) = \frac{\binom{8}{1}\binom{2}{1}}{\binom{10}{2}} = \frac{8(2)}{45} = \frac{16}{45}$$

Finally,

$$P(\text{both yellow}) = \frac{\binom{8}{0}\binom{2}{2}}{\binom{10}{2}} = \frac{1}{45}$$

Since we have considered all possible outcomes, the sum of their probabilities should and does equal unity.

EXERCISES

9-3-1 A certain university committee is to consist of 5 individuals chosen randomly from 15 faculty members and 15 students. What is the probability that it will contain no students? No faculty? Exactly one student? No more than three faculty members?

9-3-2 There are 10 faculty and 15 students available for a committee of 5 individuals to be selected at random. What is the probability that the committee will have no students? No faculty? Exactly one student? At most, four students?

9-3-3 Suppose that there are 10 faculty and 20 students available for the committee of 5. What is the probability that the committee will have no students? Exactly one student? No faculty? Exactly one faculty?

9-3-4 A bridge hand consists of 13 cards. What is the probability that it will contain no spades? Five spades? Six spades? Seven spades? Four aces? No aces? One ace?

9-3-5 Of 20 practice weldings, 5 are defective. The instructor examines a random sample of 4 weldings and is certain to detect defective ones if present. What is the probability that all will be found satisfactory? At least one defective? At most, one defective? No more than one defective?

The *sampling fraction* here is 4/20, which would seem moderate. Compute the probabilities required above by using the binomial distribution as an approximation. Compare the two sets of probabilities.

9-3-6 Suppose there are 25 defectives in a lot of 100 items. This is the same percentage defective as in the preceding exercise. Again, a sample of 4 is taken so that the sampling fraction is now a small 4/100. Compute the same probabilities as before, using both the exact hypergeometric and the approximate binomial distributions. Again compare the results.

The two exercises that follow are algebraic in nature and intended only for those students with such an interest.

9-3-7 In developing the hypergeometric distribution, we made no mention of the range of Y. For some situations, Y may be as small as zero, but in other cases this may be impossible. For example, if five males and five females are available for a committee of six persons to be drawn at random and if Y represents the number of females on the committee, then Y must necessarily have the range $1, 2, \ldots, 5$, zero not being possible.

In general, the lower value of the range of Y may be described as max $(0, v_l)$ where v_l is the lower or minimum possible value when 0 is not possible. How is v_l described in terms of N, N_1, and n?

Similarly, the upper value of the range may be described as min (v_u, n) where v_u is the upper or maximum value when n is not possible. How is v_u described in terms of N_1 and n?

9-3-8 Tables for the hypergeometric distribution are necessarily limited and must be entered with four values: N, N_1, n, and n_1. One is able to extend the availability of the tables by noting the following equation where the roles of N_1 and n are seen to be interchanged on the two sides of the equality.

$$\frac{\binom{N_1}{n_1}\binom{N - N_1}{n - n_1}}{\binom{N}{n}} = \frac{\binom{n}{n_1}\binom{N - n}{N_1 - n_1}}{\binom{N}{N_1}}$$

Prove that this equality is true.

9-4 THE POISSON DISTRIBUTION

There are many situations where we must count the number of individuals possessing a certain characteristic yet have difficulty in defining the basic experiment. In turn, it becomes difficult to say what is the sample size or what is the probability of the occurrence of a single event. For example, suppose we are inspecting a continuous sheet of glass for flaws, a dangerous intersection for accidents, or a switchboard for the number of calls it receives. It is easy to count the events, but what are the nonevents?

In situations like those just discussed, we customarily resort to specifying a unit size of glass to be examined, or a time interval in which to observe the intersection or the switchboard. We find then that we are observing counts that fluctuate around some mean value that might be defined in terms of some sort of underlying binomial parameter p and sample size n as a value of np, a product never separable into its component parts and simply given the symbol μ. Thus we observe only the number of flaws in a pane of glass of some standard size, or the number of accidents per week at the intersection, or the number of calls coming through the switchboard in a 5-minute interval.

With somewhat more formality, we assume that for a short enough unit of time or space, the probability of an event occurring is proportional to the length of time or size of the space, and that it is close enough to impossible for two such events to occur in the same unit that this can be ignored. We also assume that for nonoverlapping units, the results in one unit are of no value in predicting when or where another event will occur.

The above assumptions underlie the probability or density function given by Equation 9-4-1.

$$P(Y = y) = \frac{e^{-\mu}\mu^{y}}{y!} \qquad y = 0, 1, 2, \ldots \tag{9-4-1}$$

This is often called the Poisson distribution. Notice that the sample space is infinite. The parameter μ represents the mean value and corresponds to the np value of our earlier argument, a value actually computed when the Poisson is used as an approximation to the binomial distribution. The letter e represents a constant which is frequently encountered in mathematics, $e = 2.71828$, approximately.

We have a *Poisson distribution* when:

1 A single trial is defined by first specifying a reasonable unit of time or space; the trial is to result in an outcome that consists of a 0, 1, 2, . . . , that is, the number of times an event is observed in the unit.
2 The average number of outcomes, the parameter μ, in a trial remains constant from trial to trial, trials to be nonoverlapping. When a binomial distribution is being approximated, $\mu = np$.
3 Nonoverlapping trials are independent.

Illustration 9-4-1

A particular type of canned fruit of acceptable quality has been found to have an average of 2.5 insect particles per can of a specified size. What is the probability that a randomly selected can of this size will have five or more insect particles?

In terms of the requirements for a Poisson distribution, the single trial is defined on the basis of a specified can size which becomes the unit of space; the event involves the identification of an insect particle and a count of these is made. Quite possibly the parameter does not change from trial to trial since the canner or other agency probably recommended a spraying schedule to be followed by all growers as the crop grew and the canning process introduced a certain amount of homogeneity. Independence of trials could raise some questions and these, in turn, could reflect on the validity of the previous assumption. Here, the problem is that insects lay many eggs simultaneously, so that, as they hatch, they will be together; thus finding one will usually mean finding many, at least prior to canning. Let us assume that all requirements are appropriately met.

To answer the question about the single can, we need $P(Y = 5) + P(Y = 6) + \cdots$. We will answer it by computing $1 - P(Y = 0) - P(Y = 1) - \cdots - P(Y = 4)$.

$$\begin{aligned} P(Y = 0) &= \frac{e^{-2.5}(2.5)^0}{0!} \\ &= e^{-2.5} \qquad \text{since } 2.5^0 = 1 \text{ and } 0! = 1 \\ &= .0821 \end{aligned}$$

Values of e raised to various powers, both positive and negative, are found in many mathematical tables.

$$\begin{aligned} P(Y = 1) &= \frac{e^{-2.5}2.5}{1} = P(Y = 0)\,\frac{2.5}{1} = .2053 \\ P(Y = 2) &= \frac{e^{-2.5}(2.5)^2}{2!} = P(Y = 1)\,\frac{2.5}{2} = .2566 \\ P(Y = 3) &= \frac{e^{-2.5}(2.5)^3}{3!} = P(Y = 2)\,\frac{2.5}{3} = .2138 \\ P(Y = 4) &= \frac{e^{-2.5}(2.5)^4}{4!} = P(Y = 3)\,\frac{2.5}{4} = .1336 \end{aligned}$$

Hence,

$$\begin{aligned} P(5 \text{ or more}) &= 1 - .2053 - .2566 - .2138 - .1336 \\ &= .1086 \end{aligned}$$

The unattractive prospect does not appear to be too unlikely.

Notice how each probability may be constructed from the previous one.

Illustration 9-4-2

A manufacturing plant produces an item required in the assembly of minicomputers. The item is shipped in boxes of 100, and the plant is known to release about 3 percent defective items.

What is the probability that a box will contain no more than three defective parts?

Here a single trial is defined with reference to a box of 100 items. However, the range cannot exceed 100, so we know that we are more likely to be dealing with a binomial distribution. The average number of outcomes is $\mu = np = 100(.03) = 3$, and there is no reason to think this varies from trial to trial. Experience should provide the information relative to independence with respect to an item and, in turn, for boxes. Because we have a binomial distribution, the Poisson is serving as an approximation.

The required probability is $P(Y = 0) + P(Y = 1) + P(Y = 2) + P(Y = 3)$.

$$P(Y = 0) = \frac{e^{-3}3^0}{0!} = .0498$$

We find this by looking up e^{-3} in appropriate tables.

$$P(Y = 1) = \frac{e^{-3}3}{1!} = .0498(3) = .1494$$

$$P(Y = 2) = \frac{e^{-3}3^2}{2!} = .0498(4.5) = .2241$$

$$P(Y = 3) = \frac{e^{-3}3^3}{3!} = .0498(4.5) = .2241$$

Hence, by adding these probabilities, we find

$$P(Y \leq 3) = .6474$$

We may also observe that $P(Y > 3) = .3526$.

Reasonably extensive tables of the Poisson distribution have been prepared by Molina (1949).

EXERCISES

9-4-1 A certain fire station averages 3.8 false alarms per week. Check the Poisson assumptions to see if this distribution provides an appropriate model for computing probabilities. What is the probability that the station will get no false alarms in one week? Fewer than two? More than five?

9-4-2 A secretary claims she averages only one detectable erasure per page of typing. Is the Poisson distribution appropriate for computing probabilities of detecting various numbers of erasures? What is the probability of detecting none? One? More than four?

9-4-3 A "special service" telephone operator receives 10 calls per hour, on the average, for a certain period of the day. What is the probability that, in one hour, he will receive exactly 10 calls? No calls? Five or fewer calls?

9-4-4 A type of manufactured cloth averages .7 flaws per linear yard. A certain dress pattern requires 3 yards. What is the probability of getting a flawless piece of material? One flaw? Two flaws? More than two flaws?

9-4-5 A manufacturing process produces an item where there are, on the average, 1.5 defectives per 100. These are sold in lots of 50.

Use the Poisson distribution to determine the probability that there will be no defective items in a randomly selected lot. Exactly one. Two. Three. Four. Five. More than five.

Use the binomial distribution to compute the same probabilities.

Compare the two sets of corresponding probabilities. Which distribution has the more appropriate assumptions?

The problem that follows is for those who enjoy a little algebra.

9-4-6 Write out Poisson terms for the probabilities that Y takes the values of n and $n + 1$. From these, write $P(Y = n + 1)$ in terms of $P(Y = n)$ and any necessary multiplier. This provides a *recursion formula*, that is, one which can be used repeatedly to generate successive terms.

9-5 OTHER DISCRETE DISTRIBUTIONS

It must be clear that we have not exhausted all possibilities for discrete probability functions. For example, many classification systems allow for more than two outcomes. A voter may have decided, with respect to a certain bond issue, to vote "yes," "no," or not to vote, or still be undecided.

One option for a probability function to describe such a situation would have essentially the same sort of requirements as for a binomial distribution with the exception that there would be several possible outcomes. In turn, each outcome would have its own probability, and the sum of all these would be unity. This would be the *multinomial* rather than the binomial distribution.

The multinomial distribution can be illustrated in connection with a die problem where the outcome is any one of the numbers 1, 2, . . . , 6. This particular case is an example of the discrete *uniform probability function.*

A distribution perhaps less likely to come to mind is one where the number of trials has not been fixed but is made dependent on the number of successes. For example, suppose a party is planned and the menu is to include baked potatoes. The cook will, no doubt, have to sort the available potatoes for exactly the right number of good or uniform ones. The number sorted, not the number selected, is the sample size, and this is not known until the predetermined number of acceptable potatoes has been found.

While the example above may seem trivial, such a sampling scheme requiring statistical treatment is very realistic. When the assumptions for a binomial distribution are met, except that that of fixed sample size is replaced by the one under consideration, the probability function is called the *negative binomial.*

Chapter 10

THE NORMAL DISTRIBUTION

10-1 INTRODUCTION

The rectangular or uniform distribution was discussed in Section 8-3 as an introduction to continuous distributions. While it is important, it yields first place, in statistical theory and application, to the so-called normal distribution.

The *normal distribution*, also called the *normal law of error*, is widely used in research in the biological, physical, and social sciences. There, we find that observations are often assumed to have come from a normal population and that this is a reasonable model. As we shall see, the term "normal" has a specific interpretation and is not intended to distinguish between normal and abnormal situations. In fact, the word "anormal" is used to describe a distribution of unspecified form when it is known to be other than normal.

Historically, the normal distribution goes back to the seventeenth and eighteenth centuries and is associated with the names of DeMoivre (1667–1754), Laplace (1749–1827), and Gauss (1777–1855). At that time, it received the attention of mathematicians and natural and social scientists. Its application to biological data was pioneered at a later date by Sir Francis Galton (1822–1911).

This chapter deals with the normal distribution.

10-2 THE NORMAL DISTRIBUTION

The characteristic appearance of the *normal curve* is described as bell-shaped. It is, therefore, symmetric. Symmetry is about the mean, commonly designated by μ, which is also the median. The curve, shown in Figure 10-2-1, is concave downward in the center, but at a distance of 1 standard deviation σ on either side of the mean it changes to concave upward. The two points on the normal curve for which y equals $\mu - \sigma$ and $\mu + \sigma$, often written as $\mu \pm \sigma$, are called *points of inflection.* Both μ and σ appear as parameters in Equation 10-2-1, a mathematical description of the normal curve, and are the only parameters required to completely describe a normal distribution.

$$f(y) = \frac{1}{\sqrt{2\pi}\,\sigma} e^{-[(y-\mu)^2/2\sigma^2]} \qquad -\infty < y < \infty \qquad (10\text{-}2\text{-}1)$$

The symbols π and e that appear in Equation 10-2-1 are common numerical constants; to reasonable accuracy, $\pi = 3.14159$ and $e = 2.71828$.

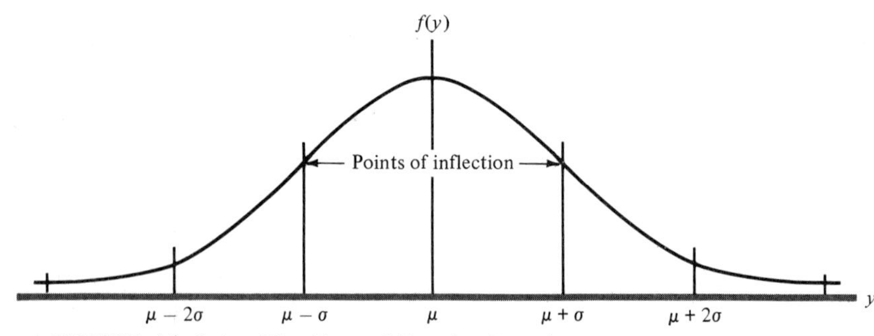

FIGURE 10-2-1 The Normal Density Function.

For a discrete variable, we were able to write $f(y) = P(Y = y)$, but this is no longer valid when we deal with a continuous variable. For a continuous variable, $F(y)$ is not a probability. Now probabilities are in terms of intervals in which the random variable may occur, so are measured as areas under the curve and above the y axis. The total area under the curve is unity.

The shape of this density curve implies that observations occur most frequently in the neighborhood of the mean and decrease in frequency as they become increasingly distant from the mean. Approximately 68 percent of the area under the curve lies within 1 standard deviation of the mean, approximately 95 percent within 2, and almost all within 3. This information is also stated in the following equations which give the probability that the random normal variable Y will fall within the three intervals specified, respectively. These equations help us in understanding the concept of a standard deviation.

$$P(\mu - \sigma \leq Y \leq \mu + \sigma) = .6827$$
$$P(\mu - 2\sigma \leq Y \leq \mu + 2\sigma) = .9545$$
$$P(\mu - 3\sigma \leq Y \leq \mu + 3\sigma) = .9973$$

Note that the range of the variable is no longer finite but that a value may be negatively infinite, $-\infty$, or positively so, $+\infty$. This symbol is not intended to represent a number but, rather, something larger than any number which you can conceive. Of course, we have already seen that values outside the interval $\mu \pm 3\sigma$ are very unusual, occurring with a probability of only $1 - .9973 = .0027$.

From Equation 10-2-1, it can be seen that $f(y)$ is a maximum when $y = \mu$. Since $e^0 = 1$, this maximum is $1/(\sqrt{2\pi}\,\sigma)$. Note that this must decrease as σ increases so that the normal curve flattens and spreads out with increasing standard deviation, just as it reasonably should. Keep in mind that the area under the curve must always equal unity.

For many continuous random variables, when a large number of observations are obtained and used to construct a histogram, it appears that they would be distributed in a manner that looks like the normal distribution if the histogram were to be based on small enough class intervals. Here we try to visualize a smoothed curve of the same area superimposed upon the histogram. Consequently, it is easy to justify use of the normal density function as a reasonable mathematical model or explanation for observations made on these continuous random variables.

It is not possible to present an equation which describes the cumulative distribution function, that is, an equation into which we can substitute a possible value of the random variable and produce the probability of a value no larger than this occurring by chance. Instead, we must rely on tables prepared by other methods. For example, Table B-1 provides areas or probabilities in the right tail of the probability function. The normal distribution is used so extensively that many other convenient tables have been prepared, tables giving areas in both tails, areas located symmetrically about the mean, and so on. One must always read table headings for the normal distribution with special attention. It is also possible, though less practical, to use a graph of the cumulative normal distribution function, such as that in Figure 10-2-2.

The left and right marginal columns of Table B-1 have values, z, to one decimal place, taken by a random variable, Z; a second decimal place is given in the top and bottom rows. It is customary to use the letter Z for a random variable from a normal density function with zero mean, $\mu = 0$, and unit standard deviation, $\sigma = 1$. The body of the table contains areas under the normal curve or probabilities associated with the right tail of the distribution, that is, to the right of the tabulated value of the normal variable, z or z_α (see Figure 10-2-3). The subscript α, Greek letter alpha, designates the probability of a larger value of the random variable

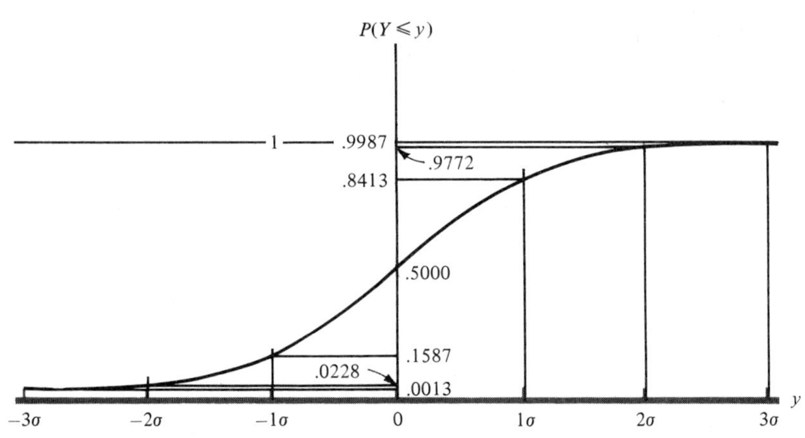

FIGURE 10-2-2 The Cumulative Normal Probability Distribution Function.

occurring by chance, that is, when z is specified, we can compute the probability, namely α, remaining in the right tail of the distribution. In the other words, α is defined by $P(Z \geq z_\alpha) = \alpha$, and the subscript on z indicates the one-sided nature of the probability. Since the normal curve is symmetric, probabilities need be given for only the right half of the distribution, as is the case here.

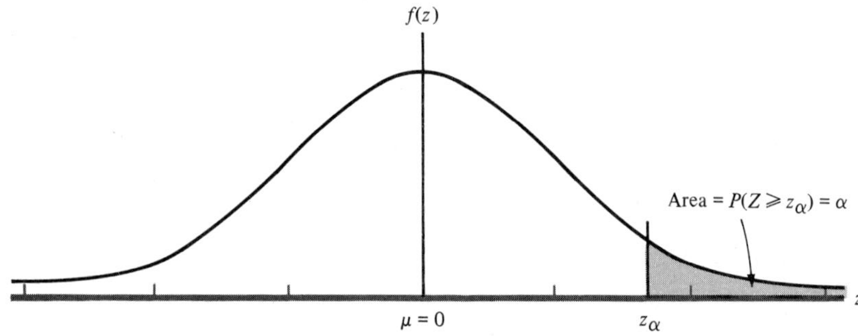

FIGURE 10-2-3 Areas Under the Normal Curve Related to Table B-1.

10-3 THE NORMAL DISTRIBUTION WITH ZERO MEAN AND UNIT VARIANCE

Table B-1 gives probabilities that a randomly selected value of the variable Z will be larger than a specified value, z or z_α, when Z is normally distributed with zero mean and unit standard deviation. This is illustrated in Figure 10-2-3. The reading of Table B-1 is shown in the following illustrations.

Illustration 10-3-1

Find $P(Z \geq 1.52)$.

The requirements of the problem are sketched in Figure 10-3-1. We enter the table by finding 1.5 in the left column and .02 across the top. This corresponds to $z = 1.52$. In the body of the table opposite 1.5 and under .02, read .0643. This is the probability that $Z \geq 1.52$. In other words, for $z = 1.52$, $P(Z \geq z) = .0643$.

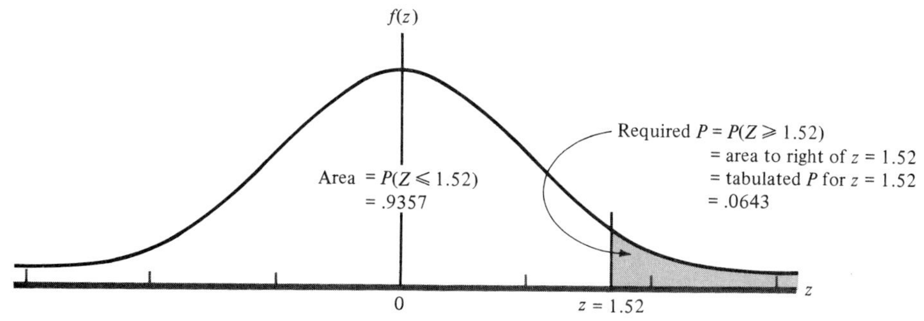

FIGURE 10-3-1 Example for a Normal Curve with Mean 0 and Variance 1.

Illustration 10-3-2

Find $P(Z \leq 1.52)$.

Again see Figure 10-3-1. Since the total area or probability is 1 and we have just found that $P(Z \geq 1.52) = .0643$, then $P(Z \leq 1.52) = 1 - .0643 = .9357$. Since there is not a discrete amount of probability at any point, we need not be especially concerned about inclusion or exclusion of the equals sign within the parentheses.

Illustration 10-3-3

Find $P(Z \leq -1.84)$.

Negative z's do not appear in the table, so we rely on the symmetric property of the normal curve. In Figure 10-3-2 and from symmetry, we observe that $P(Z \leq -1.84) = P(Z \geq 1.84)$. In turn, from Table B-1, $P(Z \geq 1.84) = .0329$, the required value.

Illustration 10-3-4

In many problems, the required probability is that of a random value falling within or outside an interval placed symmetrically about the mean. For example, we are asked to find $P(-1.27 \leq Z \leq 1.27)$

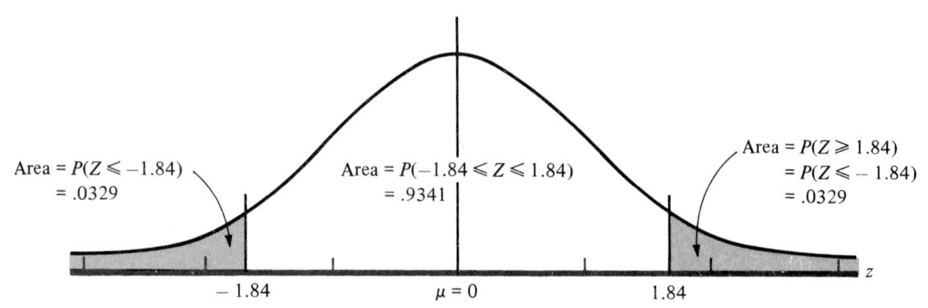

FIGURE 10-3-2 Probabilities for Negative Values of z.

$= P(|Z| \leq 1.27)$ or, on the other hand, to find $P(|Z| \geq 1.27)$. Recall that vertical bars tell us to ignore the sign of Z.

Figure 10-3-2 applies with a change of $|z| = 1.84$ to the required $|z| = 1.27$. To begin, we enter the table and find $P(Z \geq 1.27) = .1020$. Because of the symmetry of the normal curve, we have $P(|Z| \geq 1.27) = 2(.1020) = .2040$. Hence $P(|Z| \leq 1.27) = 1 - .2040 = .7960$.

Sometimes our problems will specify a P-value and ask for a corresponding value of z.

Illustration 10-3-5

Find a value of the random variable Z, say z_α, such that $P(Z \geq z_\alpha) = .1000$.

Here a figure like Figure 10-3-1 is appropriate. Table B-1 does not contain a probability of exactly .1000, the nearest value being .1003. This is opposite 1.2 and under .08. Hence the required value is approximately 1.28.

Illustration 10-3-6

Find a value of the random variable Z, say z, such that $P(Z \leq z) = .9000$.

Again, a figure like 10-3-1 applies. However, probabilities are not given for a set of points to the left of z so we look for $P(Z \geq z) = 1 - P(Z \leq z) = .1000$. Now we enter Table B-1 as in Illustration 10-3-5 to find $z = 1.28$, approximately.

Illustration 10-3-7

If our concern is with small probability values in the left tail of the distribution, we may require z such that $P(Z \leq z) = .0250$. Here the left side of Figure 10-3-2 illustrates the problem. We should enter the Table B-1 using $P(Z \geq z) = 1 - .0250 = .9750$, but such large values of P are not contained in our table and we must rely on the symmetry of the normal curve for help. Thus, since z is to the left of 0 and, consequently, negative, then $z' = -z$ is positive and to the right. Symmetry requires that $P(Z \leq z) = .0250 = P(Z \geq z')$. So, finally, we use $P(Z \geq z') = .0250$ to enter the body of the table. We find $z' = 1.96$. Our required value is then $z = -1.96$ and $P(Z \leq -1.96) = .0250$.

Sometimes our problem will concern sampling from normal distributions which are unsatisfactorily or incompletely specified. Thus, prior to the sampling, we specify a reasonable set of values of the random variable to expect if the distribution is located where we think it is. If the sample value, when found, does not lie in this interval, we may conclude that we were wrong about the location of the distribution.

Illustration 10-3-8

Suppose we have a problem where circumstances are somewhat different from those that led us to Figure 10-3-2 and think, in particular, that the mean may be different from 0. We may proceed by assuming that the mean is where past experience indicates, namely, at 0, and then see whether or not the data support such an assumption.

Our question is now, "What data will support the assumption that the mean of a distribution is zero if the standard deviation is unity?"

To answer this, we first find a reasonable or not-unusual set of values from which our value might come if the mean is 0. Such values can be defined as those toward the middle of the distribution where the density is greatest and not those in the tails where it is least.

At this point, we introduce a certain amount of arbitrariness by specifying dividing points between not-unusual and unusual. Let our definition specify that 90 percent of all possible values, say 45 percent on each side of the mean, are to be considered as not-unusual. Of course, in a real situation, we may have criteria that will help us to choose such a value.

We now need to find z such that $P(|Z| \leq z) = .9000$.

The remaining probability of $1 - .9000 = .1000$ is equally divided between the tails of the distribution. Hence we look in Table B-1 for .0500. We find $z = 1.64$ for $P = .0505$ and $z = 1.65$ for $P = .0495$. Since these values are so close together, we will assume that the probabilities change along a straight line and use $z = 1.645$. Since $P(Z \geq 1.645) = .0500$ and the normal curve is symmetric,

$P(|Z| \geq 1.645) = 2(.0500) = .1000$ and $P(|Z| \leq 1.645) = 1 - .1000 = .9000$, as was required.

Now we sample. If we observe a sample value within the range $(-1.645, 1.645)$, then we have insufficient evidence to say that $\mu \neq 0$. On the other hand, if we obtain a sample value outside this range, then we conclude that the data do not support the assumption that $\mu = 0$.

EXERCISES

10-3-1 Suppose we randomly sample a normal distribution with zero mean and unit variance. Give the probability that we will draw an observation greater than:
(a) 1.00 (b) 2.00 (c) 3.00 (d) 1.64
(e) .67 (f) 1.96 (g) 2.57

10-3-2 For the same distribution, give the probability that an observation drawn at random will be less than:
(a) −.38 (b) −.84 (c) −1.28 (d) −1.66
(e) −2.58

10-3-3 For the same distribution, give the probability that a random observation will be, in absolute value, greater than:
(a) .50 (b) 1.00 (c) 1.28 (d) 1.64
(e) 1.96 (f) 2.33

10-3-4 For the same distribution, what is the probability that a random observation will be:
(a) Less than 1.55 (b) Greater than −2.25
(c) Greater in absolute value than 1.75 (d) Less than 2.50
(e) Less than −2.25 (f) Smaller in absolute value than 1.88

10-3-5 For the same distribution, find the value of z_α such that $P(Z \geq z_\alpha)$ will equal the following values of α:
(a) .5000 (b) .4000 (c) .3000 (d) .2500
(e) .2000 (f) .1000 (g) .0500 (h) .0100

10-3-6 For the same distribution, find the value of z_α such that $P(Z \geq z_\alpha)$ will be:
(a) .9900 (b) .9500 (c) .9000 (d) .8000
(e) .7500 (f) .7000 (g) .6000 (h) .5000

10-3-7 For the same distribution, find the value of z such that $P(Z \leq z)$ will be:
(a) .99 (b) .95 (c) .90 (d) .75
(e) .50 (f) .25 (g) .10 (h) .05
(i) .01

Note: these are values of $1 - \alpha$ when the usual notation is used.

10-3-8 For the same distribution, find the value of z such that $P(|Z| \geq z)$ will be:
(a) .01 (b) .05 (c) .10 (d) .20
(e) .25 (f) .50

When α is used to designate probabilities to the right of z_α, the probabilities given in this exercise must have the values 2α. Thus we enter Table B-1 with the values $\alpha = .005$, etc.

10-3-9 For the same distribution, find the value of z such that $P(|Z| \leq z)$ will be:
(a) .99 (b) .98 (c) .95 (d) .90
(e) .75 (f) .50

10-3-10 The normal distribution with zero mean and unit variance is sampled at random and a value of 1.86 obtained. What was the probability of obtaining a larger value? What was the probability of obtaining an observation larger in absolute value? What would the probabilities have been if the sample values were .75? 1.25? 1.75? 2.22?

If your random value of z was 2.22, would you be suspicious about whether or not the sampled population had a mean of 0?

10-4 THE NORMAL DISTRIBUTION WITH ANY MEAN AND VARIANCE

Equation 10-2-1 describes the probability density of a normal random variable with mean μ and standard deviation σ and, therefore, variance σ^2. This equation contains both of these parameters. The normal random variable for which examples of probability computations were just illustrated had $\mu = 0$ and $\sigma = 1 = \sigma^2$. Probability tables for this variable are used for all normal distributions.

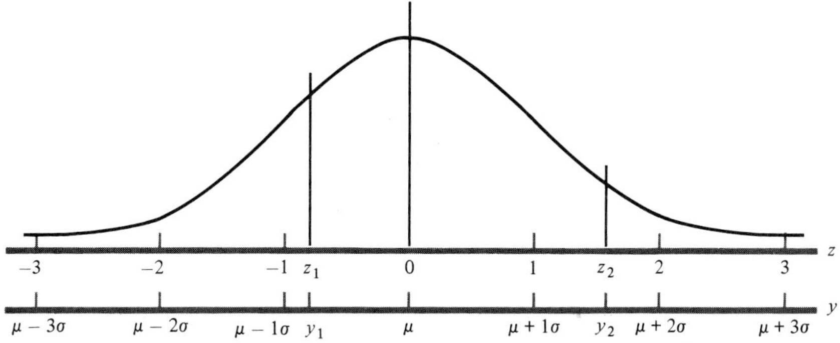

FIGURE 10-4-1 Relation Between a Normal Random Variable with $\mu = 0$ and $\sigma = 1$ and a Normal Variable with any Mean and Variance.

The relation between a normal variable, say Z, with $\mu = 0$ and $\sigma = 1$ and another normal variable, say Y, with mean μ and standard deviation σ is illustrated in Figure 10-4-1. From Figure 10-4-1, Equation 10-4-1 is seen to be valid.

$$P(y_1 < Y < y_2) = P(z_1 < Z < z_2) \qquad (10\text{-}4\text{-}1)$$

To be able to make use of Table B-1 to compute probabilities for normal distributions with means other than zero and variances other than unity, we need only the relation between y and z. An examination of the two scales in Figure 10-4-1 shows that any y-value may be measured as a distance from μ in multiples of σ, the multiple being the corresponding value of z. Thus, we have Equation 10-4-2.

$$y = \mu + z\sigma \qquad -\infty < z < \infty \qquad (10\text{-}4\text{-}2)$$

This may be solved for z to give Equation 10-4-3.

$$z = \frac{y - \mu}{\sigma} \qquad (10\text{-}4\text{-}3)$$

Finally, the random variables are related by Equation 10-4-4.

$$Z = \frac{Y - \mu}{\sigma} \tag{10-4-4}$$

The random variable Z described above is a *standard* or *standardized normal variable*. Equation 10-4-4 alone, with no requirement of an underlying normal distribution, is called simply a standard variable, as first pointed out in Section 4-3.

Probabilities associated with the random normal variable Y may be computed from Table B-1 by measuring the deviation of Y from its mean in units of its standard deviation. Such a variable is then standardized. Illustrative examples follow.

Illustration 10-4-1

A sample observation is to be drawn at random from a normal population with mean 50 and variance 9. What is the probability that the observation will be greater than 55?

We require $P(Y \geq 55)$, but must rely on Table B-1 where the population mean is 0 and the variance is unity. We must, then, standardize $y = 55$ by Equation 10-4-3. Figure 10-4-2 illustrates this situation. We compute

$$z = \frac{y - \mu}{\sigma} = \frac{55 - 50}{3} = 1.67$$

From Table B-1, we find $P(Z \geq 1.67) = .0475$. Hence $P(Y \geq 55) = 0.475$. At the same time, we note that $P(Y \leq 55) = 1 - .0475 = .9525$.

When an observation is less than the mean, we must be particularly careful about the sign of z.

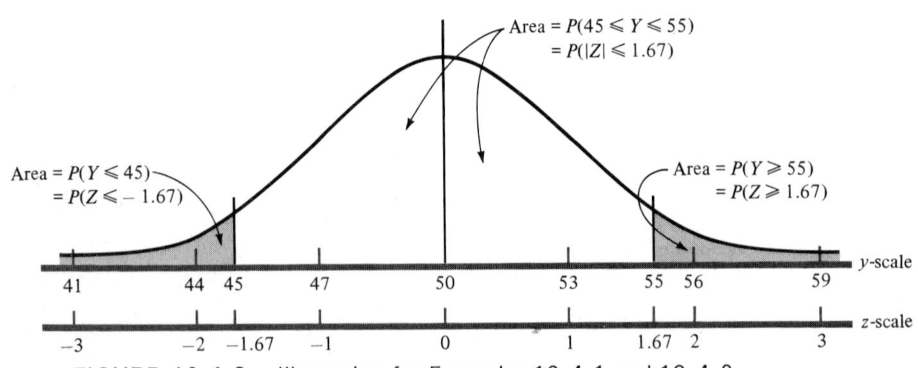

FIGURE 10-4-2 Illustration for Examples 10-4-1 and 10-4-2.

Illustration 10-4-2

Suppose we are sampling as in Illustration 10-4-1. What is the probability of observing a value less than 45?

Again, refer to Figure 10-4-2. Notice that 45 is as far to the left of $\mu = 50$ as 55 was to the right, namely, 1.67 standard deviations. We compute

$$z = \frac{y - \mu}{\sigma} = \frac{45 - 50}{3} = -1.67$$

We must find $P(Y \leq 45) = P(Z \leq -1.67)$. From symmetry, we observe that $P(Z \leq -1.67) = P(Z \geq 1.67)$. This has already been evaluated as .0475. This time we note that $P(Y \geq 45) = P(Z \geq -1.67) = 1 - P(Z \leq -1.67) = 1 - .0475 = .9525$.

We may also use the results of Illustrations 10-4-1 and 10-4-2 to conclude that

$$P(Y \leq 45 \text{ or } Y \geq 55) = P(|Z| \geq 1.67) = 2(.0475) = .0950$$

and

$$P(45 \leq Y \leq 55) = P(|Z| \leq 1.67) = 1 - .0950 = .9050$$

Illustration 10-4-3

Another sort of problem calls for us to draw a random sample and conclude whether or not the evidence is in favor of the population having a hypothesized mean. Thus, suppose we draw an observation from a population with a mean of 160 or greater and variance 81. Experience has indicated that 160 has been appropriate in the past but changing conditions suggest that, if the mean is now different, it will be higher. This is like Illustration 10-3-8 except that now we have specified the direction in which the alternative to 168 may lie.

The alternative to $\mu = 168$ suggests that we might agree to continue believing that 160 is the mean if the observed value is in the range such that $P(Y \leq y) = .90$; again, .90 is an arbitrary choice. The values in the specified range represent the sort of values to be expected if $\mu = 160$; large values of the random variable would suggest a $\mu > 160$.

We must now solve for that value of y such that smaller values occur, on the average, 9 out of 10 times when the mean is, in fact, 160. We begin with the probability statement $P(Y \geq y) = .10$ and enter the body of Table B-1 to look for .1000. The closest tabulated value is .1003, and the corresponding z is 1.28; that is, the required value on the original scale is 1.28 standard deviations beyond the mean. Hence $y = 160 + 1.28(9) = 171.5$. This is an application of Equation 10-4-2. If the observed sample value is less than 171.5, we continue to believe that the true mean is 160, because chance and a mean of 160 offer an adequate explanation. If the sample value is larger than 171.5, then chance and a true mean of 160 do not offer an explanation satisfactory by our standard, basically the .90. Rather than say chance gave us an unusual sample value, we will probably say that we hypothesized the wrong mean and that the true one must be greater than 160.

In this chapter, we have been concerned primarily with introducing the normal distribution and learning how to apply the available tables. Some illustrations have been given concerning its use. However, much has been left for Chapter 12, where its general applicability and use in statistics is further discussed.

EXERCISES

10-4-1 A random sample of one record is drawn from a population of production records for a particular industry. The population of data available appears to be normally distributed with mean $\mu = 40$ units and standard deviation $\sigma = 12$ units. Give the probability that the record drawn will be:
(a) Greater than 60 units
(b) Less than 58 units
(c) Less than 20 units
(d) Less than 35 units
(e) Less than 45 units
(f) Greater than 25 units

10-4-2 Records are available of the pulling power, in 10-pound units, exerted by a tractor working under specified experimental conditions. Suppose these data can be assumed to be normally distributed with mean $\mu = 75$ units and standard deviation $\sigma = 15$ units.

If a single random observation on pulling power is made, give the probability that it will be:
(a) Between 70 and 80 units
(b) Between 65 and 85 units
(c) Outside the range 64 to 86 units
(d) Outside the range 67 to 83 units
(e) Less than 60 units

10-4-3 Suppose soybeans are grown in pairs of plots with the rows spaced 30 inches apart in one plot and 36 inches apart in the other. The difference in yield is measured as bushels per acre; it is the signed difference of yield for the 30-inch spacing less that for the 36-inch spacing.

Assume such differences can be considered to be drawn from a normal population with mean $\mu = 0$ bushels per acre and standard deviation $\sigma = 2.8$ bushels per acre.
(a) For a randomly selected pair of plots, what positive difference must be observed before we will be able to say that larger differences are observed with (1) $P = .01$? (2) $P = .05$? (3) $P = .25$?
(b) What difference, regardless of sign, must be observed before we can say that numerically larger differences are observed with (1) $P = .01$? (2) $P = .05$? (3) $P = .10$?
(c) What difference, regardless of sign, must be observed before we can say that numerically smaller differences are observed with (1) $P = .80$? (2) $P = .90$? (3) $P = .98$?

10-4-4 Standardized scores, introduced in Section 4-3, have a mean of 500 and a standard deviation of 100. Suppose we are dealing with such scores and that, in addition, they are from a normal distribution. Where would Q_1 be located? Q_3? P_{10}? P_{90}?

How great is the range $Q_3 - Q_1$ when measured as a multiple of σ? $P_{90} - P_{10}$ as a multiple of σ?

Suppose a large sample of data had been collected and summary numbers Q_3, Q_2, Q_1, P_{90}, and P_{10} made available. How might we estimate the mean and standard deviation from these summary numbers?

Chapter 11

EXPECTATIONS FOR DISCRETE DISTRIBUTIONS

11-1 INTRODUCTION

A random experiment is such that we cannot say with certainty what the outcome will be. However, we are usually prepared to propose a model which specifies our view of the probabilistic aspect of the problem and permits us to compute the probability of any and all outcomes, though often in terms of unknown constants. This is a lot of information. In the case of a sample, we were prepared to look first at the mean and the standard deviation or variance, averages of observations or of functions of observations, as a comprehensible summary of the information.

Expectations are population averages. For expectations, we take the probabilities of the model and use these as weights to find weighted averages of the values taken by random variables. We are finding what happens *on the average* for the model involved.

This chapter is about expectations and their computation and presents some illustrations for random variables with discrete distributions.

11-2 EXPECTATIONS OF RANDOM VARIABLES

Let us first consider the random variable Y, the sum of the numbers on the down-turned faces of two tossed tetrahedra. Recall that a tetrahedron is a three-sided pyramid where each side is a triangle with equal sides; it sits on a base which is also a triangle, consisting of one side of each of the original three triangles. Consequently, it has four surfaces to which we may assign the numbers 1, 2, 3, and 4. The probability distribution, developed as Table 8-2-1, is presented here as Table 11-2-1. This table states that the value $y = 2$ is observed with probability 1/16 or,

TABLE 11-2-1 / Probability Distribution for the Sum of Two Faces on Two Tetrahedra

Sums, y	2	3	4	5	6	7	8
Probabilities, $P(Y = y)$ or $f(y)$	1/16	2/16	3/16	4/16	3/16	2/16	1/16

on the average, in 1 out of 16 trials of a random experiment repeated infinitely often, and so on for the other sums.

Illustration 11-2-1

Find the average value of Y, the sum of the hidden faces on two tossed tetrahedra.

To find this population average of the y-values, we need to have $y = 2$ in 1/16 of the cases, $y = 3$ in 2/16 of the cases, . . . , and $y = 8$ in 1/16 of the cases. In other words, each y must be present in the mean in proportion to its frequency in the population, say according to its relative frequency or probability. This calls for a weighted mean, defined in Equation 3-5-1 as $\bar{y} = \sum w_i y_i / \sum w_i$. Since we are dealing here with relative frequencies or probabilities, $\sum w_i = 1$. Thus, we find

$$\text{Average of } y\text{'s} = \frac{1}{16}(2) + \frac{2}{16}(3) + \cdots + \frac{1}{16}(8) = 5$$

In this computation, we have used a probability function, that is, the mathematical representation of a population. Consequently, we now have a value for a population. This value, 5, is called the average, mean, or *expectation* of the random variable Y. It is symbolized by $E(Y)$; thus, $E(Y) = 5$ for this problem.

In the case of the two-tetrahedra experiment, $E(Y) = 5$ is an observable value of the sum. This sort of occurrence is not necessary. For example, if an outcome must be a 1 or a 2, as would be the case if we assigned a 1 to an odd number on a single tetrahedron and a 2 to an even number, or a 1 to a head and a 0 to a tail when tossing a coin, and if they occur with equal frequency, then the expected value must be $1\frac{1}{2}$ for the tetrahedra and 1/2 for the coin, values that never occur. Also, $y = 5$ is the most probable value for the two-tetrahedra problem since $P(Y = 5) = 4/16$. Clearly this property will also not always hold; and, in particular, for a distribution such as the U-shaped one depicted in Figure 2-4-8, the expected value has to be the least frequent value, and yet one that can occur.

More generally, we define the *expectation of a random variable Y,* that takes only a finite set of values, by Equation 11-2-1 where the summation is over all values of the random variable.

$$E(Y) = \sum_{i=1}^{n} y_i P(Y = y_i) \qquad (11\text{-}2\text{-}1)$$

This definition also covers random variables that take a denumerable infinity of values, as in the case of the Poisson distribution. It is clear, however, that our definition of addition or summation needs some further explanation.

$E(Y)$ is also called the *first moment of Y about the origin.* It is a first moment because Y is to the first power and a moment about the origin because Y is measured from the origin and not as a deviation from any other point. Equation 4-3-9, $\sigma^2 = \sum (y_i - \mu)^2/N$, gives a second moment about the mean because of the second power and the $y_i - \mu$, a deviation from the population mean. Notice that $E(Y)$ and μ are symbols for the same quantity.

Illustration 11-2-2

In Illustration 9-2-4, we had a random sample of four plants classified according to the presence or absence of a genetic characteristic. The probability of obtaining a plant with the characteristic is 13/16. Let us find the average number of plants, in samples of four, having the characteristic.

The random variable must take on the values 0, 1, 2, 3, and 4. For the probability function, we use the binomial distribution with $n = 4$ and $p = 13/16$. The probabilities are

$$\left(\frac{3}{16}\right)^4, \quad \binom{4}{1}\frac{13}{16}\left(\frac{3}{16}\right)^3, \quad \binom{4}{2}\left(\frac{13}{16}\right)^2\left(\frac{3}{16}\right)^2,$$

$$\binom{4}{3}\left(\frac{13}{16}\right)^3\frac{3}{16}, \quad \text{and} \quad \left(\frac{13}{16}\right)^4$$

respectively. Hence,

$$E(Y) = 0\left(\frac{3}{16}\right)^4 + 1(4)\,\frac{13}{16}\,\frac{3^3}{16^3} + 2\,\frac{4(3)}{1(2)}\,\frac{13^2}{16^2}\,\frac{3^2}{16^2}$$

$$+ 3\,\frac{4(3)2}{1(2)3}\,\frac{13^3}{16^3}\,\frac{3}{16} + 4\,\frac{13^4}{16^4}$$

$$= \frac{0 + 1{,}404 + 18{,}252 + 79{,}092 + 114{,}244}{16^4}$$

$$= \frac{212{,}992}{65{,}536}$$

$$= 3.25 \text{ plants}$$

In view of the fact that the probability associated with finding the characteristic is 13/16, this seems a not unreasonable value.

In summary, to compute an expected value for a random variable that takes only a finite set of values, we first propose a model. The model provides a probability function which may be a listing of the values of the random variable with their probabilities or which simply may provide the information which allows us to compute probabilities for all values of the random variable. For the two-tetrahedra experiment, the model was basically for a single tetrahedron, but this allowed us to compute probabilities for a sum associated with two independent trials, as was done in Section 8-2. For Illustration 11-2-2, we had a model for a single trial and used this to find the probability function for the required $n = 4$. The probabilities for the various values of the random variable are then used as weights to find a weighted average.

The expectation of a random variable, measuring what happens on the average, may be considered as a measure of where the center of the population or of the distribution of the random variable is, a measure which, in a sense, locates it. For this reason, the expectation of a random variable may be called a measure of centrality or a measure of location of its probability function. Symbolically, we use the letter μ to represent this measure, whereas the use of E tends to imply that we are instructed to find the expectation. In other words, E is an instruction to be carried out, sometimes called an *operator*.

EXERCISES

11-2-1 In a triangle taste test, three samples are presented to a judge. Two are alike, and the judge is told to select the odd one by her sense of taste.

If the judge has no discriminatory sense of taste and only guesses, what is the probability that she will correctly identify the different flavor?

If, in the course of a day, she regularly makes a dozen such trials, what is the long-range average number of correct identifications per day?

What is the probability that she will observe this average number on a given day?

11-2-2 Suppose that for 5-pound bags of oranges, a certain market finds that 5 percent of the bags have at least one bad orange. In the course of a week, the market buys 50 bags.

Over a long period of time, what is the average number of bags per week that will contain one or more bad oranges?

What is the probability that, for any week, the number of bags with one or more bad oranges will be smaller than the expected average number per week?

11-2-3 One tetrahedron with faces numbered 1, 2, 3, and 4 is tossed, and the value on the hidden face recorded as a random variable. What is the probability function of this random variable? What is the average value of the random variable?

11-2-4 Suppose the tetrahedron of Exercise 11-2-3 is tossed twice. Let the random variable be the number of even numbers in the two-toss trial (see Exercise 8-2-1). What is the average value of the number of even numbers hidden? If the random variable is the number of divisors in the sum of the two face values (see Exercise 8-2-1), what is the average value of the number of divisors?

11-2-5 A fair die is rolled twice and the sum of the number of spots on the two trials is recorded as a random variable. What is the average value of the sum?

11-2-6 A fair die is rolled three times and the observed number of 6s recorded as a random variable (see Exercise 9-2-1). What is the average number of 6s observed in a long sequence of such trials?

11-3 EXPECTATIONS AND PARAMETERS

An expectation is seen to be a summary number derived from a model for a population. All the possible values of a random variable are used, each value in proportion to its probability. In other words, populations and probability functions contain essentially the same information, and expectations are population summary numbers just as statistics are sample summary numbers. Such population numbers are called parameters.

In general, we may begin with one random variable and its probability function and proceed to another related random variable and its expectation without ever specifically deriving the probability function of the latter. For example, a population variance is defined as $E(Y - \mu)^2$, but it is computed as $\sum (y - \mu)^2 f(y)$ where $f(y) = P(Y = y)$, the probability function of the original random variable rather than that of $(Y - \mu)^2$. Thus, suppose we have a random variable Y and its probability function $f(y)$, and define a new random variable as $g(Y)$; then we can find the average or expected value of $g(Y)$ by Equation 11-3-1.

$$E[g(Y)] = \sum g(y)f(y) \qquad (11\text{-}3\text{-}1)$$

Summation is always over all values of y.

Functions of random variables which are often of interest include the various powers of Y. Thus $E(Y^k)$ is the *kth moment about the origin*, while $E(Y - \mu)^k$ is the *kth moment about the mean.* These moments are defined by Equations 11-3-2 and 11-3-3.

$$\text{Moments about origin} = E(Y^k) = \sum y^k f(y) \tag{11-3-2}$$

$$\text{Moments about mean} = E(Y - \mu)^k = \sum (y - \mu)^k f(y) \tag{11-3-3}$$

All such moments can be considered to be parameters of the probability function. They have a number of applications in statistics.

Parameters may appear as constants in the mathematical statement of the probability function or may not, as is the case with most moments of a distribution. Sometimes they are observable, in which case there is no doubt about their values; at other times, they must be estimated or we may propose some hypothesis about the numerical value of one of them and then test this hypothesis using sample data.

As an illustration concerning models, parameters, and expectations, suppose we have a coin problem or a simple inheritance problem. We propose a model that describes a single observation as a Bernoulli or binomial trial. However, we intend to collect data on a multitrial basis and use the sum of the individual outcomes as the relevant random variable. The required distribution of the sum is given by Equation 9-2-9, namely,

$$P(Y = y) = \binom{n}{y} p^y (1 - p)^{n-y} \qquad \text{for } y = 0, \ldots, n$$

This is a general form, and the parameters that appear in the statement that allows us to compute probabilities must still be specified for the particular distribution.

A particular binomial distribution is seen to be specified by two parameters, n and p. Since n is known, it is an *observable parameter* and its value need not be estimated or hypothesized. Recall that this is not the case for the negative binomial distribution where n is, instead, a random variable. The value of p is customarily unknown and may need to be estimated or, alternatively, may be hypothesized. Thus, for a thumbtack, it would seem appropriate to estimate the parameter p, say the probability that, when tossed, it will land on its head; for a coin that could be biased, we might be more concerned with testing the hypothesis that $p = P(\text{H}) = P(\text{T}) = 1/2$, that is, testing whether or not we have a fair coin. For a binomial distribution, then, the only parameter of special concern that appears in the mathematical description or model of the population is the parameter p. Unknown and unobservable parameters, especially those which occur in the model, are a first concern in statistics.

EXERCISES

The exercises for this section are primarily algebraic. They extend the subject of expected values but the ideas developed are not used explicitly elsewhere in the text. They may be skipped by those who are more anxious to get on with applications and exercises more nearly concerned with everyday events.

11-3-1 In Section 8-2, Figure 8-2-1 and Equation 8-2-2 are concerned with what might be called the joint distribution of a variable associated with the red tetrahedron and another with the blue. Let Y_1 be the random variable that assigns the value i to the sample point (i, j) and Y_2 be another that assigns the value j to this sample point.

To find $E(Y_1 + Y_2)$ directly, we would have to carry out a summation over all points in the sample space as

$$\sum_{i,j} (y_{1i} + y_{2j})P(Y_1 + Y_2 = y_{1i} + y_{2j}) = \sum_{i,j} (y_{1i} + y_{2j})P_{ij}$$

for P_{ij} as defined by Equation 8-2-2. Find $E(Y_1 + Y_2)$. Observe that this is the same value as for Illustration 11-2-1, as it should be.

11-3-2 A coin and tetrahedron with faces numbered 1, 2, 3, and 4 are tossed simultaneously. Let Y_1 assign a 1 if the coin shows a head, 0 if a tail; let Y_2 be the number on the hidden face of the tetrahedron. Let the recorded outcome of the experiment be the ordered pair of numbers assigned by the two random variables. Construct a sample and find probability values for each sample point. Show that $E(Y_1 + Y_2) = 3$.

11-3-3 Two different sized thumbtacks, a red one and a white one, are tossed. Each may land flat on its head or on its point and the edge of its head. Let the random variable Y_1 assign a 1 if the red tack lands on its head, a 0 if on its point; let the random variable Y_2 assign a 1 if the white tack lands on its head, a 0 if on its point. Suppose also that the probability is p_1 that the red tack lands on its head and p_2 that the white one lands on its head.

(a) Show that $E(Y_1 + Y_2) = p_1 + p_2$.

(b) Consider each variable separately and show that $E(Y_1) = p_1$ and $E(Y_2) = p_2$.

Note: There is a theorem that states that when Y_1 and Y_2 are defined over the same sample space, then $E(Y_1 \pm Y_2) = E(Y_1) \pm E(Y_2)$. (The symbol $\pm$ is read as "plus or minus.")

11-3-4 Show that $E(aY) = aE(Y)$ for a a constant.

11-3-5 Show that $E(b) = b$ when b is a constant. (Think of a "random variable" defined so that it assigns a constant value to every point in the sample space and apply Equation 11-3-1.)

11-3-6 Show that $E(aY + b) = aE(Y) + b$.

11-3-7 In Exercise 11-2-3, it was shown that $E(Y) = 2.5$. Exercise 11-3-1 defined a similar experiment, but consisting of two independent trials, using two random variables Y_1 and Y_2. Show that

$$E(Y_1 Y_2) = 6.25 = 2.5^2$$

11-3-8 In Exercise 11-3-2, two independent trials constituted the experiment and there were two random variables. Show that

$$E(Y_1 Y_2) = 1.25 = .5 \times 2.5$$

11-3-9 For Exercise 11-3-3, there were two independent trials and two random variables. Show that

$$E(Y_1 Y_2) = p_1 p_2 = E(Y_1)E(Y_2)$$

Note: The statement "$E(Y_1 Y_2) = E(Y_1)E(Y_2)$ when Y_1 and Y_2 are independent random variables" is another important theorem about expectations.

11-4 THE EXPECTED VALUE OF A BINOMIAL SUM

The binomial probability function is an appropriate model for many real-world problems and so receives considerable attention in statistics. For n Bernoulli trials, a sample point is an ordered n-tuple, that is, a sequence of n numbers, consisting of the outcomes of the n trials. We usually quantify each trial by a 1 for a have, or success, and a 0 for a have-not, or failure. A random variable of considerable interest is the number of successes, that is, the sum of the results of the n trials, with possible values $0, 1, \ldots, n$. In turn, we are interested in the average number of successes when the experiment consisting of n trials is repeated indefinitely, that is, in the expected value of the sum of the quantified outcomes of the n trials.

We have already used Illustration 11-2-2 to show how to find $E(Y)$ when $n = 4$, and $p = 13/16$ for $y = 1$, and $1 - p = 3/16$ for $y = 0$. The general result when p and n are numerically unspecified parameters is sufficiently important that we will now demonstrate how to find it algebraically. If your interests are not at all in algebra, we suggest you go directly to the result, namely, Equation 11-4-1.

Suppose we have n independent Bernoulli trials where $P(\text{success}) = P(1) = p$, an unknown parameter. Consequently, $P(\text{failure}) = P(0) = 1 - p$. Let Y be the random variable equal to the number of 1s in the n trials; that is, Y is a count of the number of successes. To find the mean number of successes, namely, $\mu = E(Y)$, we apply Equation 11-2-1, using the binomial distribution as given by Equation 9-2-9.

$$
\begin{aligned}
\mu = E(Y) &= \sum_{y=0}^{n} yP(Y = y) \\
&= (0)p^0(1-p)^n + 1\binom{n}{1}p(1-p)^{n-1} + 2\binom{n}{2}p^2(1-p)^{n-2} \\
&\qquad + 3\binom{n}{3}p^3(1-p)^{n-3} + \cdots + np^n(1-p)^0 \\
&= 0 + np(1-p)^{n-1} + n(n-1)p^2(1-p)^{n-2} \\
&\qquad + \frac{n(n-1)(n-2)}{2!}p^3(1-p)^{n-3} + \cdots + np^n(1-p)^0 \\
&= np\left[(1-p)^{n-1} + \binom{n-1}{1}p(1-p)^{n-2}\right. \\
&\qquad \left. + \binom{n-1}{2}p^2(1-p)^{n-3} + \cdots + p^{n-1}\right] \\
&= np[p + (1-p)]^{n-1} \\
&= np
\end{aligned}
$$

In words, the mean or expected number of successes in n trials is n times the probability of success in a single trial or n times the proportion of successes in the population. Such a result makes good common sense.

Note that near the end of the proof there is an application of Equation 9-2-10, the binomial expansion $[p + (1 - p)]^n$, but with exponent $n - 1$.

Also, in the above proof, note that for $y = 1, 2, \ldots, n$, we have used

$$\begin{aligned} y\binom{n}{y} &= y\frac{n!}{y!\,(n-y)!} \\ &= \frac{n!}{(y-1)!\,(n-y)!} \\ &= n\frac{(n-1)!}{(y-1)!\,[(n-1)-(y-1)]!} \\ &= n\binom{n-1}{y-1} \end{aligned}$$

When $y = 0$, $y\binom{n}{y}$ is 0 because 0 is a multiplier; recall that $0! = 1$, so that we are not faced with the problem of a 0 in the denominator.

Theorem

The mean of the random variable Y in the binomial distribution, where Y is the number of successes in n trials and p is the probability of a success at each trial, is given by Equation 11-4-1.

$$\mu = E(Y) = np \qquad (11\text{-}4\text{-}1)$$

Illustration 11-4-1

In Illustration 11-2-2, we considered a binomial distribution with $n = 4$ and $p = 13/16$ for the probability of a plant having a certain characteristic. We then defined Y as the random variable giving the number of plants with the characteristic in the four trials. We asked for $E(Y)$.

To find $E(Y)$, we first generated the probability function of Y and then applied Equation 11-2-1. This was a fairly lengthy process and involved some rather large numbers even though the problem was a small one, basically.

We now see that application of Equation 11-4-1 gives a direct answer simply.

$$E(Y) = np = 4\tfrac{13}{16} = 3\tfrac{1}{4} \text{ plants}$$

EXERCISES

11-4-1 An unprepared student finds she must guess the correct option in each of 10 questions with 4 options each (see Exercise 9-2-3). What will be the average number of correctly answered questions per person if a great many students answer by guessing? What is the probability that this student will correctly answer exactly the average number?

11-4-2 In a raffle, 100 tickets are available at $1 each. One ticket will pay $50, four will pay $5 each, 5 will pay $1 each, and the remaining tickets pay nothing. If a person regularly buys a single ticket on such a raffle, in the long run how much will be the average return on $1 invested?

11-4-3 I am asked to play a game wheie three coins are tossed. I will be paid 10 cents a head so long as at least one head turns up; otherwise, I must pay 20 cents. Will it be profitable for me to play if I can stay in the game for a very long time?

11-4-4 In an inheritance problem, a geneticist is not sure whether the observed have:have-not pattern should exhibit a ratio consistent with a true 1:1 or a true 3:1 ratio. She decides to conduct an experiment and base her conclusion on the results. If she wishes the true difference between the expected values of the haves under the two hypotheses to be 60 or more, what is the minimum sample size she should choose for her experiment?

The following questions are algebraic and intended only for those who have such an interest.

11-4-5 Use the notes following the proof that $E(Y) = np$ when Y is the number of successes in n random independent binomial trials to develop a more compact presentation of this derivation, using as much summation notation as possible.

11-4-6 For some binomial trials, there is no clear argument to suggest a value of p. For example, consider tossing a tack and observing whether it comes to rest on its head or on the point and edge. Consequently, we may estimate p as the ratio of the number of times on head to the number of tosses in a sequence of n trials. Call this estimate $\hat{p}$.

What is the expected value of $\hat{p}$? Note that you must use $\hat{p}$ as the random variable and find its expectation using the true, but unknown, value of p. If $E(\hat{p}) = p$, the parameter being estimated, then $\hat{p}$ will be said to be an *unbiased estimator* of p.

11-4-7 Show that $E(Y) = \mu$ for the Poisson distribution as presented in Equation 9-4-1. (You will find some help in the development in this section and the fact that the sum of a complete set of probabilities is 1. Notice that this is a sum of a denumerable infinity of probabilities, any one of which is computable and greater than 0.)

11-5 THE POPULATION VARIANCE

While a measure of location, such as the mean, is very useful, it is even more so if we also have a measure of how closely the observations cluster about it, that is, if we have a *measure of spread* or of *dispersion*. One such measure is called the variance.

The *population variance* is symbolized by σ^2 and defined by Equations 11-5-1 for a discrete random variable with a finite number of distinct observations. Summation is over all values taken by the random variable.

$$\begin{aligned}\sigma^2 &= E(Y-\mu)^2 \\ &= \sum (y_i - \mu)^2 P(Y = y_i)\end{aligned} \qquad (11\text{-}5\text{-}1)$$

The variance is also called the *second moment* of the random variable about its mean. Together, the mean and variance or its square root, the standard deviation σ, contain a great deal of useful information concerning a population.

For the binomial distribution, we developed a general algebraic expression for $E(Y)$. We now do the same for the variance of Y in this important distribution. If your interests are primarily in results and their application, you may skip this development, including the definition of factorial moments, and go directly to Equation 11-5-4.

Consider the first of Equations 11-5-1. Expand the quantity $(Y - \mu)^2$ and then take expectations term by term. The result is simply a rearrangement of the arithmetic.

$$\begin{aligned}E(Y-\mu)^2 &= E(Y^2 - 2Y\mu + \mu^2) \\ &= E(Y^2) - E(2Y\mu) + E(\mu^2)\end{aligned}$$

For $E(\mu^2)$, we are finding the expectation of a constant (Exercise 11-3-5) and so have the following:

$$E(\mu^2) = \mu^2 P(Y = y_1) + \cdots + \mu^2 P(Y = y_n) = \mu^2 \sum P(Y = y_i) = \mu^2(1) = \mu^2$$

For $E(2Y\mu)$, 2μ is a constant in every term of the summation, and so can be factored out (Exercise 11-3-4). Hence, since $E(Y) = \mu$, we have

$$E(2Y\mu) = 2\mu E(Y) = 2\mu^2$$

Finally, substituting these results in Equation 11-5-1 leads to Equation 11-5-2.

$$\begin{aligned}E(Y-\mu)^2 &= E(Y^2) - E(2Y\mu) + E(\mu^2) \\ &= E(Y^2) - 2\mu^2 + \mu^2 \\ &= E(Y^2) - \mu^2\end{aligned} \qquad (11\text{-}5\text{-}2)$$

Equation 11-5-1 is a definition formula that has meaning in terms of the sort of quantity a variance is. Equation 11-5-2 is not much help in understanding what variance is, but it is a convenient computing formula.

For distribution where factorials appear in denominators, as for the binomial and Poisson distributions, still another computing device is helpful. If we compute $E[Y(Y-1)]$, we have an opportunity for cancellations. In turn, $E[Y(Y-1)] = E(Y^2 - Y) = E(Y^2) - E(Y)$ with the first term, $E(Y^2)$, being wanted for Equation 11-5-2. Solving the above for $E(Y^2)$, we have $E(Y^2) = E[Y(Y-1)] + E(Y)$. $E[Y(Y-1)]$ is called the *second factorial moment*.

We now apply this idea to finding the variance of a binomial distribution.

$$
\begin{aligned}
E[Y(Y-1)] &= \sum_{y=0}^{n} y(y-1)P(Y=y) \\
&= \sum_{y=2}^{n} y(y-1)P(Y=y), \\
&\qquad\qquad \text{since either } y \text{ or } y-1 \text{ is zero for } y = 0, 1 \\
&= \sum_{y=2}^{n} y(y-1)\frac{n!}{y!\,(n-y)!}p^y(1-p)^{n-y} \\
&= \sum_{y=2}^{n} \frac{n!}{(y-2)!\,(n-y)!}p^y(1-p)^{n-y} \\
&= n(n-1)p^2 \sum_{y=2}^{n} \frac{(n-2)!}{(y-2)!\,[(n-2)-(y-2)]!} \\
&\qquad\qquad \times p^{y-2}(1-p)^{(n-2)-(y-2)} \\
&= n(n-1)p^2 \sum_{y-2=0}^{n-2} \frac{(n-2)!}{(y-2)!\,[(n-2)-(y-2)]!} \\
&\qquad\qquad \times p^{y-2}(1-p)^{(n-2)-(y-2)} \\
&= n(n-1)p^2 \sum_{z=0}^{n-2} \binom{n-2}{z} p^z(1-p)^{(n-2)-z} \\
&\qquad\qquad \text{on changing the index of summation from } y-2 \text{ to } z \\
&= n(n-1)p^2(p+[1-p])^{n-2} = n(n-1)p^2
\end{aligned}
$$

Once again, by observing that some terms are 0 and that a certain amount of cancellation and factorization is possible, we produce a sum that is a binomial expansion to a power other than n, here $(n-2)$, and have the consequent simple result, namely, Equation 11-5-3.

$$E[Y(Y-1)] = n(n-1)p^2 \qquad (11\text{-}5\text{-}3)$$

In turn,

$$
\begin{aligned}
E(Y^2) &= E[Y(Y-1)] + E(Y) \\
&= n(n-1)p^2 + np
\end{aligned}
$$

Finally, substitution in Equation 11-5-2 gives us the variance of a binomial population.

Theorem

The variance for a binomial distribution is given by Equation 11-5-4.

$$
\begin{aligned}
\sigma^2 &= E(Y^2) - \mu^2 \\
&= n(n-1)p^2 + np - (np)^2 \\
&= np(1-p) \qquad (11\text{-}5\text{-}4)
\end{aligned}
$$

In the case of a variance, the unit is the square of that in which the sample observations were measured. Since the original units are more meaningful to an experimenter, one regularly takes the square root of the variance and recovers them. The resulting parameter, σ, is called the standard deviation.

Illustration 11-5-1

Find the mean, variance, and standard deviation for the number of heads observed when tossing a fair coin 20 times. The expected mean number of heads is:

$$np = 20(1/2) = 10 \text{ heads}$$

The variance is:

$$np(1 - p) = 5 \text{ (heads)}^2$$

This latter measure is not too easily visualized. However, the standard deviation is

$$\sqrt{np(1 - p)} = \sqrt{5} = 2.24 \text{ heads},$$

which is a measure in the same units as the observations and easily understood.

While we will not make much use of the standard deviation in this chapter, we will later see its use as a unit useful in deciding whether certain events are ordinary or unusual. Recall the use of standard normal variables in Section 10-4, also statements about probabilities associated with values of the normal variable at distances measured in standard deviations from the mean in Section 10-2, and similar general statements using Chebyshev's theorem in Section 4-3.

We have already seen that the probability of a specific outcome will not measure whether it is unusual or ordinary, since most sample spaces are so large that all such probabilities are necessarily small. Thus, we are led to associate probabilities with sets of outcomes or events. Standard deviations are often used to determine whether a specific outcome is included in a set of unusual outcomes.

EXERCISES

11-5-1 Find the variance of the random variable in the one-tetrahedron problem, Exercise 11-2-3.

11-5-2 Find the variance for the random variable in the two-tetrahedra problem with distribution given in Table 11-2-1.

11-5-3 Find the variance for the random variables given by (a) the number of evens and (b) the number of divisors in the sum when a tetrahedron is tossed twice. Refer to Exercise 11-2-4.

11-5-4 Find the variance for the random variables: (a) the sum of the numbers, (b) the number of evens, and (c) the number of divisors in the sum when two numbered discs are drawn as in Exercise 8-2-4b, d, and e, respectively.

11-5-5 Find the mean and variance for the random variable that is the number of spots on the upturned face of a tossed fair die.

11-5-6 Find the variance of the random variable that is the sum of the numbers observed in two tosses of a fair die as in Exercise 11-2-5.

11-5-7 Find the variance of the random variable that is the number of 6s observed in three tosses of a fair die as in Exercise 11-2-6.

11-5-8 Find the variance of the random variable that is the number of correct answers in the matching problem of Exercise 8-2-2b.

11-5-9 Find the variance of the random variable that is the number of correct answers in the multiple-choice test of Exercise 11-4-1.

11-5-10 Consider n binomial trials with $p = .5$. What is the probability that the number of successes will be farther from the mean than 2 standard deviations for $n = 5, 10, 15, 20$? Notice how the probability varies with increasing sample size.

The remaining problems are algebraic and intended only for those with such an interest.

11-5-11 What is the variance of $\hat{p}$ as defined in Exercise 11-4-6?

11-5-12 If you were successful with Exercise 11-4-7, find the variance of the variable in a Poisson distribution. Make use of the notion of factorial moments in doing so.

Chapter 12

THE NORMAL DISTRIBUTION IN STATISTICS

12-1 INTRODUCTION

For any sample, we may compute a mean, a standard deviation, or other sample quantity tending to characterize the sample in some way. All such quantities are called statistics. Repeated sampling from the same population produces more of these statistics and they, clearly, must vary from sample to sample. In fact, we can see that there has to be a population for each statistic, for example, a population of means, associated with any particular sample size. Such populations are called derived distributions or sampling distributions.

The random variable for any sampling distribution will have its own mean and variance and, more generally, its own moments. To find these, we do not necessarily need derived sampling distributions, as was pointed out in Chapter 11 in the case of the mean and variance of the number of successes in n binomial trials. However, we have not even presented working definitions for finding expectations of continuous random variables in their own parent populations and, in fact, this requires mathematics beyond the scope of our text. On the other hand, a great deal of useful algebraic manipulation relevant to expected values can be carried out without such mathematics by the use of a few simple operating rules. We will not be concerned with such manipulations here.

In statistics, probably the most useful continuous distribution is the normal distribution discussed in Chapter 10. We shall now consider its mean and variance, the related central limit theorem, and the normal approximation to the binomial distribution.

12-2 THE MEAN AND VARIANCE OF A NORMAL RANDOM VARIABLE

A normal random variable is a continuous one so that probabilities are not associated with points. Equation 11-2-1 does not, then, show us how to compute its expected value since it is applicable to discrete variables only.

For the normal distribution, two parameters with symbols μ and σ^2 appear in the mathematical description, Equation 10-2-1. In Chapter 11, the symbols μ and σ^2 were associated with expected values representing the mean and variance. As a matter of fact, the mean and variance of a normally distributed variable are the parameters μ and σ^2 of Equation 10-2-1. The choice of symbols for this equation was determined as the result of taking the expectations of Y and $(Y - \mu)^2$ for a normally distributed variable and observing the relationship between the expected values and the parameters of the equation. We will simply accept these two results.

In general, an investigator is not satisfied with a single observation. Inevitably, an investigator feels that conclusions will be better, and so can be used more effectively, as the number of sample observations increases. The limit on the number is determined only by practical considerations such as cost, time, possibly destructive sampling methods, and so on. Thus, our real statistical problems are concerned with samples of more than one observation and, in turn, with sampling distributions.

The information in a sample cannot usually be appreciated until summarized. In Section 11-4, where a sequence of Bernoulli trials was discussed, the summary was the sum of the observations. With a continuous variable, we more commonly use a mean since the result will be of the same order of magnitude as a single observation.

In either case, we begin with a *parent population*, for example, the Bernoulli trial which can result in a 1 or a 0, or the continuous distribution from which we will sample. In the former case, we proceeded to the distribution of the sum of the number of successes; we used a sum for the two tetrahedra in Illustration 8-2-1. For the continuous variable, we will want the distribution of the mean or, at least, to know some of the parameters of this distribution. These distributions, that for the sum, the mean, or any other statistic, are called *derived distributions* or *sampling distributions.*

The important theorem that follows tells us about the derived distribution of the sample mean when sampling a normally distributed random variable.

Theorem

Means of n random observations drawn from a normal population with mean μ and variance σ^2 are, in turn, normally distributed with mean μ and variance σ^2/n.

This reproductive property states that if we begin with a normally distributed variable Y whose mean and variance are μ and σ^2, take n random observations and average them to generate a new variable $\bar{Y}$, then the population of $\bar{Y}$'s will also be normally distributed. In addition, the derived population will have the same mean μ as the parent population but a smaller variance, namely, σ^2/n. In other words, means cluster about μ but more closely than do single observations.

If we use $\mu_{\bar{Y}}$ and $\sigma_{\bar{Y}}^2$ to symbolize the parameters of the derived distribution, then Equations 12-2-1 and 12-2-2 hold when a normal distribution with parameters μ and σ^2 is randomly sampled to provide means.

$$\mu_{\bar{Y}} = E(\bar{Y}) = \mu \tag{12-2-1}$$

$$\sigma_{\bar{Y}}^2 = E(\bar{Y} - \mu)^2 = \frac{\sigma^2}{n} \tag{12-2-2}$$

As a matter of fact, Equations 12-2-1 and 12-2-2 are true regardless of the form of the distribution.

Illustration 12-2-1

A population of weights of adult males, which is known to be distributed in a manner reasonably close to normal, has a mean of 150 pounds and a standard deviation of 15 pounds. What can we say about average weights for 100 randomly selected individuals from this population?

In the parent population, we have $\mu = 150$ pounds, $\sigma^2 = 225$ pounds squared, and $\sigma = 15$ pounds.

In the derived population, we have $\mu = 150$ pounds, $\sigma^2 = 225/100 = 2.25$ pounds squared, and $\sigma = 1.5$ pounds. This is regardless of the form of the parent population.

In addition, because of the normality of the parent population, the derived population is also normal. In turn,

$$P(\mu - 2\sigma_{\bar{Y}} \leq \bar{Y} \leq \mu + 2\sigma_{\bar{Y}}) = .9545$$

Since $\mu \pm 2\sigma_{\bar{Y}} = 150 \pm 2(1.5) = (147, 153)$, our sample mean should be within 3 pounds of 150 unless we have an unusual sample ($P = 1 - .9545 = .0455$), or the parent population was incorrectly characterized. For example, the mean may be other than 150 pounds.

EXERCISES

12-2-1 A normal population with mean μ and variance σ^2 is randomly sampled.

For the population of sample means from samples of 10 observations, what is the density function? What are its mean and variance, $\mu_{\bar{Y}}$ and $\sigma_{\bar{Y}}^2$, in terms of the parameters of the original population? What is the case for $n = 20$? 30? 40? 50? What is the standard deviation for each of the above populations of means?

As sample size increases, which decreases faster, the variance or the standard deviation?

12-2-2 If it is desired to have a population of means in which the variance is $\frac{1}{10}$ that of the parent population, what must the sample size be?

If it is the standard deviation of the means that is to be $\frac{1}{10}$ that in the parent population, what must the sample size be?

12-2-3 A die is rolled twice and the average number of spots recorded. Notice that the range of this random variable is $1[\frac{1}{2}]6$. What are its mean and variance? Did you need to find the distribution of the average to produce this result?

12-2-4 In Exercises 11-4-1 and 11-5-9, you found the mean and variance for the variable *number of correct guesses* by an unprepared student. Suppose a class of 25 consists entirely of "guessers." What is the mean of this variable for such a class? The variance in a population of such classes? Did your computations call for any assumptions not specified in the problem?

The remaining exercises are algebraic and need not be done by those with limited time or whose interests in statistics lie elsewhere.

12-2-5 Prove that $E(\bar{Y}) = \mu$, Equation 12-2-1.

You will need to begin with a definition of $\bar{Y}$ and recall some of the exercises you did in Section 11-3 for a discrete random variable. For example, the note at the end of Exercise 11-3-3 implies that $E(\sum Y_i) = \sum E(Y_i)$. For a continuous variable, we assume that it is all right to do this for any distribution with which we will be involved. We might call these "reasonable" distributions.

We also need to think about the meaning of $E(Y_i)$ as opposed to $E(Y)$. Since we have random samples, the order in which the observations were obtained is also random. In considering $E(Y_i)$, we are talking about the population of Y's that consist of those values appearing in the ith position of every possible random sample. It is easy to see that, for example, in repeated sampling, the parent population will eventually appear in the first position, or in any other. Hence, $E(Y_i) = E(Y) = \mu$.

12-2-6 Prove that $E(\bar{Y} - \mu)^2 = \sigma^2/n$.

As in the previous exercise, you will need to begin with a definition of $\bar{Y}$. If you then write μ as $n\mu/n$, you will see how to associate a μ with each Y_i. Now you are dealing with the expected value of a sum squared. Expand this, being sure to include all cross products. Exercise 11-3-9 will help you with the latter.

12-2-7 For a population consisting of the possible outcomes from a single Bernoulli trial, what are the mean and variance?

Suppose an experiment is defined as consisting of n trials. The number of successes will be counted and divided by n. This will give $\hat{p}$ = number of successes/n, which can serve as an estimate of p, the probability of a success in a single trial.

In a population of $\hat{p}$'s, what are the mean and variance? Did you need the distribution of $\hat{p}$ to produce this result?

12-3 THE CENTRAL LIMIT THEOREM

The theorem of Section 12-2 was concerned with means of values randomly selected from a normal distribution. While this result is interesting, it does not often apply in practice, and the question can be raised as to how means are distributed in more general situations.

Consider a single, randomly selected birth weight. We know that if we state that this particular observed value, say y, is an estimate of μ, we have followed a reasonable procedure because $E(Y) = \mu$. In other words, if we were to repeat the procedure indefinitely, the average of all our estimates would be μ. Now, intuition tells us that, to estimate μ, an average of 50 such weights is better than a single weight, and an average of 1,000 is better still. Occasional unusual values will simply lose their effect in an average. However, we are still dealing with the mean of a finite number of observations and so still have a random variable subject to sampling variation. What, if anything, can be said with certainty about the sampling distribution of means of random observations when we have no real assurance that we are dealing with a normally distributed variable?

A most important theorem concerning sampling distributions is the central limit theorem, a theorem which holds almost regardless of the form of the density function sampled. Contributions to the development of this theorem go back to DeMoivre and are still continuing.

The *central limit theorem* states that if a population has a mean μ and a variance σ^2, then the distribution of sample means derived from this distribution approaches the normal distribution with mean μ and variance σ^2/n as the sample size increases.

The theorem requires that we start with a reasonably well-behaved distribution. If the instructions for finding the variance result in a value that is infinite, then we say that the random variable has no variance and so the distribution is not well behaved. But, in almost any practical situation, the variable will have a finite range and, in turn, a finite variance.

Surprisingly enough, regardless of the form of the parent density function, the form of the derived density function approaches normality as n increases. This is the truly important part of the theorem. Thus, at some stage, means for large enough sample sizes, whether the random variable is discrete or continuous, will be approximately normally distributed. Clearly, the form of the parent density function will have some effect on the sample size required, and an asymmetric distribution will generally call for a larger n than a symmetric one. However, a sample size of 30 is sufficiently large for many applications.

The theorem also states that means are distributed about the same central value as are observations, that is, $E(\overline{Y}) = \mu = E(Y)$, and that they are distributed more closely than observations. In particular, $E(\overline{Y} - \mu)^2 = \sigma^2/n$ where $\sigma^2 = E(Y - \mu)^2$. Intuition, which favored large samples, is vindicated. These two facts about the mean and variance of the derived distribution were, of course, part of the theorem in Section 12-2, where it was stated that they did not depend on normality.

We have now seen that whereas it is often easy to justify the use of the normal distribution as a reasonable mathematical model for a real-world situation involving observations made on a continuous random variable, justification is not always necessary since normal theory can be used for a broader spectrum of density functions when means are available. Two illustrations follow.

Illustration 12-3-1

Suppose that in a large experiment concerned with the vitamin C content of lima beans, it is decided that the available resources will permit five observations to be made on each cultivar in the study. The vitamin C content of one of the cultivars of lima beans is known to be distributed about a mean of 88 milligrams per 100 grams of dry weight of beans, with a variance of 56.25 milligrams squared. Other cultivars have different means but the same variance. What can be said about the density function for means of five observations? What will be the mean and variance of this derived population? Do we need to be concerned about normality? How will things change if the means are based on 10 observations?

If the parent distribution is normal, then so is that of the means. If the parent population is not normal, we cannot make such a positive statement. However, regardless of the distribution involved, sample means based on 5 observations have a density function with mean 88 milligrams per 100 grams of dry weight of beans and a variance of $56.25/5 = 11.25$ milligrams squared.

The experiment involves many cultivars. Perhaps the one referred to is a standard and the others are new and their population means unknown. An analysis of the results will almost certainly call for comparisons of the unknown population means with the standard and for comparisons among the unknowns. Sampling variations will have to be allowed for in making statements comparing population means when the statements have to be based on sample means. To know the distribution of means is to have help in developing the theory necessary for the suggested comparisons, and, as it turns out, development is facilitated when the means are normally distributed.

Here we have biological data, and such data are known to be approximately normally distributed in many instances. The mean is not close to 0 and the standard deviation in the parent population is $\sqrt{56.25} = 7.5$ milligrams so that 2 or 3 standard deviations less than the mean is still a long way from 0. There would seem to be no reason to suspect that the parent population was asymmetric, a suspicion that would be justifiable if the mean were only 2 standard deviations above 0. Pretty clearly, means from symmetric distributions should approach normality faster than means from asymmetric distributions. We are probably reasonably safe in assuming that means of five vitamin C determinations are approximately normally distributed.

If each sample mean is based on 10 observations, then the derived distribution of such means is more nearly normal than when each was based on only 5 observations. The mean and variance are now 88 milligrams and 5.625 milligrams squared, respectively, and this is regardless of considerations of distribution.

The second illustration, which follows, concerns a discrete variable.

Illustration 12-3-2

Suppose that a sequence of clinical trials, each using 100 individuals under a specified set of conditions, shows that the number reporting quicker-than-usual recovery from common cold symptoms is distributed about a mean of 75 persons with a variance of 25 persons squared. What can be said about the density function of means of four such trials? Would it be an advantage if we knew that means were approximately normally distributed?

The density function of observations consisting of means of four such clinical trials has a mean of 75 persons as before, and a variance of $25/4 = 6.25$ persons squared. This much is certain. Observe also that while the parent population has a variable that changes by units, the derived distribution changes by quarter units. Because of this, it is easy to visualize a histogram of data consisting of a population of means as more nearly a smooth curve than a histogram of the observations. However, the central limit theorem did not restrict us to continuous variables.

In applying the central limit theorem here, our best argument is in the fact that there are 100 individuals in each trial. It is true that we are using totals rather than means; however, this will not change the basic distribution, only its mean and variance. We have simply not introduced a constant divisor into our random variable. In other words, results based on an individual clinical trial are fundamentally the equivalent of results based on a mean of 100 Bernoulli trials. Hence, in applying the theorem to a mean of four clinical trials, we are really talking about 400 observations in a statistic comparable to a mean. The distribution of such means must be very close to normality.

Is normality an advantage?

Suppose that it is decided to repeat the sequence of trials in many cities. From the normal distribution, we can compute any probabilities we wish; for example, about 68 percent of means will be within the interval $75 \pm \sqrt{6.25} = (72.5, 77.5)$ persons and 95 percent within $75 \pm 2\sqrt{6.25} = (70, 80)$ persons, and so on. This knowledge could help us detect unusual results, should they occur.

As a computing alternative, the binomial distribution is the most obvious. For a single clinical trial, the mean and variance of the number of successes are

$$\mu = np = 75$$

and $$\sigma^2 = np(1 - p) = 75(1 - .75) = 18.75 \text{ persons}^2$$

A variance, $\sigma^2 = 25$, has been provided. This is larger than the binomial variance and might suggest that perhaps not all the criteria required for a binomial distribution are met. It seems that there may be additional variability beyond that called for by the binomial distribution. Perhaps the parameter p is not constant. This could, in turn, introduce variation in addition to that of a binomial distribution, that is, variation into the distribution of means of four clinical trials. In other words, use of binomial probabilities will not necessarily provide more acceptable results than use of normal probabilities.

EXERCISES

12-3-1 A tetrahedron is tossed, with the observed random variable Y being the number on the hidden face. What are the mean and variance of Y? (See Exercise 11-5-1.) What are the mean and variance of the random variable that is the average value resulting from two tosses? Three tosses? Four tosses? Find and graph the probability distribution in each of the above cases and observe the shape as it changes with sample size. It is easy to believe that the shape is moving in the direction of normality.

12-3-2 Repeat the preceding problem but with a die so that the initial experiment has six rather than four possible outcomes.

12-4 THE NORMAL APPROXIMATION TO THE BINOMIAL DISTRIBUTION

For a binomial distribution with large n, and p not too close to 0 or 1, the normal distribution may be used to provide reasonably satisfactory cumulative probabilities. This is implicit in Illustration 12-3-2. Terms such as "large n," "not too close to 0 or 1," and "reasonably satisfactory" are, of course, relative, but studies to determine adequate values of n for various values of p have been made. For example, Cochran (1965, table 3.3) presents a table of minimum values of np for use of the normal approximation.

Consider a binomial distribution with $n = 15$ and $p = .4$. This is clearly an asymmetric distribution with mean $\mu = np = 6$, given by Equation 11-4-1, and variance $\sigma^2 = np(1 - p) = 15(.4)(.6) = 3.6$, given by Equation 11-5-4. Probabilities of individual terms computed by Equation 9-2-9, and cumulative probabilities, are given in Table 12-4-1.

TABLE 12-4-1 / Binomial Probabilities for $n = 15, p = .4$

y	$P(Y = y)$	$P(Y \leq y)$	$P(Y \geq y)$
0	.0005	.0005	1.0000
1	.0047	.0052	.9996
2	.0219	.0271	.9949
3	.0634	.0905	.9730
4	.1268	.2173	.9096
5	.1859	.4032	.7828
6	.2066	.6098	.5969
7	.1771	.7869	.3903
8	.1181	.9050	.2132
9	.0612	.9662	.0951
10	.0245	.9907	.0339
11	.0074	.9981	.0094
12	.0017	.9998	.0020
13	.0003	1.0000	.0003
14	.0000	1.0000	.0000
15	.0000	1.0000	.0000

The computation of approximate probabilities by use of the normal distribution calls for that distribution with $\mu = 6$ and $\sigma^2 = 3.6$ or $\sigma = \sqrt{3.6} = 1.90$. We then compute z-values corresponding to $y = 0[1]15$ and read the corresponding probabilities for the cumulative distribution from Table B-1. These cumulative probabilities, working from each tail, and the differences between the binomial and approximating normal-distribution probabilities are given in Table 12-4-2.

The approximate probabilities are indifferently close to the true values and least reliable for values away from the tails. Whereas a sample size of 30 was a general recommendation for normality to hold in many situations, Cochran (1965) recommends a binomial sample size of 50 when $p = .4$. Our sample size falls far short of this value, and we are not very happy with this normal approximation. Can it be improved?

Figure 12-4-1 shows the exact cumulative binomial distribution, a *step function*, and its approximating normal distribution, a smooth curve, plotted together. A step function is one that looks like treads in a stairway; the flat steps

TABLE 12-4-2 / Approximate Binomial Probabilities for $n = 15, p = .4$, computed from Normal Tables

y	z	$P(Z \leq z)$	$P_B - P_N$†	$P(Z \geq z)$	$P_B - P_N$
0	−3.16	.0008	−.0003	.9992	+.0008
1	−2.63	.0043	+.0009	.9957	+.0039
2	−2.11	.0174	+.0097	.9826	+.0123
3	−1.58	.0571	+.0334	.9429	+.0301
4	−1.05	.1469	+.0704	.8531	+.0565
5	−0.53	.2981	+.1051	.7019	+.0809
6	0.00	.5000	+.1098	.5000	+.0969
7	0.53	.7019	+.0850	.2981	+.0922
8	1.05	.8531	+.0519	.1469	+.0663
9	1.58	.9429	+.0233	.0571	+.0380
10	2.11	.9826	+.0081	.0174	+.0165
11	2.63	.9957	+.0024	.0043	+.0051
12	3.16	.9992	+.0006	.0008	+.0012
13	3.68	.9999	+.0001	.0001	+.0002
14	4.21	1.0000	.0000	.0000	.0000
15	4.74	1.0000	.0000	.0000	.0000

†P_B = probability computed from binomial distribution; P_N = approximating probability, computed from normal distribution.

are read as cumulative probabilities and rise at the values representing an increase in a discrete random variable. If n were to increase, the number of steps in the binomial distribution would increase so that the height of the jumps would decrease and the step function would tend toward a smooth curve.

The possibility of improving the approximation can be seen from Figure 12-4-1. For example, if we wish to know the exact probability that the number of observations falls in the interval 6 ± 2, that is, is one of the values 4, 5, 6, 7, or 8, we go to Table 12-4-1, reproduced in Figure 12-4-1 as the step function, to find $P(Y \leq 8) = .9050$ and $P(Y < 4) = P(Y \leq 3) = .0905$. For any step function, we have to be particularly careful with $<$ and $=$ signs when they refer to the variable. The required exact cumulative probability is then $.9050 - .0905 = .8145$, the sum of the probabilities of the five discrete events. Working directly with the numbers 4 and 8, we find the approximating normal probability as $P(Y \leq 8) - P(Y \leq 4) = P(4 \leq Y \leq 8) = .7062$. Notice that we can even drop the equality part of each sign, since individual y-values have zero probabilities.

On the other hand, it is readily seen from Figure 12-4-1 that $P(Y \leq 4)$ is too high on the approximating normal curve because we must read $P(Y \leq 3) = P(Y < 4)$ on the binomial in order to include, in the required probability, the discrete jump in the binomial distribution corresponding to $P(Y = 4)$. On the

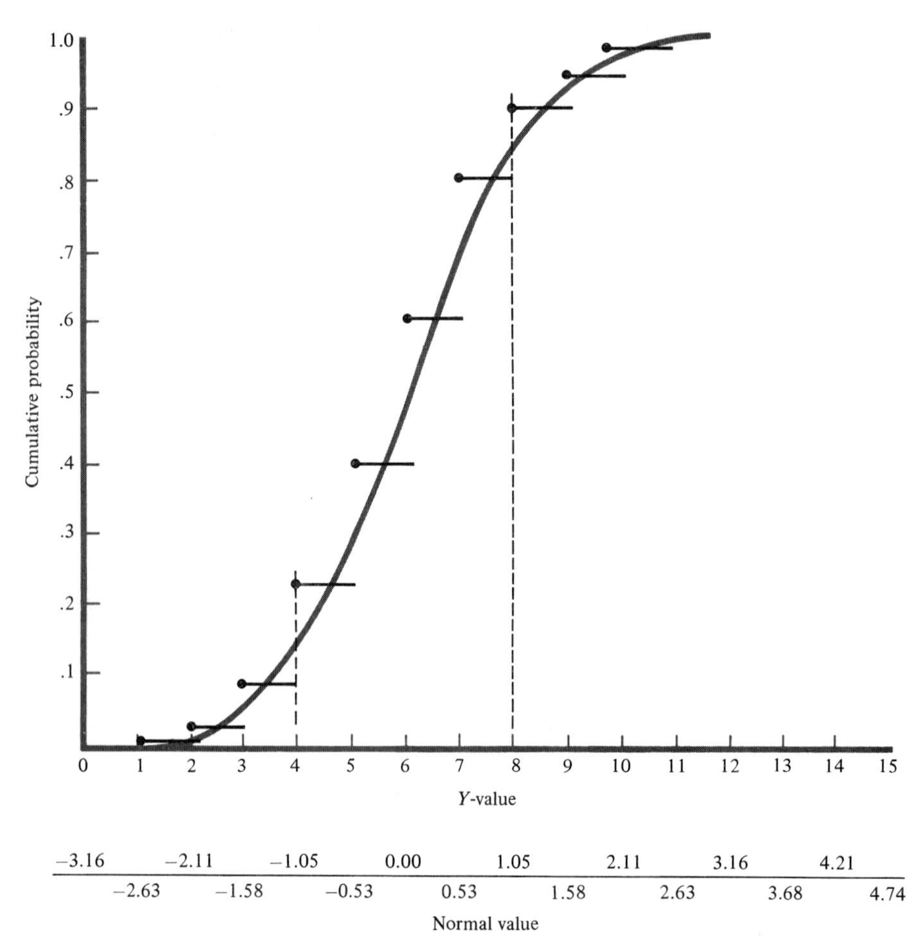

FIGURE 12-4-1 The Cumulative Binomial Distribution (Horizontal Line Segments) and an Approximating Normal One ($n = 15, p = .4$).

other hand, $P(Y \leq 8)$ is too low on the approximating normal curve because we need to read $P(Y \leq 8)$ on the step function for the binomial probability. Since we are trying to approximate a step function with a smooth curve, it is clear that we would do better to consider the integers as midpoints of intervals and compute probabilities based on endpoints chosen to include appropriate midpoints. Thus, in our case, we would begin at 3.5 and end at 8.5. Using these values and Table B-1, we find $P(3.5 \leq X \leq 8.5) = P(-1.32 \leq Z \leq 1.32) = .8132$, a real improvement over $P = .7062$ and quite close to the exact probability of .8145.

In spite of the improvement that is possible in approximating cumulative binomial probabilities by normal probabilities, it is not uncommon to ignore it. This comes from the fact that good tables of the binomial distribution are readily available so that there is no need for an approximation until n is quite large. For example, there are the tables of the National Bureau of Standards (1950). Outside the range of values given there, any large sample approximation is likely to be satisfactory for most purposes.

EXERCISES

12-4-1 Consider a binomial distribution with $p = .5$. Graph the density function of $\hat{p}$, the ratio of the number of times the event occurred $[P(E) = .5]$ to the total number of trials made, for $n = 1, 2, 3, 4, 5$, and 6, respectively. What are the mean and variance of each distribution? Equation 12-2-2 is appropriate for finding the variance. Observe how the shape of the distributions changes as n increases.

12-4-2 Consider a binomial distribution with $p = \frac{1}{3}$. Graph the density function of $\hat{p}$ for $n = 1[1]6$, respectively. What are the mean and variance of each distribution? Observe that, though the distributions are not symmetric, the skewness tends to decrease as n increases.

12-4-3 For a binomial random variable with $p = .4$ and $n = 15$, compute an approximating probability distribution based on the normal distribution. For this, use the suggestion that integral values of the variable should be midpoints of intervals rather than endpoints, as used in Table 12-4-2.

Using your computations, construct a table like Table 12-4-2 and a graph like Figure 12-4-1 so that you can see how use of the suggestion improves the approximating probabilities. Be sure to notice that your z-values lie between corresponding integral values.

12-4-4 A sack of 100 pennies is dumped on a table. Use a cumulative normal distribution to approximate the probabilities of the following events:

(a) 50 or fewer heads.

(b) No more than 45 heads. No more than 40. No more than 25.

(c) Number of heads will be no less than 30 and no more than 50.

12-4-5 Use the suggestion given in the text for improving the approximation to recompute the probabilities in Exercise 12-4-4a and c.

12-4-6 A die is rolled 300 times and the number of times each face appears is recorded. What is the probability that the combined number of 1s and 2s will be no less than 90 or more than 110? No less than 80 or more than 120?

12-4-7 Compute the probabilities required in Exercise 12-4-6 using the improved approximation of the section.

12-4-8 About 15 percent of the items produced by a certain machine are defective. A buyer has purchased 800 items. Use the normal distribution to approximate the probabilities of the following events.

(a) Less than 25 percent of the items are defective. Less than 15 percent. Less than 10 percent.

(b) The number of defectives is not less than 14 or more than 16 percent. Not less than 13 or more than 17 percent.

If the probability is .90 that the number of defectives is no more than k, what is the value of k?

12-4-9 A college basketball player sinks 75 percent of his foul shots in the long run. If he gets 120 foul shots in a season, what is the probability that he will sink 100 or more? (Use normal approximation.)

12-4-10 In a classic genetic study, Gregor Mendel (1948) observed plant-to-plant variation. One of the characteristics observed was seed form as either round or angular. In one sample, there were 88 round seeds and 24 angular ones. Approximate the probability of finding 24 angular seeds, the number observed, or fewer if seeds should be in the theoretical ratio of 3 round to 1 angular.

12-4-11 A test consists of 100 questions where the true answer for each is one of 4 given options. Thus, if a person selects an option at random, there is a probability of .25 that she will answer the question correctly. For what value of k is the probability .90 that she will answer no more than k questions correctly?

12-5 SAMPLING DISTRIBUTIONS

We have already discussed a number of distributions from which samples might be drawn. These often provide a mathematically tractable and realistic formulation of real-world problems. For example, the uniform distribution was used for dice and wheels-of-chance problems, the binomial for coins and genetics, and the normal as an approximation to many other real-world problems. Probabilities were considered from the point of view of a sample of one observation but, to compute them, it was necessary to specify certain parameters of the distribution. When we had both the form of the distribution and the parameters, we had a completely specified model or a parent distribution.

A sample improves as the number of observations increases; that is, it contains more information relevant to the parent distribution. However, it becomes necessary to summarize the information in a sample if we are to have any real hope of drawing inferences. This fact led us to a consideration of the sample mean, called the sample proportion when the random variable was binomial, and its distribution or a reasonable approximation.

The sample mean is a statistic, as is any random variable associated with a sample, and its distribution is called a sampling distribution or a derived distribution. In fact, it will usually be derived as a consequence of knowledge of the parent distribution.

Many sampling distributions have been derived from the normal distribution but are used in dealing with samples of observations which may or may not be normally distributed. How good such derived distributions are for nonnormal data is necessarily an area of statistical research.

It is easy enough to look ahead and see at least some of the statistics for which it will be necessary to have sampling distributions. For example, the sample variance s^2 is clearly a very important statistic. Because of its use in measuring spread, we need to know something about its behavior in repeated sampling, that is, about its distribution. This is available when the parent distribution is normal.

Again, we have seen that we often use σ as a unit of measurement. For example, we enter normal tables with values of the random variable $Z = (Y - \mu)/\sigma$ or, more often, with $Z = (\bar{Y} - \mu)/(\sigma/\sqrt{n})$. Since σ is rarely known, how much better it would be if we could use the statistic $(\bar{Y} - \mu)/(s/\sqrt{n})$. This too has its sampling distribution and it is not that of Z, even when the observations are normally distributed, the case for which the distribution is available.

Many other sampling distributions, especially ones concerned with the comparison of several means, have been derived for observations from the normal distribution. For samples from other distributions, statistics are available that compare the sample distributions to see if they may have come from a common parent, without specifying what the form is. This latter type of comparison is in the class of so-called *distribution-free* or *nonparametric methods.*

The following chapters take a more careful look at some sampling distributions and their application to inductive inference.

Chapter 13

ESTIMATION OF PARAMETERS: KNOWN DISTRIBUTIONS

13-1 INTRODUCTION

Since we cannot examine an entire population, we must be content with a sample. This applies whether we have conducted an experiment or made a sample survey. Our interest, then, is to make use of the sample information to draw some inference concerning the population. To do this involves an inductive process where we reason from a part, the sample, to the whole, the population. Consequently, we conclude with an uncertain inference. However, by applying the laws of probability, we can place a precise measurement on the uncertainty of inferences in general.

The process of drawing inferences generally takes one of two forms, namely, the estimation of population parameters or the testing of hypotheses which specify the values of parameters. In estimation, we are interested in answers to questions such as, "What is the average blood pressure of male and of female adults of the same age?" On the other hand, in testing hypotheses, we are interested in answers to questions such as ,"Is there any difference in the average blood pressure of male and female adults of the same age?"

The two inferential processes are not independent of one another. In order to test hypotheses, it is first necessary to estimate the relevant parameter or parameters before we can test whether one equals a specified value or whether or not two are equal.

In this chapter, we consider some of the principles involved and procedures used to estimate parameters when certain rather strong assumptions are made, that is, when we specify the form of the sampled population. Two types of populations will be examined, the binomial and normal. We also give some attention to the problem of deciding upon an adequate sample size.

13-2 ESTIMATION AND INFERENCE

To estimate a population parameter, we use a sample statistic. In other words, any estimator of a parameter is computed from sample data by means of a formula or estimator. The estimator is, then, a statistic and subject to sampling variation. More formally,

---

An *estimator* is a random variable or statistic used to estimate a parameter. The number resulting from applying an estimator to the data of a particular sample is an *estimate*.

It follows that the random variable or statistic that is the sample mean $\bar{Y}$ is an estimator of the population mean μ, while the numerical value that is a particular sample mean $\bar{y}$ is an estimate of μ. Such an estimate is called a *point estimate*. Many point estimators were given in Chapters 3 and 4.

Random samples drawn from the same population differ, as do point estimates based upon them. Thus, we attach no certainty but only uncertainty to any estimate; for example, we can reasonably say no more than that the population mean should be somewhere near our $\bar{y}$. Accordingly, the use of an *interval estimate* within which we might state that the parameter probably lies would be a more realistic estimation procedure. Such a procedure is available, and, in addition, we find that a specific numerical value can be assigned to the adverb "probably." Thus, for any parameter to be estimated, two sample values, say L_1 and L_2, can be defined computationally and will determine an interval such that, before the sample is drawn, the probability is $1 - \alpha$ that the parameter will lie in the interval from L_1 to L_2. This interval, a random variable, is called a *confidence interval*; $1 - \alpha$ is the *confidence coefficient*, and we are $(1 - \alpha)100$ percent confident that the interval will contain the parameter. The value $1 - \alpha$ is chosen more or less arbitrarily.

In earlier chapters, we began with populations and made probability statements about events that could occur when we would sample. Now, our approach has turned completely around. We are beginning with samples and drawing inferences about populations. We are in the field of statistics rather than probability and the term probability does not quite apply. Consequently, "confidence" as an idea and a term has been introduced.

Once the sample is drawn and one computes the endpoints of the interval, it either contains or does not contain the parameter being estimated. Thus, it is incorrect to say, "The probability that the parameter lies in a particular interval is $1 - \alpha$." Since we do not know the parameter, we are never certain that a particular interval does or does not contain it. However, we do know that the procedure will lead us, on the average, to make $(1 - \alpha)100$ percent correct statements, in the long run, when we say that the parameter does lie in the interval. Thus, for a confidence interval determined for a parameter and computed from a randomly selected sample, we may go so far as to make the probability a personal one applicable to a particular inference and say that we are $(1 - \alpha)100$ percent confident that the computed interval contains the parameter. Failure to distinguish between statements made before and after sampling is equivalent to tossing a fair coin, observing the face showing, say, a head, and stating that the probability of that face is .5 when, in fact, once a coin has been tossed and the face observed, probability is not involved. Before the coin was tossed, there was a probability of .5 of a head appearing, but not after the result is observed.

While we intend to discuss the mechanics of confidence-interval estimation in this chapter, let us first try to clarify some of our ideas by an illustration. For this, we have drawn 20 samples of 10 observations each from a population with $\mu = 40$. We pretended not to know the value of μ and, for each sample, estimated it by means of a 95 percent confidence interval. The results are shown graphically in Figure 13-2-1.

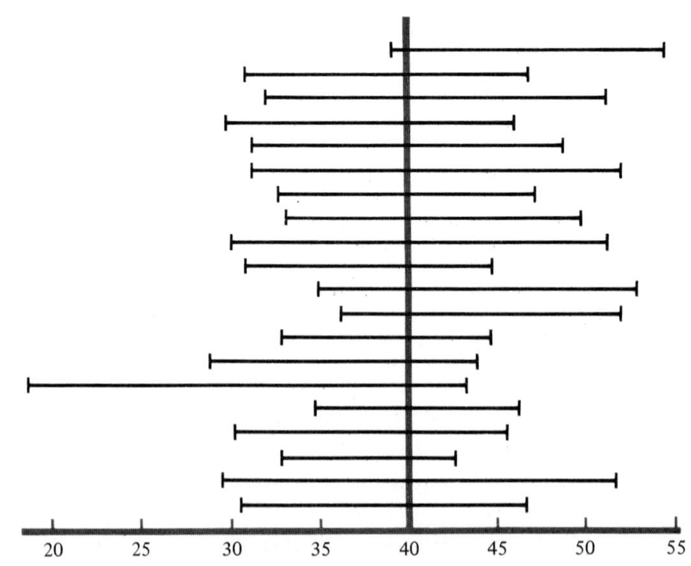

FIGURE 13-2-1 Twenty 95 percent Confidence Intervals Estimating $\mu = 40$.

We see that the lower values of the confidence intervals vary, as do the upper ones and the interval lengths. Each of these three quantities is a random variable, and so we expect this.

We have associated a 95 percent confidence coefficient with our interval-estimation procedure, and so would expect that, on the average, 19 out of 20 intervals would include the population parameter, in this case, $\mu = 40$. Actually, all 20 of our intervals contain the parameter. However, we could not know this sort of information in a real-world situation, and, moreover, we would likely be making only one interval estimate. In that case, we would have assumed that our interval contained the parameter being estimated unless an unusual sampling situation, one unrecognized by us, had occurred. By unusual, we mean a sampling situation likely to occur only 1 time in 20, on the average, since we expect 95 percent of our confidence intervals to contain the estimated parameter.

Now we need to decide what statistic computed from our sample data will best estimate a particular unknown population parameter of interest. In other words, we now need some sort of criteria specifying what constitutes a good estimator. This need was also discussed in Section 3-9 in connection with selecting a measure of central tendency.

EXERCISE

13-2-1 What is the probability that, for 20 computed 95 percent confidence intervals, all will contain the estimated parameter? Exactly one will not contain the parameter? What model did you use in computing your probabilities?

13-3 PROPERTIES OF ESTIMATORS

In Chapters 3 and 4 we discovered that a parameter may have more than one estimator. Thus, for the normal distribution, μ may be estimated by the mean, median, or mode. How good are these different estimators of the parameter?

Since an estimator is a function of the sample, it is a random variable possessing a sampling distribution; one must, consequently, be content in knowing about its performance in repeated sampling rather than at each and every use. Thus, one needs to consider some of the characteristics of the sampling distribution involved to decide what is a good estimator of a given parameter, and what makes a particular one "best."

For example, consider three different estimators of the parameter μ, denoted by $\hat{\mu}_1$, $\hat{\mu}_2$, $\hat{\mu}_3$, where the $\hat{}$, call it "hat," indicates an estimator of μ. All possible values of each of $\hat{\mu}_1$, $\hat{\mu}_2$, and $\hat{\mu}_3$ might generate the density functions of Figure 13-3-1. Observe that $\hat{\mu}_1$ gives values not centered on μ but fairly consistent, that is, with relatively small variance; $\hat{\mu}_2$ gives values that are centered on μ and are also consistent; $\hat{\mu}_3$ gives values centered on μ but with considerable variance. The problem is to choose the best estimator where "best" must be defined. To this

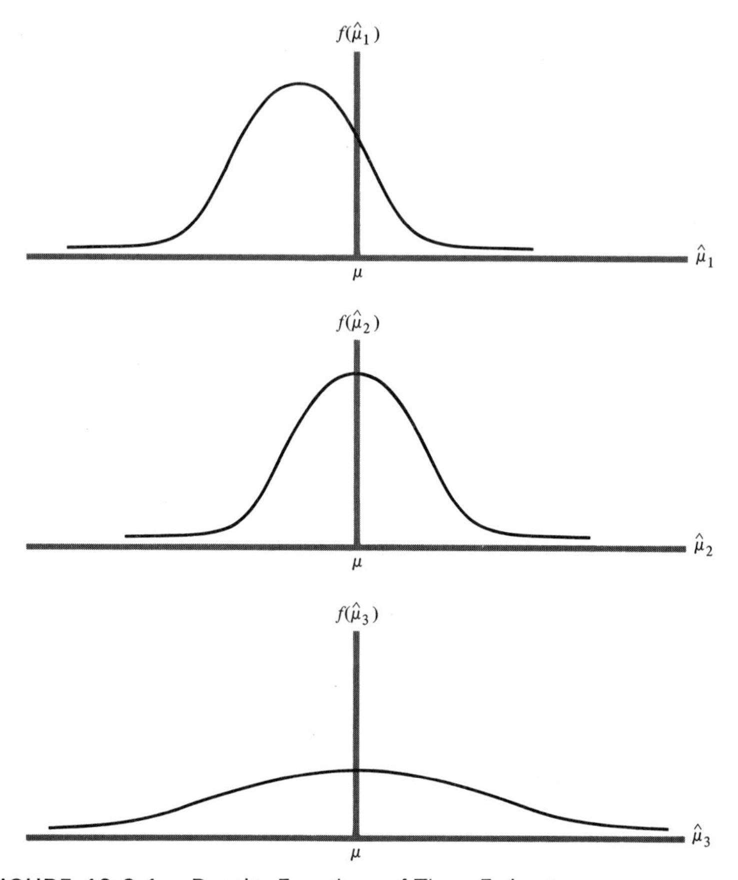

FIGURE 13-3-1 Density Functions of Three Estimators.

end, we might consider a number of desirable properties and then choose an estimator which possesses as many of these as possible. Unbiasedness and minimum variance are two such properties of special concern.

An *unbiased estimator* is one such that its expected value is the parameter being estimated.

Recall that an estimator is a random variable and that its expected value is the mean of the population of values it takes in repeated sampling. Then if this mean is the parameter being estimated, the estimator is said to be unbiased. For example, $\overline{Y}$ is an estimator of μ. We have $E(\overline{Y}) = \mu$. Since μ is the parameter being estimated by $\overline{Y}$, $\overline{Y}$ is said to be an unbiased estimator of μ.

The term mean-unbiased estimator is also used in this case to be more specific. Alternatively, we might ask that the median of all estimates be the parameter being estimated and so obtain a median-unbiased estimator.

In Figure 13-3-1, both $\hat{\mu}_2$ and $\hat{\mu}_3$ are meant to be unbiased estimators of μ, whereas $\hat{\mu}_1$ is biased.

Since an estimator generates a population of values, it has a variance as well as a mean. If we have several estimators, then that with the smallest variance is the best by a variance criterion. Thus,

---

An estimator of a parameter is the *minimum variance estimator* if its variance is less than that of any other estimator for samples of the same size.

The *efficiency* of an estimator is related to its variance, being inversely proportional to it; that is, a small variance implies an efficient estimator. In Figure 13-3-1, $\hat{\mu}_1$ and $\hat{\mu}_2$ are more efficient than $\hat{\mu}_3$ as estimators of μ. The *relative efficiency* of two estimators is found by examining the ratio of the reciprocals of their variances. This is conveniently expressed as a percentage. Thus, for $\hat{\mu}_2$ and $\hat{\mu}_3$, unbiased estimators of μ, the efficiency of $\hat{\mu}_2$ relative to $\hat{\mu}_3$ is given by Equation 13-3-1.

$$\text{Relative efficiency of } \hat{\mu}_2 \text{ to } \hat{\mu}_3 = \frac{\text{variance of } \hat{\mu}_3}{\text{variance of } \hat{\mu}_2} \, 100 \text{ percent} \qquad (13\text{-}3\text{-}1)$$

Equation 12-2-1 states that the sample mean is always an unbiased estimator of the population mean and, in particular, is so for a normal population. The median Q_2 is also an unbiased estimator of μ for a normal population. The variance of $\bar{Y}$ is σ^2/n and that of Q_2 is $1.57\sigma^2/n$ for samples of the same size. Thus, the efficiency of the median relative to the mean is

$$\frac{\sigma^2/n}{1.57\sigma^2/n} \, 100 \text{ percent} = 64 \text{ percent}$$

This shows that the mean, to have the same efficiency as the median as an estimator of μ for a normal population, requires a sample only 64 percent as large as needed for the median.

Unbiasedness and minimum variance are not the only desirable properties that estimators may possess, and, in fact, we do not always ask that an estimator be unbiased. The sample standard deviation, for example, is not an unbiased estimator of σ. For the time being, we will not consider other properties but simply try to keep these two in mind.

13-4 ESTIMATION OF PROPORTIONS

A binomial population is defined as one with two possible outcomes or classes, often referred to as success and failure. The probability of a random observation belonging to the first class is p and to the second class is $1 - p$. Repeated trials are to be independent, which is like sampling a finite population with replacement.

To symbolize the estimate of the parameter p, we use $\hat{p}$, the proportion of the sample observed in the first class. The mean of all possible $\hat{p}$ values is p (Exercise 11-4-6); that is, $\hat{p}$ is an unbiased estimator of p. The variance of the statistic $\hat{p}$ is $p(1 - p)/n$ (Exercise 11-5-11), where n is the total number of observations in the sample; it is estimated by $\hat{p}(1 - \hat{p})/n$ when necessary.

Extensive tables of probabilities for the binomial distribution are available and can be used in the construction of confidence intervals for the binomial parameter p. From such tables, further tables and charts have been prepared for finding such confidence intervals more conveniently. However, we will proceed by using the normal distribution as an approximation to the binomial distribution. This use was discussed in Section 12-4 for the number of successes rather than the proportion.

If Y, the number of successes in n binomial trials, is converted to a proportion $\hat{p}$, and this in turn to standard units by Equation 10-4-3, then the resulting random variable Z can be referred to normal probability tables and any required probabilities can be computed. Equation 10-4-3 is now written as 13-4-1.

$$Z = \frac{\hat{p} - p}{\sqrt{p(1 - p)/n}} \tag{13-4-1}$$

where $\hat{p} = Y/n$ is the point estimate of p; $\hat{p}$ proceeds by steps of $1/n$ from 0 to 1.

Since we are estimating p, we know neither its value nor that of the variance. We will estimate the latter by $\hat{p}(1 - \hat{p})/n$ and use Equation 13-4-2.

$$Z = \frac{\hat{p} - p}{\sqrt{\hat{p}(1 - \hat{p})/n}} \tag{13-4-2}$$

A probability statement about Z can now lead us to a confidence interval for p. Write

$$P(z_{1-\alpha/2} < Z < z_{\alpha/2}) = P(-z_{\alpha/2} < Z < z_{\alpha/2}) = 1 - \alpha$$

Replace Z by its definition from Equation 13-4-2 and rework the probability statement as one about a random interval covering p. Recall that the subscript on z tells us how much probability lies to the right. Consequently, we have assigned equal probabilities, namely $\alpha/2$, to each tail. Thus, we begin with a statement about the random variable $\hat{p}$.

$$P\left(-z_{\alpha/2} < \frac{\hat{p} - p}{\sqrt{\hat{p}(1 - \hat{p})/n}} < z_{\alpha/2}\right) = 1 - \alpha$$

Multiplying the inequality in the parentheses by the standard deviation, we get

$$P\left(-z_{\alpha/2}\sqrt{\frac{\hat{p}(1 - \hat{p})}{n}} < \hat{p} - p < z_{\alpha/2}\sqrt{\frac{\hat{p}(1 - \hat{p})}{n}}\right) = 1 - \alpha$$

Subtracting $\hat{p}$ from each term in the parentheses gives

$$P\left(-\hat{p} - z_{\alpha/2}\sqrt{\frac{\hat{p}(1 - \hat{p})}{n}} < -p < -\hat{p} + z_{\alpha/2}\sqrt{\frac{\hat{p}(1 - \hat{p})}{n}}\right) = 1 - \alpha$$

Since p must lie between 0 and 1, multiply within the parentheses by -1, consequently reversing the direction of the inequality. Now rearrange the terms so that the smallest is on the left, and so obtain Equation 13-4-3.

$$P\left(\hat{p} - z_{\alpha/2}\sqrt{\frac{\hat{p}(1 - \hat{p})}{n}} < p < \hat{p} + z_{\alpha/2}\sqrt{\frac{\hat{p}(1 - \hat{p})}{n}}\right) = 1 - \alpha \qquad (13\text{-}4\text{-}3)$$

Thus, the $(1 - \alpha)100$ percent confidence limits for p are approximated by Equations 13-4-4.

$$L_1 = \hat{p} - z_{\alpha/2}\sqrt{\frac{\hat{p}(1 - \hat{p})}{n}}$$

$$L_2 = \hat{p} + z_{\alpha/2}\sqrt{\frac{\hat{p}(1 - \hat{p})}{n}} \qquad (13\text{-}4\text{-}4)$$

In review, we began with a valid probability statement about a normal random variable Z. We converted this to an approximate one about a binomial variable and, after some manipulation, to Equation 13-4-3. This is a statement about the random interval (L_1, L_2) and its placement about the parameter p, which is a constant. Since p is either in or not in the interval, Equation 13-4-3 is not really a probability statement about p, but only appears to be, so we call it a *confidence-interval statement*.

We may now state that each computed interval contains the estimated parameter. In this case, the confidence coefficient measures the proportion, $1 - \alpha$, of statements that will be true in the long run. On the other hand, we may say of each interval that it contains the parameter unless we have a sample so unusual that it is in a class whose members occur only a small proportion, α, of the time.

The choice of a confidence coefficient is somewhat arbitrary. We could choose $1 - \alpha = .99$ and have a large proportion of our intervals contain p. However, our intervals would necessarily be large also and, so to speak, give us little to be confident about since increasing length implies increasing vagueness as to where p is. On the other hand, we could choose $1 - \alpha = .50$ and have much shorter intervals but less confidence in them since only about 50 percent would contain p. The choice of $1 - \alpha = .95$ has served as a very satisfactory general value with departures depending upon special circumstances.

Illustration 13-4-1

Suppose 400 persons are selected at random and questioned about their preference for sports cars A and B; 224 prefer A. Construct a 95 percent confidence interval for the population proportion of persons preferring car A.

Here $n = 400$, $y = 224$, $\hat{p} = 224/400 = .56$, and $1 - \alpha = .95$. Consequently $\alpha = .05$ and $\alpha/2 = .025$. We refer to Table B-1 and find $z_{.025} = 1.96$.

Now, since

$$s_{\hat{p}} = \sqrt{\frac{\hat{p}(1-\hat{p})}{n}} = \sqrt{\frac{.56(.44)}{400}} = .0248$$

we find

$$l_1 = .56 - 1.96(.0248) = .51$$
$$l_2 = .56 + 1.96(.0248) = .61$$

Again, we have used lowercase letters to indicate sample values. These values are the endpoints of a confidence interval with confidence coefficient of .95. We know that in the long run, 95 percent of the intervals so computed will contain the parameter. Thus .95 is a measure of our confidence in the estimation procedure, although some may wish to call it a personal probability, measuring their confidence in the single statement that the sample interval contains the parameter being estimated.

Suppose that the sampled population contains 100,000 people. The sample of 400 is so small relative to the population size that no adjustment to our interval estimate is needed. In fact, we can expand our sample result to estimate the number in the population of 100,000 who prefer car A. We can say, with approximately 95 percent confidence in our procedure, that the number of individuals who prefer car A lies between 51,000 and 61,000.

The reasoning that led to the confidence limits of Equation 13-4-4 can be extended to estimates of the difference between the parameters of two binomial proportions.

Let the parameters of the two populations be p_1 and p_2. Independent random samples of n_1 and n_2 observations are drawn from the two populations and p_1 and p_2 estimated by $\hat{p}_1$ and $\hat{p}_2$, respectively.

The point estimator of $p_1 - p_2$ is $\hat{p}_1 - \hat{p}_2$, an unbiased estimator (Exercise 11-3-3). Its standard deviation is found by taking the square root of the sum of the variances as in Equation 13-4-5.

$$\sigma_{\hat{p}_1-\hat{p}_2} = \sqrt{\frac{p_1(1-p_1)}{n_1} + \frac{p_2(1-p_2)}{n_2}} \tag{13-4-5}$$

This population variance is estimated by the sample value given in Equation 13-4-6.

$$s_{\hat{p}_1-\hat{p}_2} = \sqrt{\frac{\hat{p}_1(1-\hat{p}_1)}{n_1} + \frac{\hat{p}_2(1-\hat{p}_2)}{n_2}} \tag{13-4-6}$$

Thus, the interval estimate of $p_1 - p_2$, with confidence coefficient $1 - \alpha$, has endpoints given by Equation 13-4-7.

$$(L_1, L_2) = (\hat{p}_1 - \hat{p}_2) \pm z_{\alpha/2} s_{\hat{p}_1 - \hat{p}_2} \tag{13-4-7}$$

Illustration 13-4-2

Consider two pain-relieving drugs compared on independent samples of 1,000 individuals. Suppose 750 of those individuals receiving drug 1 and 800 of those receiving drug 2 reported some pain relief. Construct a 95 percent confidence interval for the difference between population proportions.

The point estimate of $p_1 - p_2$ is $\hat{p}_1 - \hat{p}_2 = .80 - .75 = .05$. The interval estimate with confidence coefficient $1 - \alpha = .95$ is

$$\begin{aligned}(l_1, l_2) &= (\hat{p}_1 - \hat{p}_2) \pm z_{\alpha/2} \sqrt{\frac{\hat{p}_1(1 - \hat{p}_1)}{n_1} + \frac{\hat{p}_2(1 - \hat{p}_2)}{n_2}} \\ &= (.80 - .75) \pm 1.96 \sqrt{\frac{.80(.20)}{1000} + \frac{.75(.25)}{1000}} \\ &= .05 \pm .026 = (.024, .076)\end{aligned}$$

The true difference in the percent relief given by the two drugs favors drug 2 and should lie in the interval 2.4 to 7.6 percent. We know that in repeated sampling approximately 95 percent of such intervals will include the true difference $p_1 - p_2$.

EXERCISES

13-4-1 When a population is large relative to sample size, the error in using the binomial is negligible. Consider a population of 100,000 individuals to be sampled for preference for two kinds of cars. Suppose that 55,000 of the individuals prefer car *A* and 45,000 car *B*.
(a) What is the probability that the first randomly selected person prefers car *A*?

The probability that the second person in the random sample prefers *A* now depends upon the preference of the first individual.
(b) What is the probability that the second person prefers car *A* if the first one did also?
(c) What is the probability that the second person prefers car *A* if the first one preferred car *B*?

Note: Observe the closeness of all three computed probabilities. In practice, we often assume independence of probabilities under such circumstances and use the binomial distribution because the result will be very close to the true situation.

13-4-2 Was the population sampled for Illustration 13-4-1 on confidence-interval estimation a finite one? Illustration 13-4-2?

13-4-3 In a random sample of 250 voters, 150 preferred candidate Jones and 100 candidate Smith. Construct 90 and 95 percent confidence intervals for the proportion of voters preferring candidate Jones. What will be the intervals in Smith's case? What is the relationship between the intervals for the two candidates? Is the sampled population finite or infinite?

13-4-4 A random sample of 400 light bulbs contained 10 that were defective. Determine the point estimate and 95 percent confidence interval for p, the percentage of defective light bulbs.

13-4-5 A random sample of 5,000 consumers indicated that 3,500 preferred a colored refrigerator over a white one. Construct 95 and 99 percent confidence intervals for the proportion of consumers preferring the colored product.

13-4-6 A random sample of 1,600 Chicago families indicated that 200 intended to purchase a new car during the next 12 months. Construct a 95 percent confidence interval for the proportion of families that expect to purchase a new car during the next 12 months.

13-4-7 In a poll of college students in a large state university, 300 of 400 students living in dormitories approved a certain course of action, whereas 200 of 300 students not living in dormitories approved it. Estimate the difference in the proportions favoring the course of action and compute 90 and 95 percent confidence intervals for it.

13-5 ESTIMATION OF MEANS IN THE NORMAL DISTRIBUTION

If the sampled population is normal, then we would expect to find an interval estimate of the mean by proceeding, as was done for proportions in Section 13-4, from a probability statement about the random variable Z. However, the true standard deviation is rarely known, so that we would necessarily substitute a sample value, as was done in Equation 13-4-2, and conclude with what would obviously be an approximation even if the assumption of normality is valid. The approximation would be in the confidence coefficient. Fortunately, the derived distribution of the random variable which measures a distance from μ in units of the sample standard deviation is known, so that such an approximation is not required. This distribution is called Student's t-distribution. Equation 13-5-1 defines Student's t and Table B-2 provides the associated probabilities.

Student's $$t = \frac{\overline{Y} - \mu}{s_{\overline{Y}}} \qquad (13\text{-}5\text{-}1)$$

For *interval estimation* of the mean, by use of Student's t, to be valid, the following assumptions must be true:

1. Each random observation consists of the fixed population mean plus a random deviation. Formally, we wrote $Y_i = \mu + \varepsilon_i$ as the model equation describing an observation and read ε_i as epsilon-sub-i.
2. The deviations are random, independent, and from a normal distribution with mean 0 and common, but unknown, variance.

Statement 2 completes the model because it tells us about the probabilities involved. It says, essentially, that all observations come from the same normal population and are drawn at random and, hence, independently of one another. In other words, this statement is no different in content than previous ones, only somewhat more formal.

To use Student's t, we first compute $\bar{Y} = \sum Y_i/n$. Next, we need the variance of $\bar{Y}$. The true variance of sample means is $\sigma_{\bar{Y}}^2 = \sigma^2/n$, which is estimated by $s_{\bar{Y}}^2 = s^2/n$. Finally, to estimate the standard deviation of the parent population and of sample means, σ and $\sigma_{\bar{Y}}$, respectively, we take the square root of the corresponding variance estimators. Thus,

$$s = \sqrt{s^2} \qquad \text{and} \qquad s_{\bar{Y}} = \sqrt{s_{\bar{Y}}^2} = \sqrt{\frac{s^2}{n}}$$

In Equation 13-5-1, the only remaining unknown value is μ, the parameter for which we require an interval estimate.

The t-distribution is symmetric about $\mu_t = 0$ and depends upon a single parameter, the degrees of freedom, abbreviated df. Thus it is many distributions. This parameter, df, was discussed in Section 4-3 where it was given the value $n - 1$ when s^2 was computed from a single random sample. Table B-2 gives values of t such that α is the probability of a larger value of t, sign ignored; that is, Equations 13-5-2 hold.

$$\begin{aligned} P(t < -t_{\alpha/2} \text{ or } t > t_{\alpha/2}) &= \alpha \\ P(-t_{\alpha/2} < t < t_{\alpha/2}) &= 1 - \alpha \end{aligned} \qquad (13\text{-}5\text{-}2)$$

Notice that a capital letter is not used to indicate the random variable. However, the subscript $\alpha/2$ tells us that $t_{\alpha/2}$ is a tabulated and, therefore, nonrandom, value. Also, a probability of $\alpha/2$ has been assigned to each tail of the distribution.

In the inequalities within the parentheses of Equations 13-5-2, the second one in particular, we may now substitute the definition of t and proceed as in Section 13-4 until we have an equation that says something about the probability of a random interval covering μ, namely Equation 13-5-3. The interval is symmetric about $\bar{Y}$.

$$P(\bar{Y} - t_{\alpha/2}s_{\bar{Y}} < \mu < \bar{Y} + t_{\alpha/2}s_{\bar{Y}}) = 1 - \alpha \qquad (13\text{-}5\text{-}3)$$

We can now say that we are $(1 - \alpha)100$ percent confident that μ will lie between the two sample quantities presented as random variables in the following formulas, where $t_{\alpha/2}$ has $n - 1$ degrees of freedom.

$$\begin{aligned} L_1 &= \bar{Y} - t_{\alpha/2}s_{\bar{Y}} \\ L_2 &= \bar{Y} + t_{\alpha/2}s_{\bar{Y}} \end{aligned}$$

Alternatively, we may say that μ will lie in the interval (L_1, L_2) unless we have an unusual sample, unusual being measured by saying that such samples occur only $\alpha 100$ percent of the time, on the average.

If σ is known, essentially the same formulas can be applied except that σ replaces s and the degrees of freedom for t become infinite so that we are back to the standard normal variable Z. This is seen by comparing the last line of the t-table with values of z for corresponding probabilities. Note also that the length of the interval estimate will not be a random variable, although its location is, but will have a fixed length.

Illustration 13-5-1

Exercise 3-6-3 contains heart rates of 13 cats that had received 10 milligrams of procaine. Let us construct a 95 percent confidence interval for the mean of the population of such data.

The point estimate of μ is $\bar{y} = \sum y_i/n = 1{,}919/13 = 147.6$ heartbeats. The sample variance is $s^2 = \sum (y_i - \bar{y})^2/(n-1) = 9{,}049/12$, with $13 - 1 = 12$ degrees of freedom, and the standard deviation of the mean is $s_{\bar{y}} = \sqrt{s^2/n} = \sqrt{9{,}049/(12)(13)} = 7.62$ heartbeats. This is a measure of the variability in a population of heartbeat means based on random samples of 13 cats.

From Table B-2, we find the value of $t_{\alpha/2}$ for 12 degrees of freedom, namely $t_{.025} = 2.179$. The interval estimate is then

$$l_1 = \bar{y} - t_{.025}s_{\bar{y}} = 147.6 - 2.179(7.62) = 131.0 \text{ heartbeats}$$
$$l_2 = \bar{y} + t_{.025}s_{\bar{y}} = 147.6 + 2.179(7.62) = 164.2 \text{ heartbeats}$$

We are now able to state that the true mean lies in the interval (131.0, 164.2) unless we have an unusual ($\alpha = .05$) sample. Alternatively, we might say that we are 95 percent confident that the interval contains the parameter because, in repeated sampling, approximately 95 percent of such intervals will.

The above reasoning can also be used to provide a procedure to estimate the difference between the means of two normal populations just as was done for two binomial parameters.

Let the means of two normal populations be μ_1 and μ_2. Independent random samples of n_1 and n_2 observations are drawn from the populations, and μ_1 and μ_2 estimated by $\bar{Y}_1$ and $\bar{Y}_2$, respectively. The point estimate of $\mu_1 - \mu_2$ is $\bar{Y}_1 - \bar{Y}_2$. Denote its standard deviation by $\sigma_{\bar{Y}_1 - \bar{Y}_2}$. This, or an estimate, will be necessary in order to allow construction of a criterion that measures the distance between $\bar{Y}_1 - \bar{Y}_2$ and its corresponding population parameter in standard deviation units.

If we assume ${\sigma_1}^2 = {\sigma_2}^2 = \sigma^2$, say, then the variance of $\bar{Y}_1 - \bar{Y}_2$ is $\sigma^2(1/n_1 + 1/n_2)$. We must estimate σ^2 by using all the available information and do so by pooling the sums of squares and degrees of freedom from the two samples as in Equation 13-5-4.

$$s^2 = \frac{\sum (Y_{1i} - \bar{Y}_1)^2 + \sum (Y_{2i} - \bar{Y}_2)^2}{(n_1 - 1) + (n_2 - 1)} \tag{13-5-4}$$

This estimate has $(n_1 - 1) + (n_2 - 1)$ degrees of freedom. Notice that we have introduced a second subscript, the 1 and the 2, to keep our two samples distinct. Also, notice that s^2 is a weighted average of two independent estimates of σ^2, namely ${s_1}^2$ and ${s_2}^2$. The weights are the degrees of freedom rather than the sample sizes. Equation 13-5-5 is equivalent to 13-5-4.

$$s^2 = \frac{(n_1 - 1){s_1}^2 + (n_2 - 1){s_2}^2}{(n_1 - 1) + (n_2 - 1)} \tag{13-5-5}$$

We now estimate $\sigma_{\bar{Y}_1 - \bar{Y}_2}$ using Equation 13-5-6.

$$s_{\bar{Y}_1 - \bar{Y}_2} = \sqrt{s^2 \left(\frac{1}{n_1} + \frac{1}{n_2} \right)} \tag{13-5-6}$$

Note that this is simply the square root of the sum of the variances of the two means, that is, $\sqrt{s_{\bar{Y}_1}^2 + s_{\bar{Y}_2}^2}$.

We can now construct a sample t-value with $\mu_1 - \mu_2$ as the only unknown, make a probability statement about a random t, and rearrange this to give a confidence interval statement about $\mu_1 - \mu_2$. The endpoints are given by Equation 13-5-7.

$$(L_1, L_2) = [(\bar{Y}_1 - \bar{Y}_2) \pm t_{\alpha/2} s_{\bar{Y}_1 - \bar{Y}_2}] \quad \text{for df} = n_1 + n_2 - 2 \tag{13-5-7}$$

Illustration 13-5-2

In a study by Watson et al. (1949), two lots of Holstein heifers were each fed different rations. The average gain in weight for the 15 heifers of lot 1 was 235.9 pounds and for the 15 heifers of lot 2 was 187.6 pounds. The pooled variance s^2, an estimate of a common σ^2, was 2,199 pounds squared. We require a 95 percent confidence interval on the difference between the population means.

We will assume that these are random samples. Certainly, the available animals were assigned to the two lots at random. Definition of the populations must, of course, be restricted to specify the conditions under which the animals were acquired and kept.

The 95 percent sample confidence limits are determined by Equation 13-5-7 and computed below.

$$\begin{aligned}(l_1, l_2) &= (\bar{y}_1 - \bar{y}_2) \pm t_{.025} s_{\bar{y}_1 - \bar{y}_2} \qquad \text{for df} = n_1 + n_2 - 2 \\ &= 235.9 - 187.6 \pm 2.056 \sqrt{\frac{2(2{,}199)}{14}} \end{aligned} \tag{13-5-7}$$

Hence, $l_1 = 11.9$ pounds and $l_2 = 84.7$ pounds.

In repeated sampling approximately 95 percent of such intervals will contain the true difference.

Two other cases calling for the estimation of a difference between population means are of interest.

The first arises when our samples are not independent, as when we have n pairs of individuals available and expect the members of a pair to respond very similarly if treated alike. Such would be the case if we had sets of twins. We then assign the individuals of a pair at random, one to each treatment, expecting the difference between members of a pair to be attributable largely to the treatment difference if such exists. Here, we take the difference, $D_i = Y_{1i} - Y_{2i}$, for each pair and treat these differences as our sample. The mean of these differences,

$\bar{D} = \bar{Y}_1 - \bar{Y}_2$, is an estimate of $\mu_1 - \mu_2$, and the population standard deviation of this estimate is, in turn, estimated by $s_{\bar{D}} = s_D/\sqrt{n}$, with $n - 1$ degrees of freedom, where s_D is computed using the n differences. This case is considered in Chapter 17.

The other case arises when we do not have pairing and are not justified in assuming homogeneity of variance. This is a more complicated situation and solutions are only approximations (see Steel and Torrie 1960, chapter 5).

EXERCISES

13-5-1 Use the data of Exercises 3-6-1, 3-6-2, 3-6-5, 3-6-6, and 3-6-7 (three samples) to construct 95 percent confidence intervals for the appropriate population means.

13-5-2 What do you think about the validity of the assumptions you have made in connection with the data of Exercises 3-6-6 and 3-6-7?

13-5-3 Construct 99 percent confidence intervals for some of the population means for which you have just computed 95 percent intervals. Notice the change in length as the confidence coefficient changes.

13-5-4 Using the data in Exercise 3-6-4, construct a 95 percent confidence interval for the difference between population means. Be sure to note that these are before and after observations and, consequently, paired.

13-5-5 Monarch butterflies acquire cardiac glucosides which make them unpalatable to birds. In a quantitative evaluation of amounts, suppose it is found that 35 males have a mean concentration of .195 units and a variance of .0100 units squared, while 40 females have a mean of .205 units with a variance of .0144 units squared.

Construct a 95 percent confidence interval for the difference between population means for males and females, assuming the two sample variances are estimates of a common value.

13-5-6 Special experiments are required to show that silicon is required for normal growth in chicks. Suppose that a study of this is made with supplemented and unsupplemented diets being given two randomly selected groups and that average daily weight gains are measured for each chick over a suitable period of time. Let the sample means for 36 chicks in each group be 3.10 and 2.37 grams and sample variances be .3600 and .4356 grams squared for the supplemented and unsupplemented groups, respectively.

Construct a 95 percent confidence interval for the difference between population means for the two groups, assuming that the sample variances are estimates of a common value.

13-5-7 What is the length of a $(1 - \alpha)100$ percent confidence interval when the population variance is unknown? Does it vary from sample to sample of the same size?

13-5-8 What is the length of a $(1 - \alpha)100$ percent confidence interval when the population variance is known? Is this length a random variable when n is fixed?

13-6 SAMPLING FROM SMALL POPULATIONS

The estimation procedures discussed so far in this chapter are based on the assumption that random sampling is from a very large population. When a sample consists of an appreciable portion of a finite population, 5 percent or greater, modification in the formulas is needed for the standard error of the statistic being estimated. For both means and proportions, the modification consists of multiplying the standard error by a *finite population correction, fpc,* given by Equation 13-6-1.

$$\text{fpc} = \sqrt{\frac{N - n}{N - 1}} \tag{13-6-1}$$

where N and n are the numbers of observations in the population and sample, respectively.

Thus, the formulas for the population standard error of the mean and its estimate, $\sigma_{\bar{Y}}$ and $s_{\bar{Y}}$, are given by Equations 13-6-2.

$$\begin{aligned} \sigma_{\bar{Y}} &= \frac{\sigma}{\sqrt{n}} \sqrt{\frac{N - n}{N - 1}} \\ s_{\bar{Y}} &= \frac{s}{\sqrt{n}} \sqrt{\frac{N - n}{N - 1}} \end{aligned} \tag{13-6-2}$$

For proportions, the formulas for $\sigma_{\hat{p}}$ and $s_{\hat{p}}$ are given by Equations 13-6-3. Notice that $\sigma_{\hat{p}}$ is a parameter, whereas $s_{\hat{p}}$ is an estimate; the notation $\hat{\sigma}_{\hat{p}}$ could just as well be used in the latter case.

$$\begin{aligned} \sigma_{\hat{p}} &= \sqrt{\frac{p(1 - p)}{n}} \sqrt{\frac{N - n}{n - 1}} \\ s_{\hat{p}} &= \sqrt{\frac{\hat{p}(1 - \hat{p})}{n}} \sqrt{\frac{N - n}{n - 1}} \end{aligned} \tag{13-6-3}$$

13-7 ESTIMATION OF SAMPLE SIZE

At the planning stage of an experiment or survey, it is desirable to know how large a sample will be necessary to provide reasonable assurance that an estimate of a mean or a proportion will have some stated precision such as a specified width of confidence interval. Alternatively stated, we may wish to know how large a sample will be required in order for us to have reasonable assurance that a sample mean or proportion will differ from the population parameter by no more than some specified amount.

Suppose we are to sample from a normal distribution with a known variance σ^2. A $(1 - \alpha)100$ percent confidence interval is computed as $\bar{Y} \pm z_{\alpha/2}\sigma/\sqrt{n}$ to give an interval of length $2z_{\alpha/2}\sigma/\sqrt{n}$. If we specify the desired maximum length as d, then $d = 2z_{\alpha/2}\sigma/\sqrt{n}$. We can now square and solve for n to get Equation 13-7-1.

$$n = \frac{4z_{\alpha/2}{}^2\sigma^2}{d^2} \tag{13-7-1}$$

If we are prepared to specify the length d as a multiple of the population standard deviation, then Equation 13-7-1 may be written as Equation 13-7-2. This is a form which is often more useful than that of the original equation.

$$n = 4{z_{\alpha/2}}^2 \left(\frac{\sigma}{d}\right)^2 \qquad (13\text{-}7\text{-}2)$$

We can now guarantee that a confidence interval with confidence coefficient $1 - \alpha$ will have the required length. Alternatively, since $(1 - \alpha)100$ percent of random $\bar{Y}$'s lie within a distance of $z_{\alpha/2}\sigma/\sqrt{n}$ from μ, we are guaranteed that $(1 - \alpha)100$ percent of $\bar{Y}$'s for the given sample size will lie within a distance $d/2$ of μ.

When σ^2 is unknown and we must rely upon estimates, we cannot provide the guarantees mentioned but must settle for reasonable assurances. However, these can be provided as is done with confidence coefficients.

We now illustrate the computation of sample size for a binomial proportion using the normal approximation.

Illustration 13-7-1

A person to whom a certain type of drug is administered responds by improving or not improving. Previous experience leads us to believe that about 70 percent of individuals will improve. The problem is to choose a sample size to estimate the true probability that a randomly selected individual will show improvement. Let us require that the sample size be such that 95 percent of estimates based on this sample size will lie within a distance of .06 of the true probability. In other words, we want a 95 percent assurance that we will be within the specified distance from μ.

For this problem, $d/2 = .06$, so that $d = .12$; also, $1 - \alpha = .95$, and, consequently, $z_{\alpha/2} = 1.96$. The parent binomial variance is estimated from the 70 percent figure as $\sigma^2 = .7(.3) = .21$. We now apply Equation 13-7-1.

$$n = \frac{4(1.96)^2 .21}{.12^2} = 224$$

Observe that the half interval is of the required length.

$$z_{\alpha/2}\sqrt{\frac{p(1-p)}{n}} = 1.96\sqrt{\frac{.7(.3)}{224}} = .06$$

Note also that the sample size appears to be large enough to justify the normal approximation.

Illustration 13-7-2

Hydrogen sulfide is produced in the anaerobic fermentation of sewage and measured in parts per million (ppm) after 42 hours at 37°F. Suppose we want to estimate μ by a 95 percent confidence interval that will not exceed a length of $\frac{4}{5} = .8$ standard deviations and are prepared to assume that the underlying distribution is approximately normal.

Here $1 - \alpha = .95$, calling for $z_{.025} = 1.96$. Also $d/\sigma = .8$. We use Equation 13-7-2 to find n.

$$n = 4(1.96)^2 \left(\frac{1}{.8}\right)^2 = 24$$

Notice that we have not been required to specify a value for σ. However, we have implied that we know this value since our procedure for finding n was based on constructing an interval using the parameter σ and the variable Z. We have oversimplified our problem and would normally have to provide a prior estimate of σ or wait until we had our sample.

Since our procedure for determining sample size can never guarantee the length of the interval estimate without knowledge of σ, it becomes desirable to introduce a measure that specifies the probability that a $(1 - \alpha)100$ percent confidence interval will be no more than the desired length. Now, we see that we have two requirements, the probability that the interval estimate will cover μ and the probability that it will be no more than the required length. In our illustrations, we have specified the former at $1 - \alpha$ and left the other open; it is approximately .5 for our procedure. More refined techniques are available for both proportions and means but are beyond the scope of our text.

EXERCISES

13-7-1 In a public opinion poll, how large a sample is necessary so that one can state that the estimated proportion of votes received by candidate A will not differ by more than 3 percent from the true proportion of votes she receives, unless the sample is so unusual as to be in a class that occurs only about one time in 20, on the average? The candidate's campaign manager admits privately that the favorable vote could be anywhere from 30 to 70 percent.

13-7-2 How large a sample would be necessary if the probability were lowered from .95 to .90? Raised from .95 to .99?

13-7-3 How many fuses need to be examined if we wish to state with 95 percent assurance that the true proportion will not differ from the sample proportion by more than 0.02? Past experience has indicated that the true percentage of rejects is about 5 percent.

13-7-4 From a normal population with a variance of 50 units squared, how large a sample need be taken so that the sample mean will not differ from the true mean by more than 2 units with 95 percent assurance.

13-7-5 An experimenter desires to estimate the difference in yield measured as the response to two different fertilizers when applied to corn, to within 5 bushels per acre with probability of 95 percent. Previous tests have indicated that the standard deviation in the case of a single fertilizer will be about 8 bushels. How many observations should he make per group, if the groups are to be of equal size?

13-7-6 Suppose that in Exercise 13-4-7 one wishes to determine how large a sample will be necessary to estimate the difference in the two proportions to within a value of .05. Assume that samples of the same size will be taken from both groups and that $p = .7$ will suffice as an approximation for both proportions.

13-8 ESTIMATION OF σ^2 IN THE NORMAL POPULATION

When the parent population randomly sampled is normally distributed, then the chi-square (χ^2) distribution can be used to construct a confidence interval for σ^2.

The χ^2 distribution, like the t-distribution, is many distributions. By definition, the sum of squares of n normally and independently distributed quantities with zero means and unit variances, that is, standardized normal variables, is distributed as χ^2 with n degrees of freedom. While this seems to require a knowledge of μ, it was pointed out in Section 4-3 that $(n - 1)s^2$ is the sum of squares of $n - 1$ quantities that truly differ, that is, that are independent. These are the $(Y_i - \bar{Y})$s, which clearly have zero mean. This fact will allow us to estimate σ^2 by means of a confidence interval. We begin with the following definition.

Chi-square. For Y a normal variable with mean μ and variance σ^2, a random sample of n independently drawn values may be used to define the random variable χ^2 by Equation 13-8-1.

$$\chi^2 = \frac{\sum_{i=1}^{n} (Y_i - \bar{Y})^2}{\sigma^2} \qquad n - 1 \text{ df} \tag{13-8-1}$$

From Table B-3, the χ^2 table, we find two values such that Equation 13-8-2 holds. The table is entered using $n - 1$ degrees of freedom.

$$P(\chi^2_{1-\alpha/2} < \chi^2 < \chi^2_{\alpha/2}) = 1 - \alpha \tag{13-8-2}$$

In Equation 13-8-2, we now replace the random variable χ^2 by definition 13-8-1 and carry out the necessary algebraic manipulations to obtain Equation 13-8-3, an equation about a random interval containing σ^2.

$$P\left[\frac{\sum (Y_i - \bar{Y})^2}{\chi^2_{\alpha/2}} < \sigma^2 < \frac{\sum (Y_i - \bar{Y})^2}{\chi^2_{1-\alpha/2}}\right] = 1 - \alpha \tag{13-8-3}$$

This interval about χ^2 is based on two tabulated χ^2 values, which are associated with the same probability, $\alpha/2$, in each tail. This equiprobability property was used as a matter of convenience and does not lead to the shortest possible interval because of the asymmetric nature of the χ^2 distribution.

Illustration 13-8-1

In Illustration 13-5-1, the following heartbeat data were given: $n = 14$, $\bar{y} = 147.6$ heartbeats, and $s^2 = 9{,}049/13 = 696.1$ heartbeats squared, with 13 degrees of freedom. Let us compute a 95 percent confidence interval for σ^2, using Equation 13-8-3.

$$l_1 = \frac{\sum (y_i - \bar{y})^2}{\chi^2_{.025}(13 \text{ df})} = \frac{9{,}049}{24.7} = 366.4$$

$$l_2 = \frac{\sum (y_i - \bar{y})^2}{\chi^2_{.975}(13 \text{ df})} = \frac{9{,}049}{5.01} = 1{,}806.2$$

Notice that this interval is not symmetric about s^2, the unbiased point estimate of σ^2. We can now say that σ^2 lies between 366.4 and 1,806.2 heartbeats squared unless an unusual random sample has been obtained, one in a class likely to occur only once in 20 times on the average.

EXERCISES

13-8-1 In Exercises 4-3-1 to 4-3-3, variances were computed for the data of Exercises 3-6-1 to 3-6-5, inclusive. Construct 90 percent confidence intervals for the true population σ^2 in the case of several of the populations sampled. Construct some 95 percent confidence intervals for σ^2. Construct some 99 percent confidence intervals for σ^2.

13-8-2 For several of the sets of data referred to in Exercise 13-8-1, compute 90, 95, and 99 percent confidence intervals for σ^2. Observe how the length of the interval increases as the confidence coefficient increases. Notice how both left and right endpoints are affected.

13-8-3 From the heartbeat data last used in Illustration 13-8-1, we might conclude that 700 is a reasonable and convenient estimate of σ^2. Using this value as σ^2, find the sample size necessary to give a 95 percent confidence interval for μ that is no larger than $\sigma/2$.

We have seen that σ^2 may, in fact, be reasonably expected to lie anywhere between 366 and 1,806. Find sample sizes necessary for a 95 percent confidence interval for μ that is no larger than $\sigma/2$, using each of these extreme values as though it were, in fact, σ^2.

Notice how widely the estimated value of n varies when we do not know the true value of σ^2.

Chapter 14
NONPARAMETRIC ESTIMATION OF LOCATION

14-1 INTRODUCTION

In Chapter 13, we discussed the estimation of parameters when strong assumptions were valid, assumptions concerning the distribution of the parent population. In particular, estimation for binomial and normal populations was presented. Since we were explicit about the form of the distribution, our concern was with only those few parameters that completed the specification and so, in turn, was with parametric statistics and procedures.

For much data, the form of the population distribution is not readily assumable. Quantiles will then often serve as substitute parameters, since we need procedures which are not too dependent upon specified parent distributions. In other words, we need distribution-free statistical procedures. These usually compare distributions without specifying their form and also estimate their quantiles. Distribution-free procedures are also said to be nonparametric. Most of them apply to a large class of distributions rather than to all possible ones.

Nonparametric procedures are often very simple to apply, using counts and ranks, but they have disadvantages. Probably the most important of these is that, if the parent distribution is known to be reasonably close to a distribution for which there is applicable theory, or if the data can be transformed to another scale of measurement so that such is the case, then nonparametric procedures do not ordinarily extract as much information from the data as do appropriate parametric procedures.

This chapter is concerned primarily with methods of estimating the median, a quantile which is also the mean for a symmetric distribution. For those with interests solely in the more traditional procedures that depend upon the binomial or normal distribution, this chapter may be omitted.

14-2 ESTIMATION OF MEDIAN; SIGN-TEST PROCEDURE

Let us assume that we cannot make strong assumptions concerning the form or shape of the population sampled. We will, then, concern ourselves with estimating the population median, or second quartile, Q_2, to locate the population. The procedure relies on the definition of Q_2, Equation 14-2-1.

$$P(Y < Q_2) = \tfrac{1}{2} = P(Y > Q_2) \tag{14-2-1}$$

For a median, each sample observation may be considered to be simply a sign, a plus if above or a minus if below. In testing hypotheses about the location of the median, the procedure calls for a count of the pluses or minuses and is called the sign test. Since testing of hypotheses about medians preceded their estimation, we sometimes say the latter is a procedure for estimating a confidence interval by use of the sign test.

The assumptions required for the *sign-test estimation* procedure to be valid are:

1. Each observation consists of the population median plus a deviation. Formally $Y_i = Q_2 + \varepsilon_i$
2. The deviations are random, independent, and from a continuous distribution with median zero.

We are now asking for less than when we used t, Equation 13-5-1, where normality was required; we need only a single continuous population with random and independent observations. Obviously these are quite weak assumptions, and the sign-test procedure will be attractive if stronger ones are not justified. Any weakness of the procedure lies in its generality, for if assumptions not required by the procedure do apply to the sampled population, then a presumably better procedure could be tailored to them.

Suppose the above assumptions are valid and we are to estimate a population median from a random sample of data, $Y_1, \ldots, Y_n$. First, we order the data from smallest to largest, writing $Y_{(1)}, \ldots, Y_{(n)}$. The obvious point estimate of Q_2 will be the middle ordered sample value if n is odd or a value between the two center ones if n is even.

For an interval estimate of Q_2, two values need to be selected, each at some distance on opposite sides of the sample median. Observe the heartbeat data bordering Table 14-2-1, assume that the necessary assumptions are valid, and consider how this estimation might be done.

TABLE 14-2-1 / Averages of All Pairs of 13 Cats for Heart Rate per Minute after Receiving 10 Milligrams of Procaine

	105	120	123	126	135	138	140	150	160	168	170	186	198
105	105.0	112.5	114.0	115.5	120.0	121.5	122.5	127.5	132.5	136.5	137.5	145.5	151.5
120		120.0	121.5	123.0	127.5	129.0	130.0	135.0	140.0	144.0	145.0	153.0	159.0
123			123.0	124.5	129.0	130.5	131.5	136.5	141.5	145.5	146.5	154.5	160.5
126				126.0	130.5	132.0	133.0	138.0	143.0	147.0	148.0	156.0	162.0
135					135.0	136.5	137.5	142.5	147.5	151.5	152.5	160.5	166.5
138						138.0	139.0	144.0	149.0	153.0	154.0	162.0	168.0
140							140.0	145.0	150.0	154.0	155.0	163.0	169.0
150								150.0	155.0	159.0	160.0	168.0	174.0
160									160.0	164.0	165.0	173.0	179,0
168										168.0	169.0	177.0	183.0
170											170.0	178.0	184.0
186												186.0	192.0
198													198.0

The specific problem is to find a set of values for Q_2, the population parameter, such that chance, as measured by a confidence coefficient, say $1 - \alpha = .95$, and any member of the set can adequately explain the actual observations. The minimum and maximum values of the set will be the endpoints of the confidence interval.

We begin by asking the question, "Can the median be to the left of $y_{(1)} = 105$?" If so, then all 13 observations fall to the right of the median. Equation 14-2-1 gives the probability of a single random observation being to the right of Q_2 as 1/2. Equation 9-2-9 shows us how to compute the probability that any number of random, independent observations will fall to the right of Q_2. In particular, where E_1 is the outcome of 13 falling to the right of Q_2, we have $P(E_1) = (1/2)^{13} = .00012$. Clearly E_1 is most improbable if the median is as proposed. Values less than 105 are, therefore, unacceptable as candidates for Q_2.

Now try a Q_2 value between $y_{(1)} = 105$ and $y_{(2)} = 120$. The observed outcome, E_2, is that of 1 observation to the left of the median and 12 to the right. Equation 9-2-9 gives

$$P(E_2) = \binom{13}{1}\left(\frac{1}{2}\right)^{13} = .00159$$

To make the decision as to whether or not a median between 105 and 120 and chance adequately explain the data, it is not sufficient to look at the probability of this single outcome. If the sample size were large enough, all outcomes would have very small probabilities. Instead, we look at the observed outcome and anything more extreme. Here, we have observed a 1:12 split but would have observed a 0:13 split if all observations had been greater than 120. The cumulative probability is

$$\begin{aligned} P(1\!:\!12 \text{ split}) + P(0\!:\!13 \text{ split}) &= P(E_1 \cup E_2) \\ &= P(E_1) + P(E_2) \\ &= .00012 + .00159 = .00171 \end{aligned}$$

Hence, Q_2 between 105 and 120 is still an unlikely choice for median.

To continue searching for the minimum acceptable value of Q_2 as we are doing, we will want a probability of .025 associated with the set of extreme values and the same probability when looking for a maximum acceptable Q_2.

Our next trial value of Q_2 lies between 120 and 123. Now we have a 2:11 split, and the probability for this outcome is

$$P(E_3) = \binom{13}{2}\left(\frac{1}{2}\right)^{13} = .00952$$

The probability of an outcome at least as extreme as that observed if Q_2 lies between 120 and 123 is

$$P(E_1 \cup E_2 \cup E_3) = P(E_1) + P(E_2) + P(E_3) = .01123$$

We are still short of $\alpha/2 = .025$.

For a trial Q_2 between 123 and 126, we find

$$P(E_4) = \binom{13}{3}\left(\frac{1}{2}\right)^{13} = .03491 \quad \text{and} \quad P(E_1 \cup \cdots \cup E_4) = \sum_{i=1}^{4} P(E_i) = .04614$$

We have gone too far if we wish our confidence interval to be symmetric.

We now proceed to look for a maximum acceptable value for Q_2. The procedure is really the same, as are the computed probabilities, because of the symmetry resulting from the value $P = 1/2$.

For the confidence interval, we might then include all values between 123 and 170, representing splits of 10:3 through 3:10. Since the probability associated with the excluded splits is 2(.01123), the confidence coefficient for this interval is $1 - 2(.01123) = .97754$. On the other hand, if we choose the interval (126, 168), including splits of 9:4 through 4:9, then the confidence coefficient is $1 - 2(.04614) = .90772$. The natural discreteness of our process makes the computations simple enough but the results somewhat unattractive. We would like an interval with lower endpoint between 123 and 126, with upper endpoint between 168 and 170, and with a confidence coefficient of .95.

To circumvent the problem of discreteness, we take the observations 123 and 126 with true $\alpha/2$ equal to .01123 and .04614 and interpolate to the required

$\alpha/2 = .025$. The same is also done for the upper values of 168 and 170. The interpolation procedure is simplified by use of Table B-4 and does not involve a linear relation with the probabilities but assumes that endpoints of the interval are linearly related to the number of standard deviations needed to include a probability equal to half the confidence coefficient. Consequently, endpoints computed by the procedure below are linearly related to z-values of the normal distribution and so are an approximation.

The *sign-test procedure*, using Table B-4, for the interval estimation of the median with confidence coefficient $1 - \alpha$ is:

1 Arrange the data from smallest to largest.
2 Select the value $c = c_{\alpha,n}$ from Table B-4, where α is the probability for a two-sided rejection rate and n is the number of observations.
3 If c is a whole number, the lower endpoint of the confidence interval is the observation $Y_{(c+1)}$. Interpolation is needed if c is a fraction of the form $a.b$, read "a-decimal-b," in which case the lower endpoint is $Y_{(a+1)} + 0.b[Y_{(a+2)} - Y_{(a+1)}]$. For the upper endpoint, use $Y_{(n-c+1)}$ if c is an integer. If c is of the form $a.b$, use $Y_{(n-a)} - 0.b[Y_{(n-a)} - Y_{(n-a-1)}]$.

Illustration 14-2-1

The data bordering Table 14-2-1 are used to illustrate the above procedure by constructing a 95 percent confidence interval estimate of Q_2. Thus:

1 The heart-rate data in ascending order are: 105, 120, 123, 126, 135, 138, 140, 150, 160, 168, 170, 186, 198.
2 Here $n = 13$, $1 - \alpha = .95$, so that $\alpha = .05$, and, from Table B-4, $c_{(\alpha,n)} = c_{(.05,13)} = 2.5 = a.b$, so that $a = 2$ and $0.b = 0.5$.
3 Since $c_{(.05,13)}$ is not a whole number, interpolation is necessary. Observations $y_{(a+1)}$ and $y_{(a+2)}$ are $y_{(3)} = 123$ and $y_{(4)} = 126$. Consequently, the lower endpoint

$$y_{(a+1)} + 0.b(y_{(a+2)} - y_{(a+1)}) = 123 + .5(126 - 123) = 124.5$$

Similarly, the upper endpoint is $170 - 0.5(170 - 168) = 169.0$.

We may say we are 95 percent confident that the true median lies in the interval 124.5 to 169.0 heartbeats per minute, since in repeated samplings approximately 95 percent of such intervals will contain the true median.

The interpolation problem is solved largely by Table B-4. Without the table, one would proceed as follows. The y-values, 123 and 126, are associated with the exact probabilities, .0112 and .0461, and in turn with corresponding z-values, 2.284 and 1.684, from the

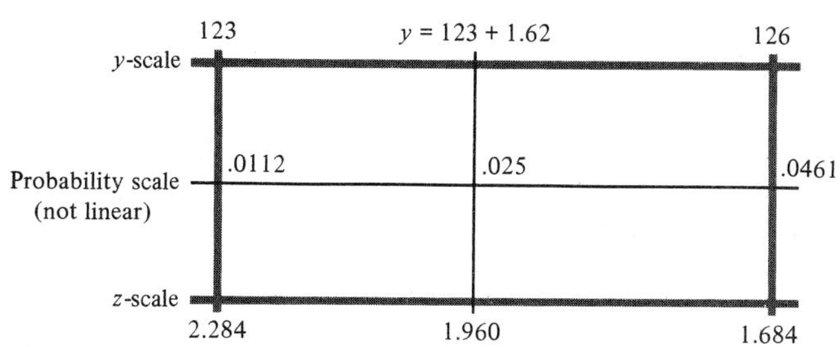

FIGURE 14-2-1 Interpolation for Confidence-Interval Endpoints.

normal distribution. All are shown in Figure 14-2-1. For the required probability of .025, the z-value is 1.960, and we must find the corresponding value on the y-scale. For this, we find what fraction the part (2.284 − 1.960) is of the total distance (2.284 − 1.684). We find (2.284 − 1.960)/(2.284 − 1.684) = .54 on the z-scale. We must now move the same fractional distance along the y-scale, that is, a distance of .54(126 − 123) = 1.62 beyond 123. Thus, the interpolated confidence interval endpoint is 124.62 heartbeats per minute.

Table B-4 tabulates only the .5 of .54, which accounts for the discrepancy between 124.5 and 124.62.

The sign-test procedure for confidence-interval estimation can be as low as 64 percent as efficient as the t-procedure if the data are, in fact, from a normally distributed population.

EXERCISES

14-2-1 Use the data of Exercises 3-6-1, 3-6-2, and 3-6-5 to construct *nearly* 95 percent confidence intervals for the population medians, using the sign-test estimation procedure. "Nearly" means you will not make use of Table B-4. Specify what the confidence coefficients really are.

14-2-2 What do you think about the validity of the assumptions you would make if you were to apply the procedure used in the previous exercise to the data of Exercises 3-6-6 and 3-6-7?

14-2-3 For each confidence interval constructed in Exercise 14-2-1, construct one that is more nearly a 95 percent confidence interval by using the interpolation values provided in Table B-4.

14-2-4 Construct a 95 percent confidence interval for the population median of differences using the sample values available in Exercise 3-6-4. Be sure to use the paired sample differences as your observations.

In doing this, find a confidence interval without benefit of Table B-4 and specify what the confidence coefficient is. Also, find a confidence interval using Table B-4 and with a confidence coefficient of $1 - \alpha = .95$.

14-3 ESTIMATION OF MEDIAN: SIGNED–RANK-SUM TEST PROCEDURE

The sign-test procedure for estimating a population median uses very little of the information in an observation. An adaptation of the signed–rank-sum test is also available. This test determines whether or not hypothesized values of the population median offer acceptable explanations of the data. It makes some use of the magnitude of the observation but is still a distribution-free procedure.

The test procedure requires all deviations from a hypothesized median. To use it then, we proceed as indicated in Figure 14-3-1. The observations are first ordered by magnitude and with the original scale preserved. For the heart rates given in Table 14-2-1, a hypothesized median of 125 heartbeats is marked. All deviations from this value are then computed. They are: -20, -5, -2, 1, 10, 13, 15, 25, 35, 43, 45, 61, and 73. These are ordered by magnitude with the sign ignored and assigned ranks from smallest to largest. Thus rank 1 goes to deviation $+1$, rank 2 to -2, 3 to -5, 4 to $+10, \ldots$, 13 to $+73$. The sign of the deviation is next assigned to the rank; those to the left of the median receive negative signs, while those to the right receive positive ones. Finally, we sum the ranks of the same sign, either the positive or negative ranks, compare the result with tabulated values, and decide whether or not the hypothesized value of $Q_2 = 125$ is an acceptable choice for the population median. The test procedure is fully illustrated in Chapter 18.

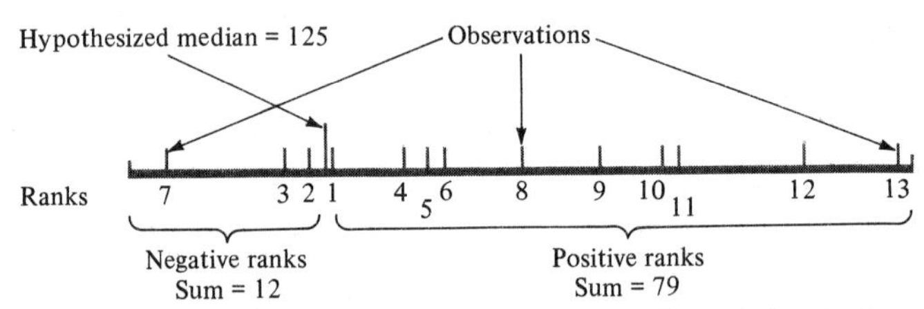

FIGURE 14-3-1 Heart-Beat Data Used for Signed–Rank-Sum Test Estimation Procedure.

We proceed to test other hypothesized values of Q_2, looking in particular for the endpoints, until we have a set of only acceptable values with a confidence coefficient of $1 - \alpha$.

Notice that the big deviations above and below the hypothesized median are assigned greater weights, by means of their greater ranks, than are small deviations. This is the advantage this procedure has over the sign-test procedure, namely that some use is made of the magnitudes of the observations.

The testing procedure as carried out does not bear an immediately obvious resemblance to the estimation procedure. However, this estimation procedure lends itself readily to an interpolation method like that used for the sign-test estimation procedure and gives results that are identical with the more lengthy and obviously related procedure indicated above.

In addition to the assumptions required for validity of the sign-test procedure, the signed–rank-sum procedure requires symmetry in the parent distribution. This means that observations equidistant from the true median, and consequently assigned equal ranks, will be associated with the same probability, regardless of sign. Note also that, with symmetry, the mean and median must be the same value.

To be valid, the signed–rank-sum test estimation procedure requires that the following assumptions be met.

1 Each original observation consists of the population median or mean and a deviation. Formally,

$$Y_i = Q_2 + \varepsilon_i = \mu + \varepsilon_i$$

2 The deviations are to be random, independent, and from a continuous symmetric distribution with the median equal to zero.

The signed–rank-sum test procedure, using Table B-5, for the interval estimation of the median or mean of a symmetric distribution, follows. The confidence coefficient is $1 - \alpha$.

1 Compute the average of all possible pairs of observations including each with itself. There will be $m = n(n + 1)/2$ averages.
2 Arrange these in order from $Y_{(1)}$ to $Y_{(m)}$, $Y_{(1)}$ being the smallest average and $Y_{(m)}$ the largest. Note that these Y's are not the original observations.
3 Select the critical value $s = s_{(\alpha,n)}$ from Table B-5.
4 The lower confidence interval limit is $Y_{(s+1)}$ when s is an integer. Interpolation is needed if s is a fraction of the form $a.b$, a-decimal-b, when

$$Y_{(s+1)} = Y_{(a+1)} + 0.b[Y_{(a+2)} - Y_{(a+1)}]$$

5 To determine the upper limit, arrange the data from largest to smallest and proceed as above.

Illustration 14-3-1

We now use the above procedure to compute a 95 percent confidence interval for the population median of the sample heartbeat data in Table 14-2-1.

1 The average of all possible pairs of observations including each with itself is given in this conveniently arranged table. Since $n = 13$, there are $m = n(n + 1)/2 = 13(14)/2 = 91$ averages.
2 All 91 ranks are assigned to the averages, from smallest to largest, with tied averages being given consecutive ranks.
3 Since $1 - \alpha = .95$ or $\alpha = .05$ and $n = 13$, then $s_{(\alpha,n)} = s_{(.05,13)} = 17.2$. Consequently $a = 17$ and $.b = .2$.

4 Since $y_{(a+1)} = y_{(18)} = 130$ and $y_{(a+2)} = y_{(19)} = 130.5$ are not the same value, interpolation is required. We find

$$\begin{aligned} y_{(s+1)} &= y_{(a+1)} + .b[y_{(a+2)} - y_{(a+1)}] \\ &= 130 + .2(130.5 - 130) \\ &= 130.1 \end{aligned}$$

5 For the upper limit, we rank the averages from large to small. However, the sum of the two ranks assigned to a number in the forward and reverse rankings must be $m + 1$. Consequently, ranks 18 and 19 become $92 - 18 = 74$ and $92 - 19 = 73$, respectively, in the reverse ranking. Hence, for the upper limit,

$$y_{(s+1)} = 165 + .2(164 - 165) = 164.8$$

We may now say that we are 95 percent confident that the true median or mean lies in the interval 130.1 to 164.8 heartbeats per minute since in repeated samplings approximately 95 percent of such intervals will contain the true value.

When several values are the same, they are assigned the set of appropriate ranks. For example, 120.0 occurs twice, and the assigned ranks are 5 and 6. If $s_{a.b}$ had been 5.3, between 5 and 6, no interpolation would have been necessary and the endpoint would simply have been 120.0.

This procedure is approximately 95 percent as efficient as the *t*-procedure in obtaining confidence limits if the assumption that the data are from a normally distributed population is valid.

It is interesting to compare the confidence limits obtained by the two nonparametric procedures of this chapter with those computed in Chapter 13 and based upon the normal theory. The same set of data was used in all cases. The 95 percent confidence limits for the cat data of Table 14-2-1 are:

Sign test = 124.6–168.9 heartbeats/minute
Rank-sum test = 130.1–164.8 heartbeats/minute
Normal theory = 131.0–164.2 heartbeats/minute

It is not surprising that the confidence interval based on normal theory is the shortest of the three since the assumptions here were the most restrictive. However, the rank-sum procedure gave an interval which differed only slightly from it. On the other hand, the sign-test procedure gave the greatest interval but is the one requiring the least constraining assumptions for its validity.

EXERCISES

14-3-1 Use the data of Exercises 3-6-1, 3-6-2, and 3-6-5 to construct 95 percent confidence intervals, by the signed–rank-sum test method, for the corresponding population medians.
14-3-2 Construct a 95 percent confidence interval for the population median of differences

using the sample values available in Exercise 3-6-4. Once again note that pairing has enabled us to use a one-sample procedure to answer a two-sample problem.

14-3-3 Compare your 95 percent confidence intervals computed using normal theory, the sign-test estimation procedure, and the signed–rank-sum test estimation procedure wherever you have used the same data. What is your conclusion?

14-3-4 In describing the estimation procedure based on the signed–rank-sum test, it is stated that $m = n(n + 1)/2$ means of pairs of observations are possible. How is this figure arrived at?

Chapter 15

TESTS OF HYPOTHESES I: BINOMIAL DISTRIBUTION

15-1 INTRODUCTION

Decisions are required of all of us. We must decide what clothes to wear, what college to attend, how to spend our lives. For every decision, data will be collected. We observe the weather and listen to the weather report before deciding whether or not to carry an umbrella; we review our high school performance and appraise college curricula before completing a college application; we try to assess the challenges and rewards of a job before committing appreciable parts of our lives.

All decisions made under realistic circumstances involve risk. Thus, we may decide to go to a movie only to find that it is poor entertainment; or we may decide against a play only to be advised that it is superior. In either case, we have made a "wrong" decision in the sense that the result could have been better. On the other hand, we may decide to go to a movie and then thoroughly enjoy it; or we may decide to stay away from a play and, later, hear only disparaging reviews, In these cases, our decisions appear to have been right. These ideas of risk, decision, and outcome are summarized in Table 15-1-1.

TABLE 15-1-1 / The Appropriateness of a Decision

Better Choice	Decision	
	To Do It	Not To Do It
To Have Done It	Right	Wrong
Not To Have Done It	Wrong	Right

The investigator must make decisions, too. Thus a factory operator may have data on two machines of different manufacture and must decide which one will be better as an addition to the operation. Experimental data might consist of the number of defectives and nondefectives and the total production for each machine over a period of time. Presumably there will be a certain amount of variation in the output of essentially identical machines, so that data for any two machines under consideration will almost certainly differ. The best decision is not immediately obvious. If the operator decides on machine A and it is superior, profit is made; if it is not superior, profit will be less. The same sort of argument holds if it is decided to use machine B.

Again, a farmer with a new cultivar of a crop must remove acreage from production of a cultivar whose quality is known to produce the new one whose performance has been measured on an experimental rather than a market basis. The farmer's decision is no easier than that of the factory operator, and the outcome is subject to the same risks.

While economics and other criteria necessarily affect decision making, we will not now concern ourselves with this aspect of the problem. Instead, we will consider decision making from the point of view of tests of hypotheses where we must decide whether one model or another offers a better explanation of the observed data. For example, a physician might compare two medical treatments for a certain ailment and measure the time required to reach some specified degree of recovery. One hypothesis might be that the times to recovery are the same, with the alternative hypothesis being that they are different.

This chapter deals with tests of hypotheses applicable to discrete data when a binomial model is reasonable.

EXERCISES

15-1-1 Prepare a table such as 15-1-1 for the factory operator. Show the two possible decisions and the correctness of each according to which machine is truly the better one.

15-1-2 Prepare a table such as 15-1-1 for the doctor. (The rightness or wrongness of the decision regarding which drug to use depends on whether the recovery times are the same or different.)

15-2 HYPOTHESIS TESTING: DISCRETE DATA

In this section, hypothesis testing is considered as a general problem for data from a binomial distribution. Vocabulary and definitions are introduced and discussed in conjunction with a continuing illustration not formally designated as such. More concisely presented illustrations are given in Section 15-3.

Consider experimentation when pairs of individuals are available.

Pairing is said to be *meaningful pairing* when individuals from the same pair will perform more nearly alike than individuals from different pairs. For example, we may have swimmers paired by age, height, weight, and previous training and find that neither member of a pair has performed with consistent superiority for 200 yards of butterfly. On the other hand, two randomly chosen swimmers may well have records to show that one has been consistently faster. Or we may have pairs of plants grown from similar cuttings from the same plant. They would be expected to grow more nearly at the same rate than cuttings from different plants.

The obvious plan of an experiment designed to assess two different future training programs, called *treatments*, is to randomly assign a different swimmer from each pair to each of the two programs for a period of time prior to competition. For an experiment assessing the effect of length of day on plant growth, we would randomly assign the members of each pair to two daylight lengths for 1 month or so prior to measuring growth. Since pairing has been used, all but the smallest of observed differences are now expected to be attributable to treatment effects. With pairing, true treatment effects should be detectable with a smaller experiment than is necessary without meaningful pairing.

For the simplest assessment of the relative value of the treatments, it is not necessary to measure swimming times or plant growth. The observations needed are a record, for each pair, of which treatment resulted in the better time or the greater stem growth. These data alone can be used to ascertain if there is sufficient consistency to suggest that one treatment is better than the other.

The random assignment is necessary in order that the laws of chance may apply and that no bias, either intentional or unintentional, favoring one treatment be introduced. It is accomplished by tossing a coin, using a similar mechanical procedure, or using a random number table.

Since the individual outcomes are, basically, success or failure, we may quantify them as 1s or 0s. The outcome of an experiment is, then, a sequence of 1s and 0s. If the appropriate assumptions are valid, we have a binomial model with the parameter unspecified.

When there are n pairs and, so, n trials or repetitions, the sample space is n-dimensional, and so not ordinarily real, with two possible values, 0 or 1, on each axis. A sample space for $n = 3$, a case of a real space, is illustrated in Figure 15-2-1. In all, there are 2^n possible outcomes for the experiment, although only one occurs. Each outcome corresponds to a sample point which may always be represented by an ordered n-tuple. Our immediate problem is to find a decision

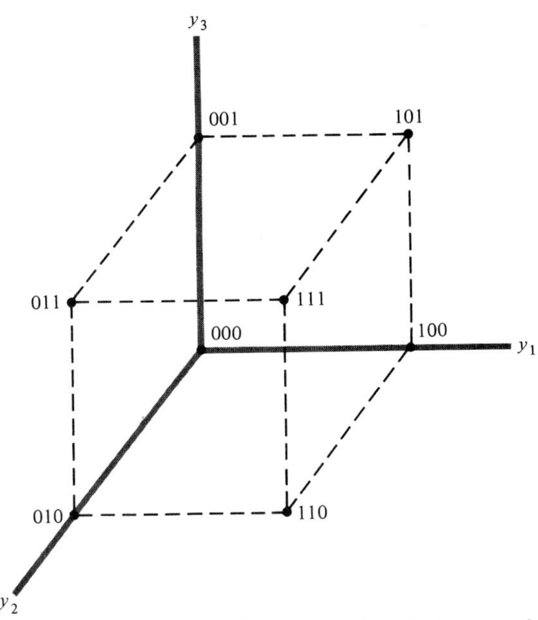

FIGURE 15-2-1 Sample Space for a Binomial Model, $n = 3$.

rule that can be applied to each point, or outcome, in the sample space and that will let us infer whether or not the observed result is attributable to chance.

In order to compute the probability to be assigned to each sample point, information which can help us to choose a decision rule, it is necessary to specify the parameter p. If the two treatments in either experiment are of equal value, then the parameter is clearly $p = .5$. With this parameter and independent trials, the probability to be assigned to each point of the sample space is $(1/2)^n$.

We have now hypothesized a parameter $p = .5$ for our model. Our decision rule must lead us to conclude, from the result of any random experiment, one of the following: either the parameter $p = .5$ is appropriate, and there is no real difference between the responses to the two treatments but only chance variation; or $p \neq .5$, and there is a real difference.

A *null hypothesis* is proposed when it specifies, for a model, a parameter such that a probability can be computed for each and every sample point. We write $H_0{:}p = p_0$ for a binomial model.

An *alternative hypothesis* is one such that we can conclude in its favor if the sample evidence does not support the null hypothesis. For a binomial model, we will write one of

$$H_1{:}p \neq p_0 \qquad \text{or} \qquad H_1{:}p > p_0 \qquad \text{or} \qquad H_1{:}p < p_0$$

A null hypothesis is always one for which we can compute probabilities of outcomes or points in the sample space or of sets of points which have meaning in terms of our decision rule. It is often a hypothesis of no difference between or among treatments, although this is not required. An alternative hypothesis is usually a set of alternatives, for example, that a parameter is different from that specified by the null hypothesis. Sometimes we may be able to specify the direction of the difference. Under the alternative hypothesis, it will not generally be possible to compute a unique set of probabilities for the sample points but rather a set for any particular option included in the alternative.

Given a null hypothesis, we must next decide which sample outcomes tend to support it and which tend to deny it. If about half the swimmers on one training program win and half lose, most of us would agree that there is little evidence to deny $H_0:p = .5$; that is, we decide that the programs are of about equal value where swimming speed is the criterion. However, if most of the swimmers on one program swim faster than their matched pair-mate, then the null hypothesis is not supported and we decide in favor of $H_1:p \neq .5$. In other words, our decision rule would say that if a sample point consists of roughly equal numbers of 1s and 0s, decide in favor of H_0; on the other hand, if a sample point is mostly 1s or mostly 0s, decide in favor of H_1.

A decision rule always makes use of the alternative hypothesis. For example, if the null hypothesis is that the training programs are of equal value but the alternative is that a specific one is better, then we would not accept the alternative if all the swimmers on this program lose, even though the 1s and 0s do not occur anywhere nearly the same number of times. The original discussion involved so-called *two-tailed* alternatives; in the present example, we have *one-tailed* alternatives.

In general, a reasonable decision rule is seen to be based on the number of 1s or 0s. Thus, suppose we have 20 pairs of cuttings in our plant-growth experiment and assign 1 to a pair where the A plant exceeds the B plant in height and assign a 0 otherwise. If we have a null hypothesis of no difference due to treatments and an alternative of a real difference, then we might decide to accept H_0 if the number of cuttings on treatment A that exceed in height their pair-mate on treatment B lies in the range of 7 to 13 inclusive, and otherwise to reject H_0 in favor of H_1. The test criterion is to be the number of times the response to treatment A exceeds that to B; that is, it is a count of the number of 1s among the 20 responses.

A *test criterion* or *test statistic* is a random variable used in deciding between a null hypothesis and an alternative hypothesis.

Once we have made our decision as to cutoff points, we can look at every point in the sample space and say whether it calls for a decision in favor of H_0 or against H_0. The sample space has been divided into an acceptance region for H_0 and a rejection region or critical region for H_0. The cutoff points separating the regions are called critical values and are a part of the critical region. Our critical values are 6 and 14.

The *acceptance region for the null hypothesis* H_0 consists of the set of values of the test criterion for which H_0 will be accepted. Similarly, the *rejection* or *critical region for* H_0 is the set of values of the test criterion for which H_0 will be rejected.

Acceptance and rejection regions, though generally specified in terms of the test criterion, correspond to points in the sample space for which the same decision would be made. For example, in testing $H_0:p = .5$ for a binomial model against $H_1:p \neq .5$, using a random sample of n trials, we accept H_0 when there are not too many of either 1s or 0s and reject H_0 when there are. In other words, our test statistic is the number of 1s, and the acceptance and rejection regions are defined in terms of this criterion. However, each value of the test criterion corresponds to one or more sample points so that we can equally well consider that the sample space has been partitioned into acceptance and rejection regions. Thus if $n = 3$ as in Figure 15-2-1, we might decide to reject $H_0:p = .5$ if we observe three 1s or three 0s. In terms of the test criterion, say c, we reject H_0 when $c = 0$ or $c = 3$, the rejection region, and accept H_0 when $c = 1$ or $c = 2$, the acceptance region. In terms of the sample space, $c = 1$ corresponds to the points 100, 010, and 001, while $c = 2$ corresponds to 110, 101, and 011. These six points in the sample space correspond to the acceptance region. For $c = 0$, there is the single sample point 000 and for $c = 3$, there is 111. These two points constitute the rejection region.

The *critical value* of the test criterion is the boundary value (or values) that separates its range into acceptance and rejection regions. It is considered to be in the critical region.

In the above illustration, $c = 0$ and $c = 1$ are critical values and, at the same time, are the critical region.

To construct and evaluate the performance of a particular decision rule, consider the plant experiment but with only 10 pairs of plants. This is a small experiment but one for which it is easy to compute and present all the probabilities involved. Day lengths A and B are to be treatments. For now, assume no ties occur. The outcome of this experiment may be represented as a 10-place ordered sequence with 1 or 0 in each place. Each possible outcome now corresponds to a point in a 10-dimensional sample space. Since there are two possible outcomes for each place, there are 2^{10} possible points in the sample space. For the test criterion, add the entries in the sequence to give the number of 1s in the outcome of the experiment, that is, the number of times the response to A exceeds that to B.

Table 15-2-1 will help us construct and evaluate the performance of any decision rule. The null hypothesis is to be $H_0:p = .5$ and the alternative,

TABLE 15-2-1 / Probability of y Successes, or 1s, in 10 Binomial Trials for Given Parameter p

y \ p	.5	.4	.3	.2	.1	
0	.001	.006	.028	.107	.349	10
1	.010	.040	.121	.269	.387	9
2	.044	.121	.234	.302	.194	8
3	.117	.215	.267	.201	.057	7
4	.205	.251	.200	.088	.011	6
5	.246	.201	.103	.026	.002	5
6	.205	.111	.037	.006	.000	4
7	.117	.042	.009	.001	.000	3
8	.044	.011	.001	.000	.000	2
9	.010	.002	.000	.000	.000	1
10	.001	.000	.000	.000	.000	0
	.5	.6	.7	.8	.9	p \ y

$H_1: p \neq .5$. Sample-space points for which test criterion values are $y = 0$ and 10 are the first to be assigned to the rejection region. These points are more likely associated with an alternative hypothesis, since the corresponding outcomes are so one-sided. When $H_0: p = .5$ is true, Table 15-2-1 shows that these values of y have the smallest and equal probabilities, namely, .001, each value arising from a single point in the sample space.

The next sample points for logical assignment to the rejection region are associated with test-criterion values $y = 1$ and 9. Now the two equal probabilities computed under $H_0: p = .5$ are larger than for $y = 0$ and 10 but smaller than any others. Each of the values 1 and 9 is based on 10 points from the sample space, but the remaining values of y all contain more.

Let us go one step further and add to the critical region those points for which $y = 2$ and 8. The probabilities and numbers of sample points increase but are still smaller than for any y-values not yet included.

Now, the critical region consists of sample points for which $y = 0, 1, 2, 8, 9$, and 10. Here we reject the null hypothesis of $p = .5$. The acceptance region consists of the remaining points where $y = 3, 4, 5, 6$, and 7. Here we accept $H_0: p = .5$. Our decision rule tells us what to do for every point in the sample space.

Stopping at test-criterion values of 2 and 8 has been arbitrary. What we really need to determine satisfactory critical values is some input as to what sort of difficulty we may be creating for ourselves. This may be seen from a consideration of the kinds of error that may be committed. Thus we can obviously get, by chance, an experiment with ten 1s when treatment A is really no different from B, just as we can toss a fair coin and observe 10 heads. However, by the decision rule, we would wrongly reject a true null hypothesis if this were the experimental outcome. On the other hand, we could observe a 5:5 split, by chance, even if $H_1: p = .1$ is true. This time, we would make a different kind of error, concluding that $H_0: p = .5$ is true when such is not the case.

An *error of the first kind*, or a *Type I error*, is made when a true null hypothesis is rejected. The probability of a Type I error is given the symbol α. This value is also called the *significance level* of the test or the *size* of the test.

Table 15-2-1 shows that $\alpha = 2(.001 + .010 + .044) = .110$ for the paired-plants experiment with the decision rule discussed. If the experiment gives a value of the test criterion of 1, then the rule says to reject the null hypothesis. We may also say that the result is significant at the .110 level. As we have seen, the value $1 - \alpha$ is called the confidence coefficient when the problem is one of estimation. Here, $1 - \alpha = .890$.

The performance of the rule seems realistic provided $p = .5$ always. On the average, we correctly accept H_0 about 9 times out of 10 and falsely reject it the other time. In the latter case, we make a Type I error. Remember that we are simply hypothesizing a p-value and are provided with only a sample on which to base our decision about the value of p. We cannot always be right.

Next, let us see how the decision rule performs if the null hypothesis is false and the alternative true. Since the alternative hypothesis includes all values of p that are different from .5, we will observe performance at each of several p-values. To help us, we construct Table 15-2-2 from information provided in Table 15-2-1. From Table 15-2-2, it is apparent that as the true parameter gets farther from $p = .5$, the probability of accepting H_0 decreases and we are increasingly likely to decide in favor of the true alternative. This is a desirable property for a decision rule because as p increases or decreases, it is because the plants subjected to one or the other of the two day lengths are demonstrating greater growth with increasing likelihood, presumably as a consequence of the superiority of that treatment. However, we do note that the true probability of observing superiority in a single

TABLE 15-2-2 / Probabilities Associated with Acceptance and Critical Regions for a Particular Decision Rule

	True Binomial Parameter is				
Probability of	.5	.4 or .6	.3 or .7	.2 or .8	.1 or .9
Accepting H_0	.890	.820	.616	.322	.070
Rejecting H_0	.110	.180	.384	.678	.930

trial will have to be greater than .7 or less than .3 before we have a better than even chance of rejecting H_0 and so of detecting the true alternative with only 10 paired comparisons and the present decision rule. Thus, it will be easy to claim that $H_0:p = .5$ is true when, in fact, it is the alternative that is true. This error is a different kind from the Type I error.

An *error of the second kind*, or a *Type II error*, is made when a true alternative hypothesis is rejected. Here we accept the null hypothesis when it is false. The probability of a Type II error is given the symbol β.

The line "accepting H_0" in Table 15-2-2 is a set of β-values except for the entry under $p = .5$. The line "rejecting H_0" is a set of $1 - \beta$ values and measures the ability of the test to reject the null hypothesis and so accept the alternative, when the alternative is true. In other words, it tells us how powerful the test is.

The *power* of a test is the probability with which it will reject H_0 when H_1 is true. In rejecting H_0, we accept H_1. This measure is given by $1 - \beta$.

The bottom line of Table 15-2-2 gives the power of our test, for the paired-plants experiment, for specific values of p as alternatives to $p = .5$. We may plot these values as a *power (function) curve*. If the power is presented as a function of the parameter under test, then we have the power function. The power curve for the plant-growth experiment is plotted in Figure 15-2-2. From such a curve, we are able to read the probability of detecting a true binomial parameter p of any value between 0 and 1, using a sample of size 10.

The *operating characteristic* and its curve are concerned with the probability of a Type II error rather than the power of the test. Thus the *OC-curve* plots β against p rather than $1 - \beta$ against p.

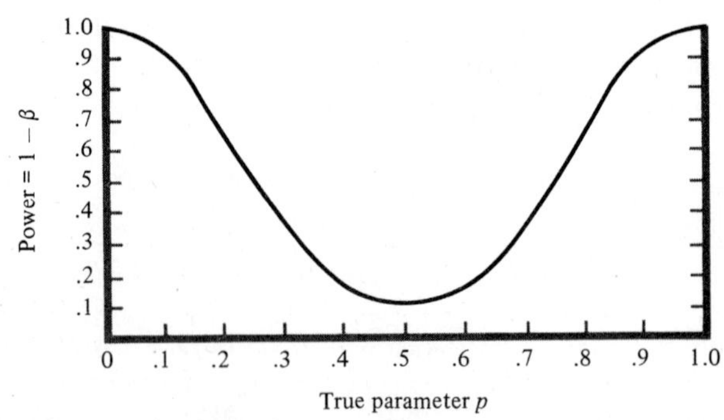

FIGURE 15-2-2 The Power of a Test for a Binomial Distribution Problem, $n = 10$.

Much of the discussion concerning the plant-growth experiment, test procedure, decision rule, and terms used is summarized in Table 15-2-3, a table basically like Table 15-1-1. At any one time, we can be in only one of the four cells of the table. For example, we can reject H_0 when it is false and be in the lower right cell; consequently, we have made a correct decision. Prior to conducting the experiment, the probability of this outcome is $1 - \beta$, necessarily computed under the assumption that H_1 is true.

TABLE 15-2-3 / The Correctness of a Decision, with Terms and Symbols Used

	Decision	
The Truth	Accept H_0, and Reject H_1	Reject H_0, and Accept H_1
H_0 Is True and H_1 Is False	Correct decision $P = 1 - \alpha$ Confidence coefficient	Wrong decision Type I error $P = \alpha$ Error rate, significance level, or size of test
H_0 Is False and H_1 Is True	Wrong decision Type II error $P = \beta$	Correct decision $P = 1 - \beta$ Power of test

In constructing the decision rule, we have assigned to the critical region those points in the sample space for which corresponding values of the test criterion have the lowest probabilities when H_0 is true. Ideally, they have probabilities which are greatest when H_1 is true, thus maximizing the power of the test and minimizing the probability of a Type II error. Again Table 15-2-1 provides the relevant information.

In Table 15-2-1, observe the probabilities for that part of the critical region consisting of the points for which $y = 8, 9$, and 10. $P(y = 8, 9, \text{ or } 10 \mid p = .5) = .044 + .010 + .001 = .055$. However, the probability that $y = 8, 9$, or 10 for any $p < .5$ is even smaller and is decreasing as p gets farther from $p = .5$. Thus we are accomplishing nothing by including the points 8, 9, and 10 in our critical region as far as the power of the test against small values of p is concerned, and are, in fact, increasing the probability of making a Type II error, relative to what it could be if we used other options for points to be included in the critical region. This suggests that we leave $y = 8, 9$, and 10 in the acceptance region and put $y = 3$ in our critical region, which includes $y = 0, 1$, and 2, to improve the power against small p alternatives. Unfortunately, $P(y = 3 \mid p = .5) = .117$, whereas $P(y = 8, 9, \text{ or } 10 \mid p = .5) = .055$, and so the size α of the test changes and we cannot fairly compare the two suggested critical regions. More particularly, note that the suggested change adversely affects the power of the test if the true p should be greater than .5. If we could have a test of the same size and of the suggested type, then the power curve would lie above that shown in Figure 15-2-2 for values of $p < .5$ but would not go even as high as .110, α for the test, for $p > .5$.

Incidentally, we can construct a test of almost the same size by excluding 8, 9, and 10 from the critical region and tossing a coin when $y = 3$ comes up for the purpose of deciding whether or not to put it in the acceptance region. It is because $(1/2)P(y = 3 \mid p = .5) = .0585$ is close to $P(y = 8, 9, \text{ or } 10 \mid p = .5) = .055$ that the size is nearly the same.

We see, then, that in the case of our two-tailed alternative, we have to compromise somewhat in selecting a critical region since the true parameter may be at either end of the scale. Ideally, we would like to have a critical region of only large values of y when $H_1{:}p < .5$ is true and of only small values of y when $H_1{:}p > .5$ is true. Of course, this is impossible.

If we can partition our sample space so that our test is more powerful than any other test, then we have *the most powerful test* for that alternative. If the test is most powerful, or at least as powerful as any other test, for all alternatives, it is a *uniformly most powerful test*. Unfortunately, such tests do not always exist.

Two questions are suggested from the previous discussion. First, what if the only reasonable alternative hypothesis calls for $H_1{:}p < .5$? In this case, we would do about what was suggested in the previous paragraphs, namely drop $y = 8, 9$, and 10 from the critical region and possibly include $y = 3$. With or without $y = 3$, the size α of the test is changed. This new test would be a one-tailed test or a test against one-sided alternatives. A similar one-tailed test can be proposed if the only reasonable alternatives are $H_1{:}p > .5$. Unfortunately, we cannot observe the data before deciding on the alternatives. This would lead to a result-guided test for which we would not know the true α-value. One must decide at the planning stage of an experiment whether to use a one-tailed or a two-tailed test.

The other question concerns the choice of α and β. How do we choose? The answer is that α is often set arbitrarily and β is left uncontrolled. Values of $\alpha = .05$ and .01 were used for many years in agricultural research experiments and seem to have proved satisfactory. Later, the same values were carried over to other areas of research. More recently, the .001 level is being used in agricultural experiments where recommendations are to be made over a large area such as a state and Type II errors are not considered serious. Also, it has been suggested that $\alpha = .10$ be used when an experiment is small. When α is set, it is obviously to the investigator's advantage to have some knowledge of the power of the test being used. In any case, as α decreases as a necessary consequence of including more points in the acceptance region, β will increase because it must include the probabilities, computed under any particular alternative hypothesis, associated with these additional points. Thus, for fixed sample size, the choice of a low α is at the cost of a high β, whereas a high α leads to a relatively low β.

When an investigator is able, in some way, to measure the seriousness of a wrong decision, he may be in a position to suggest working values of α and β. To satisfy both these values simultaneously, it is necessary to be able to choose the sample size. Discussion of this problem was begun in Section 13-7 for confidence-interval estimation and without reference to Type II errors.

In summary, the procedure for testing a statistical hypothesis in the general case is as follows:

STEP 1:

Propose a null and an alternative hypothesis, H_0 and H_1, respectively, such that an eventual decision rule will let us choose one over the other.

STEP 2:

Fix a significance level or a size α for the test and a sample size n.

STEP 3:

Choose a test statistic with a probability distribution known under H_0.

STEP 4:

With the value of α in mind, decide upon those values of the test criterion that will call for rejection of H_0 and those that will call for acceptance.

STEP 5:

Collect the data and compute the value of the test statistic.

STEP 6:

On the basis of the value of the test statistic, decide that either H_0 or H_1 is true.

At step 2, we are accepting the fact that we cannot guarantee a correct decision and so, with some notion of the seriousness of a wrong decision, are setting a numerical value on $\alpha = P(\text{rejecting } H_0 \mid H_0 \text{ true})$. Simultaneously we fix the sample size according to our overall resources and with some regard to the seriousness of the other possible type of error. Hopefully, we also have knowledge of the magnitude of this error, that is, of $\beta = P(\text{rejecting } H_1 \mid H_1 \text{ true})$, for at least some specific alternatives included in H_1.

The statistic chosen at step 3 is our test criterion for testing H_0 versus H_1. Its probability distribution is computed under the assumption that H_0 is true, and this permits the desired control over α. In the illustration, the test criterion was the number of pairs of plants for which treatment A gave the greater response.

At step 4, we partition the sample space into an acceptance and a rejection region. For every sample point, a decision is now possible. This partitioning is determined largely by H_1 since a powerful test is desired.

It is essential at step 5 that randomness be employed in the sampling scheme so that a random value of the test statistic is obtained and the laws of chance apply.

Finally, we decide which hypothesis is to be accepted. If we accept the null hypothesis, we are saying that chance and this hypothesis offer an adequate explanation of the data. There are those who feel that the words "accept the null hypothesis" are too strong; these people prefer to say that there is no evidence to reject the null hypothesis. When the null hypothesis is rejected, it is because chance and the null hypothesis do not adequately explain the data. We generally prefer to think that the null hypothesis, rather than chance, is at fault and so accept the alternative. Accepting the alternative is not considered to involve too strong a statement since the alternative is almost invariably a collection of alternatives as opposed to a unique value of a parameter or something equally restricted.

Special attention is needed in the case of choosing between only two hypotheses, a problem suggested in Exercise 9-2-9. It is clear that either may serve as the null hypothesis and, in either case, we need a one-tailed test. The following approach may be taken for choosing a decision rule.

Suppose we consider alternative rules that are reasonable for deciding between $p = .5$ and $p = .2$ on the basis of a random sample of size $n = 10$. If $y = 10, 9, 8, 7, 6$, or 5, we will obviously decide in favor of $p = .5$. If $y = 0, 1$, or 2, we will obviously decide in favor of $p = .2$. Hence, three reasonable rules to consider are the following.

RULE 1:

Decide on $p = .5$ if $y = 10, 9, \ldots, 5$ and on $p = .2$ if $y = 4, 3, \ldots, 0$.

RULE 2:

Decide on $p = .5$ if $y = 10, 9, \ldots, 4$ and on $p = .2$ if $y = 3, 2, 1, 0$.

RULE 3:

Decide on $p = .5$ if $y = 10, 9, \ldots, 3$ and on $p = .2$ if $y = 2, 1, 0$.

For each rule, and with the help of Table 15-2-1, we can compute the probability of making a wrong decision when $p = .5$ and when $p = .2$. These are presented in Table 15-2-4. For each rule, the larger probability is underlined. Among the larger probabilities, there is one which is smallest, namely $p = .172$. The rule associated with this minimum of the maximum probabilities of error is the *minimax* solution to the problem of choosing an appropriate rule. Thus, if

TABLE 15-2-4 / Probabilities of Wrong Decisions: Determining a Minimax Solution

Rule	True p	Decision Wrong When $y =$	P(Wrong Decision | p)
1	.5	0, 1, . . . , 4	.377
	.2	5, 6, . . . , 10	.033
2	.5	0, 1, 2, 3	.172
	.2	4, 5, . . . , 10	.121
3	.5	0, 1, 2	.055
	.2	3, 4, . . . , 10	.322

our criterion says to choose the least disastrous of rules in a world of worst possible decisions, we choose rule 2. Basically, we are still involved with α and β and we can examine similar rules for larger sample sizes until we find a sample size which gives adequate control of both types of error.

EXERCISES

15-2-1 You are considering buying a new car but are somewhat concerned by a recent news report about a serious traffic accident resulting from the failure of the steering mechanism in a new car. The salesperson assures you that the steering mechanism of any car sold you will not fail in traffic. You decide to consider the truth of this statement to be hypothetical. Construct a table like Table 15-2-3 but especially applicable to your problem. Indicate the seriousness of any wrong decision.

15-2-2 There is a very old overhead traffic bridge over the railroad tracks on a certain campus. Suppose that the city engineer has stated that the bridge is unsafe for any vehicle over 5 tons. Your vehicle is over 5 tons, but you are tempted to use the bridge to save time and distance on a particular occasion. You decide to consider that the city engineer has simply formulated a hypothesis. Construct a table like Table 15-2-3 but especially for this problem. Indicate just how serious any wrong decision can be.

The exercises that follow tend not to be of an applied nature and may be omitted. However, working them through thoughtfully should contribute to your understanding of the content of this section. Applied problems are to be found following Section 15-3.

15-2-3 In this section, there was a discussion concerning placing $y = 8, 9$, and 10 in the critical region, versus other options, and the consequent effect on the power of the test against alternatives calling for small values of p. Suppose that the critical region can be:

(1) $y = 0, 1$, and 2 only
(2) $y = 0, 1, 2$ and 8, 9, 10
(3) $y = 0, 1, 2$, and 3

(a) Compute the power of the test against alternatives that $p = .4$ and .2 for the three choices of critical region.
(b) What is the probability of a Type II error in each case?
(c) What is the size of the test in each case?

15-2-4 The illustration given in the text was concerned mainly with a binomial population with parameter $p = .5$ and an alternative $p \neq .5$. Suppose we now consider testing the null hypothesis $p = .5$ against alternatives $p < .5$. Let the test procedure be to accept H_0 if $y = 3, 4, \ldots, 10$ and to reject H_0 if $y = 0, 1,$ or 2.
(a) What is the value of α?
(b) Plot a power-function curve for the test by first computing the probabilities of rejecting H_0 if, in fact, $p = .4, .3, .2, .1$ and $.6, .7, .8, .9$. Graph your results and compare with the power function when the alternatives are $p \neq .5$.
15-2-5 Geneticists often observe characteristics for which there are only two categories, for example, male and female. At times, they may have to distinguish between true population ratios of 1:1 and 3:1 for these characteristics in what we will call the first and second categories, respectively. Suppose a certain geneticist has to make a decision between the two ratios on the basis of 20 observations.
(a) By rule 1, he will test $H_0{:}p = .5$ versus $H_1{:}p = .75$ by including in his rejection region the two-category tables 20-0, 19-1, 18-2, etc., until the probability is as close to $\alpha = .05$ as possible.

Exactly what does his decision rule turn out to be?
What is the true value of α?
What is the true value of the power for this test?

(b) By rule 2, he will test $H_0{:}p = .75$ versus $H_1{:}p = .5$ by including in his rejection region the two-category tables 0-20, 1-19, 2-18, etc., until the probability is as close to $\alpha = .05$ as possible.

Exactly what does his decision rule turn out to be?
What is the true value of α?
What is the true value of the power of this test?

(c) By rule 3, he will decide in favor of $p = .5$ if the first class contains $0, 1, \ldots, 12$ and in favor of $p = .75$ if the first class contains $13, 14, \ldots, 20$.

What is the probability of accepting $p = .5$ if it is the true ratio? Of rejecting it?
What is the probability of accepting $p = .75$ if it is the true ratio? Of rejecting it?

15-3 HYPOTHESIS TESTING—ILLUSTRATIONS

Testing a hypothesis using discrete data was discussed in Section 15-2 for both one-sided and two-sided alternatives. The plant-growth experiment used for illustration called for a hypothesis of $p = .5$, a hypothesis where the probabilities of individual test criterion values were symmetrical. We now review the procedure using two new illustrations. Both are such that each observation falls in either one or another of two categories, giving rise to data in a two-category or two-cell table.

Illustration 15-3-1

Some years ago, silver coins of the United States were replaced by so-called "sandwich" coins, easily recognized by the coppery color observed when one looked at their milled edge. Suppose we have heard that a sandwich quarter, if spun on edge rather than tossed in air, does not fall with equal probabilities for the two sides. We plan an experiment to provide data appropriate to a consideration of the problem.

STEP 1:

Propose the hypotheses $H_0{:}p = .5$ and $H_1{:}p \neq .5$.

STEP 2:

Choose $\alpha = .05$ or a value reasonably close since, for discrete data, probabilities will have finite jumps. Let $n = 25$.

STEP 3:

Let the sample statistic be the random variable Y = number of heads in the 25 independent trials. Its distribution is readily computed.

STEP 4:

We will want very large and very small numbers of heads in a two-tailed rejection region since such values tend to support a parameter in H_1 rather than H_0. For $n = 25$ and $p = .5$, we find $P(y = 0, 1, \ldots, 7) = .02165 = P(y = 18, 19, \ldots, 25)$. Thus for the implied critical region, $\alpha = 2(.02165) = .0433$. If we include $y = 8$ and $y = 17$, then α becomes $.0433 + 2(.03223) = .1078$. Hence we use a critical region of $y = 0, 1, \ldots, 7$ and $18, 19, \ldots, 25$ and have $\alpha = .0433$. The acceptance region will be $y = 8, 9, \ldots, 17$.

STEP 5:

The experiment is conducted and 11 heads and 14 tails observed. The value of the test criterion is 11.

STEP 6:

Since the observed value of the test criterion is in the acceptance region, we accept $H_0{:}p = .5$. Chance and the binomial model with $p = .5$ give an adequate explanation of the data.

Illustration 15-3-2

Gregor Mendel (1948) in his classic genetic study with pea plants, observed plant-to-plant variation for a number of characteristics. In one experiment, he observed the numbers of seeds with yellow and green albumen. According to the particulate theory of inheritance, one would expect a true ratio of 3 yellow to 1 green albumen; but chance is involved, and so observations fluctuate about the true ratio.

Suppose our class project is to obtain Mendel's original data, and each person is to test the implied hypothesis for one plant.

STEP 1:

Propose the hypotheses $H_0{:}p = .75$, for yellow, and $H_1{:}p \neq .75$.

STEP 2:

Choose $\alpha = .05$ or a value reasonably close. Here n is up to the plant. For illustration and computing convenience, we use $n = 20$, even though most of Mendel's plants produced more than 30 seeds.

STEP 3:

Let the sample statistic be the random variable Y = number of seeds with yellow albumen.

STEP 4:

We will want very large and very small values of Y in the critical region, but we are not dealing with symmetric probabilities as when $p = .5$.

Let us proceed by trying to associate about half of α with each tail of the distribution. For $n = 20$ and $p = .75$, $P(y = 19 \text{ or } 20) = .0243$ and we obviously will not wish to include $y = 18$.

At the other end, $P(y = 0, 1, \ldots, 10) = .0139$ and $P(y = 0, 1, \ldots, 11) = .0409$. We can stop with 10 and have $\alpha = .0382 < .05$ or include 11 and have $\alpha = .0652 > .05$. Since our sample size is small for detecting alternatives, let us arbitrarily have the critical region include 0, 1, . . . , 11 and 19 and 20. Hence $\alpha = .0652$.

STEP 5:

Suppose the experiment is completed and the data show 14 yellow to 6 green. The value of the test criterion is 14.

STEP 6:

The observed value of the test criterion is in the acceptance region, so we accept $H_0{:}p = .75$. Chance and a binomial model with $p = .75$ offer an adequate explanation of the data.

Step 4 in both illustrations calls for us to determine acceptance and critical regions. Clearly there are many possible choices. Thus, in Illustration 15-3-2, we might have a critical region consisting of the points 11 and 19, which would give $\alpha = .048$, but that does not make much sense; or we could use the value 12 and have $\alpha = .0609$, but that makes even less sense.

It is fairly clear that an alternative like $H_1 : p \neq .75$ calls for a two-tailed critical region, whereas an alternative like $p < .75$ (or $p > .75$) calls for a one-tailed critical region. In the latter case, we can and do choose the critical region so as to minimize the probability of a Type II error.

For two-tailed critical regions, we suggested that each tail have an associated probability of $\alpha/2$, as nearly as possible. An obvious way to assign values to the critical region, in general, is to begin with the y-value that has the least probability, add in the one with the next smallest, and so on until we have a α close to the suggested value. In Illustration 15-3-2, the y's would be included in the following order: $y = 0, \ldots, 8, 20, 9, 10, 19$, and finally 11. The value of $\alpha = .0652$ has been divided between the tails of the distribution when $p = .75$ as .0243 and .0409, not a very equal division. However, this general approach seems reasonable.

EXERCISES

15-3-1 Consider an experiment to test the hypothesis that a sandwich quarter is as likely to show a head as a tail if spun on edge. Plan to use $n = 20$ and α close to .10 since this is a relatively small sample size. After the planning is complete, that is, steps 1 to 4, conduct the experiment and test the hypothesis. Drawing a conclusion is required.

15-3-2 Repeat the above experiment at least 25 times. This is easily done as a class exercise with each student contributing one or more trials. There may be some coin-to-coin variation, but it is likely to be negligible. You now know how many times the hypothesis $p = .5$ was accepted and how many times rejected.

What is the expected value of the number of rejections? How does the proportion of rejections compare with your α?

Your 25 or more experiments constitute a sample from a binomial population with $p = \alpha \approx .10$; that is, α equals approximately .10. The true hypothesis should be that $p = \alpha$ where α is determined by your choice of acceptance and rejection regions.

Test the hypothesis that $p = .10$, not α, for your sample of $n = 25$ against the alternative $p > .10$, since you have probably set $\alpha > .10$ in Exercise 15-3-1.

15-3-3 If you reject a null hypothesis concerning a binomial parameter, you will normally conclude that you have hypothesized the wrong value of p. There is also, however, the possibility of an unusual random sample or even of a nonrandom sample. What other possible decision outside of our testing procedure could be made?

15-3-4 In a sweet-clover experiment, a certain progeny was examined to see whether the growth habit could be described as annual or biennial. Genetic considerations suggested that the progeny should be present in the ratio of three annual to one biennial.

If the observed ratio is 13 annual to 2 biennial, what will you conclude about the genetic analysis of the situation?

15-4 THE NORMAL APPROXIMATION FOR HYPOTHESIS TESTING

The use of the exact test for binomial data clearly requires considerable computation or extensive probability tables. A reduction in tables can be made if only critical values for selected α-levels are tabulated, but many (p, n)-pairs are still involved. Thus, it seems advisable to look for some approximation for which critical values are available or easily computed. The central limit theorem is an obvious place to start.

Section 13-4 provides all the technique necessary to test a hypothesis concerning the parameter of a binomial distribution. There, we saw that a sample proportion had mean $\mu = p$ and variance $\sigma^2 = p(1 - p)/n$, that an estimate of p was available from the data, and that this led, with the aid of the central limit theorem, to a confidence-interval statement about p, Equation 13-4-3.

Illustration 15-4-1

Let us apply that procedure to the coin example of the preceding section where we tested $H_0{:}p = .5$ against $H_1{:}p \neq .5$, using $n = 25$ and $\alpha = .0433$. For a two-sided confidence interval with $\alpha = .0433$, we find $z = 2.02$. We observed $y = 11$ heads and so find the following confidence interval for p.

$$\begin{aligned} \text{CI} = \hat{p} \pm z_{\alpha/2}\sqrt{\frac{\hat{p}(1-\hat{p})}{n}} &= \frac{11}{25} \pm 2.02\sqrt{\frac{11}{25} \times \frac{14}{25} \times \frac{1}{25}} \\ &= .44 \pm .20 \\ &= (.24, .64) \end{aligned}$$

Chance and any value of p between .24 and .64 are adequate to explain the data. Thus $p = .5$, being in the interval, must be accepted if proposed as a null hypothesis. It is apparent that confidence-interval estimation of a parameter has a built-in test procedure.

Alternatively, we may use a somewhat more direct test if we standardize the variable by Equation 15-4-1 and proceed as in Illustration 15-4-2.

$$Z = \frac{\hat{p} - p}{\sqrt{p(1 - p)/n}} \qquad (15\text{-}4\text{-}1)$$

Illustration 15-4-2

For the preceding data, $z = (11/25 - .5)/\sqrt{.5(.5)/25} = -.6$. The value is an observation made on a standardized near-normal variable. We must now observe whether the probability of a more extreme value, regardless of the sign, is large or small. In particular, we compare $|z| = .6$ with the tabulated $z = 2.02$ for $\alpha = .0433$. Since the observed value is smaller than the critical value, we accept the null hypothesis. Somewhat less simply but possibly more informatively, we observe that $P(|Z| > .6) = .5486$. This means that the probability of a more extreme value than that observed is .5486, nearly a 50:50 chance. Hence, the observed z-value is not to be considered unusual if $p = .5$.

The value $\alpha = .0433$ is unusual. Here, it is used because it was the exact probability computed by use of the binomial distribution for a critical region consisting of points in the true sample space. Ordinarily, if we use the normal approximation, we will also use the more familiar $\alpha = .05$.

This test procedure does not give precisely the same result as the confidence-interval approach. The slight difference is attributable to the fact that the confidence-interval approach uses only the data, whereas hypothesis testing requires a specific value of the parameter to be tested and this is used in computing the standard deviation.

The following computational technique gives the same answer as Equation 15-4-1. It may be more desirable in that rounding errors are introduced at a later stage and thus are less likely to affect the final computed value. Theorems 11-4-1 and 11-5-4 state that the number of successes in n binomial trials has mean $\mu = np$ and variance $\sigma^2 = np(1 - p)$. This leads to Equation 15-4-2 and the accompanying computations.

$$Z = \frac{y - np}{\sqrt{np(1 - p)}} \tag{15-4-2}$$

For the data, $z = [11 - .5(25)]/\sqrt{25(.5).5} = -.6$. Clearly the numerical results and conclusions must be the same.

When we assume that a normal approximation is appropriate, we act as though we are dealing with a symmetric distribution. This is not the case for a binomial distribution with $p \neq .5$. Thus, for a reasonable approximation to hold, it must be based on sample sizes that increase as the true value of p moves away from $p = .5$ and a formula or table is needed to provide guidelines as to an acceptable n for values of either p or $\hat{p}$. Cochran (1965, chapter 3) gives such a table.

EXERCISES

15-4-1 Test the hypothesis that $p = .75$, using z as a test criterion, for Illustration 15-3-2. How do the results compare?

15-4-2 In a study of the inheritance of sex, a random sample showed 35 males and 46 females. Test the null hypothesis that inheritance is according to a 1:1 ratio.

15-4-3 In a study of the inheritance of awns (attachments to the seed of certain grains) in barley, a random sample might show a ratio of 625 long awns to 200 short awns. Test the null hypothesis that inheritance is according to a 3 long:1 short ratio.

15-4-4 It has been suggested that women who take birth control pills are more likely to have girls than boys when they do give birth. A newspaper report provided the data used here.
(a) A random sample of nonpill mothers gave birth to 24 boys and 38 girls. Test the null hypothesis that the true ratio is 1:1 against the alternative that it is not.
(b) A random sample of pill mothers gave birth to 7 boys and 23 girls. Test the null hypothesis that the true ratio is 1:1 against the one-tailed alternative that has been suggested. Do the same using the exact binomial procedure of Section 15-3 and compare the results.
15-4-5 Additional binomial data are given in Exercises 15-5-3 to 15-5-5. Use these data to test the null hypotheses proposed there.

15-5 THE χ^2 TEST CRITERION

A test criterion that gives the same result for the problems being discussed as does the normal approximation of Equations 15-4-1 and 15-4-2 is the χ^2 test criterion with 1 degree of freedom.

The square of a normally distributed variable with zero mean and unit variance is called a chi square, χ^2, variable with 1 degree of freedom, that is, $\chi^2 = Z^2$. Thus for Y normally distributed with mean μ and variance σ^2, Equation 15-5-1 holds.

$$\chi^2 = \left(\frac{Y - \mu}{\sigma}\right)^2 = \frac{(Y - \mu)^2}{\sigma^2} \qquad 1 \text{ df} \qquad (15\text{-}5\text{-}1)$$

Large values of this test criterion tend to deny the null hypothesis since they indicate large departures of an observation from the hypothesized mean. Notice also that, in squaring Z, we lose the sign of the deviation so that large values of Z, regardless of sign, become large values of χ^2, always positive. This means we are testing against two-sided alternatives for μ although we are using only one tail of the χ^2 distribution.

If, for a binomial model, we are prepared to say that the number of successes in n trials is approximately normal, as was done in Equation 15-4-2, then the square of this z-variable is distributed as χ^2 with 1 degree of freedom. This gives us Equation 15-5-2.

$$\chi^2 = \frac{(Y - np)^2}{np(1 - p)} \qquad 1 \text{ df} \qquad (15\text{-}5\text{-}2)$$

Illustration 15-5-1

For the null hypothesis and data of the coin-tossing example of Illustrations 15-4-1 and 15-4-2, we compute

$$\chi^2 = \frac{(11 - 25/2)^2}{25(1/2)(1/2)} = .36 \qquad 1 \text{ df}$$

This value of the test criterion is referred to the distribution of χ^2 with 1 degree of freedom, a single line in Table B-3. The observed value, namely .36, is seen to be such that larger values occur randomly with a probability between .750, for $\chi^2 = .102$, and .500, for $\chi^2 = .455$. Chance and the null hypothesis offer an adequate explanation of the data so the hypothesis is not rejected. Using Z, we found the corresponding probability to be .5486.

The χ^2 test criterion for binomial model problems is usually stated in a more general form than that of Equation 15-5-2, namely as Equation 15-5-3, which is equivalent under these circumstances.

$$\chi^2 = \sum \frac{(\text{observed} - \text{expected})^2}{\text{expected}} = \sum \frac{(O - E)^2}{E} \tag{15-5-3}$$

The observed data are the numbers of observations falling in various categories or cells of a table that need not be two-celled or even one-dimensional. Summation is over all the cells, and the degrees of freedom depend upon the number of cells, the restrictions imposed on the sampling, and the independent parameters that must be estimated. Determination of degrees of freedom will be discussed in Chapter 16. This general form has many uses in applied statistics. We shall see but a few.

Illustration 15-5-2

For our coin example, the observed values are 11 and 14, and the expected values are np and $n(1 - p)$, both being $25(1/2) = 12.5$. Thus, we have

$$\chi^2 = \frac{(11 - 12.5)^2}{12.5} + \frac{(14 - 12.5)^2}{12.5} = .36 \qquad 1 \text{ df}$$

No parameters need to be estimated, but sampling has been restricted to $n = 25$ tosses. Hence, the two cells provide only a single degree of freedom. Notice that the two differences between observed and expected values are equal but of opposite sign. In general, $\sum (O - E) = 0$.

A great deal of discrete data fall in *one-way classifications* with two or more cells. For example, we might test the fairness of a die by tossing it 100 times and observing the frequency with which each of the values 1, 2, . . . , 6 occurs. For the underlying uniform distribution, all cell probabilities are 1/6 and the expected values for each cell are 100/6. Since no parameters are estimated, χ^2 will have $6 - 1 = 5$ degrees of freedom. More generally, cell probabilities will not be equal and we will have simply a multinomial distribution. In many cases, all cell probabilities will be given rather than estimated. These are used to compute expected values and, in turn, a χ^2 value by Equation 15-5-3. This χ^2 will have degrees of freedom equal to a value one less than the number of cells.

EXERCISES

15-5-1 Test the hypothesis that $p = .75$, using the χ^2 test criterion, for the data of Illustration 15-3-2. Compare your result with that obtained in Exercise 15-4-1.

15-5-2 For the data and the hypotheses of Exercises 15-4-2, 15-4-3, and 15-4-4*a*, test the proposed null hypotheses. Compare these results with those obtained earlier.

Notice that Exercise 15-4-4*b* calls for a test against a one-sided alternative. Can you see any way to test the null hypothesis against this alternative using χ^2?

15-5-3 In studying pubescence types in plants, geneticists observe whether they are glabrous, that is, with no hairs on the stems, or pubescent, that is, with hairs on the stems. A study of soybeans of a particular generation showed 320 glabrous plants and 115 pubescent (Bernard and Singh, 1969).

Test the null hypothesis of a 3 glabrous:1 pubescent population ratio.

A different cross produced 306 glabrous and 123 pubescent plants. Again test the null hypothesis of a 3:1 ratio.

15-5-4 In their studies, geneticists observe seeds to see if they are well filled, that is, normal or plump, or if they are shrunken, sometimes called dent. The observation is made on the endosperm, that is, the tissue that produces the new plant if the seed is planted.

In a study of dent times normal endosperm crosses in sorghum, a particular generation gave 686 plump:247 dent, 215:86, and 242:75 in three trials. Test the null hypothesis that the true segregation ratio is 2 plump:1 dent for each cross (Gorbet and Weibel, 1962).

15-5-5 Suppose a sample of 100 wild rabbits consists of 40 males and 60 females. Use both z and χ^2 to test the null hypothesis that the true sex ratio is 1:1.

Suppose a second sample consists of 1,000 wild rabbits with 400 males and 600 females, the same ratio as before. Again use both z and χ^2 to test the null hypothesis that the true sex ratio is 1:1.

Did you accept H_0 for both samples? Do both tests agree? What effect does sample size have on the test procedure? In which of our several equations defining a test criterion can you best see this effect?

15-5-6 A retailer places an order for 400 recapped automobile tires with a supplier who claims that no more than 5 percent of his output is ever returned as unsatisfactory. In time, 31 of the 400 tires are returned as unsatisfactory. Should the retailer continue to trust this supplier's word as to the rate of returns?

15-5-7 (a) In the same study as that which provided the data for Exercise 15-5-3 but using another cross, that is, another breeding, the authors also observed puberulent or minutely pubescent plants. They hypothesized that the true ratio of types was 12 glabrous: 3 normal pubescent:1 puberulent. The corresponding numbers of plants observed were 239:51:21 (Bernard and Singh, 1969).

Test the implied null hypothesis using χ^2 with 2 degrees of freedom as a test criterion. Do you see any way to use z as a test criterion? If you tried to use a trinomial instead of a binomial as the true distribution to give an exact test of significance, what sort of problem would you run into?

(b) Still another cross produced 189 normal plants, 66 curly, and 78 puberulent where the true ratio was hypothesized to be 9:3:4. Test this null hypothesis using χ^2 with 2 degrees of freedom as test criterion. Find the probability of a larger value of χ^2 being obtained at random.

(c) Still another cross produced 195 dense or heavily pubescent plants, 56 normal, and

80 puberulent where the true ratio was hypothesized to be 9:3:4. Test this null hypothesis using χ^2 with 2 degrees of freedom as test criterion. Find the probability of a larger value of χ^2 being obtained at random.
(d) Still another cross produced 193 sparse plants, 70 normal, and 77 puberulent where the true ratio was hypothesized to be 9:3:4. Test this null hypothesis using χ^2 with 2 degrees of freedom as test criterion. Find the probability of a larger value of χ^2 being found at random.

The remaining exercises are algebraic.

15-5-8 Show that Equations 15-4-1, 15-4-2, and 15-5-2 are algebraically equivalent, the last being simply the square of the others.

15-5-9 Show that Equations 15-5-2 and 15-5-3 are algebraically equivalent for a binomial problem.

15-5-10 Suppose that a random sample of n binomial observations results in n_1 individuals in one category and n_2 in the other. Show that, to test $H_0: p = .5$, we may use $\chi^2 = (n_1 - n_2)^2/(n_1 + n_2)$.

Chapter 16

TESTS OF HYPOTHESES II: DISCRETE DATA IN TWO-WAY TABLES

16-1 INTRODUCTION

Discrete data are frequently presented as counts in two-way tables, one example of the presentation of *categorical data.* Thus, a completely random sample of 1,000 individuals might be classified simultaneously by sex and voting habits to give data like that in Table 16-1-1. Only the total sample size has been fixed so that we have a single random sample of size 1,000.

TABLE 16-1-1 / Voting Habit—Sex Classification of 1,000 Voters

	Sex		
Action	Male	Female	Total
Participant	490	340	830
Nonparticipant	90	80	170
Total	580	420	1,000

Individuals within a population may have different characteristics, such as socioeconomic group, rural or urban home, and so on, where we have good information as to the proportions of individuals in each classification, and we may wish to carry these proportions over to our sampling. In particular, we might have data to show that voting males and females are about equal in numbers, and so decide to draw equal-sized samples from the two kinds of eligible voters to get data like that in Table 16-1-2. The responses for the two independent random samples are placed side by side or one above the other for comparison.

TABLE 16-1-2 / Voting Habits of Males and Females

	Action		
Sex	Participant	Nonparticipant	Total
Male	420	80	500
Female	360	140	500
Total	780	220	1,000

On the other hand, two independent random samples may come from the same population but be treated differently, as when they are given different remedies for the purpose of comparing treatment effectiveness in alleviating discomfort. The presentation of the data is, again, as in Table 16-1-2.

In either of the latter two cases, we may refer to the sampling scheme as *stratified random sampling*, although the term stratum goes more naturally with the first of these two examples. In both cases, the distinguishing characteristic is that the samples are independent.

Still other sampling schemes also lead to data presented in two-way tables. This chapter will consider some methods of analysis for such data.

16-2 TESTING THE EQUALITY OF BINOMIAL PARAMETERS: INDEPENDENT SAMPLES

In 1966, a newspaper reported a study of the merits of a long-term prophylactic anticoagulant therapy for treatment of patients after acute heart attacks. In a 10-year period, 32 of 88 patients who took the anticoagulant had a subsequent heart attack, while 28 of 80 untreated patients had one. The data are conveniently presented in Table 16-2-1.

The difference between the two percentages of subsequent heart attack, namely $(32/88)(100) = 36.36$ percent and $(28/80)(100) = 35.00$ percent, is less than two points, but the base numbers are moderately large. Consequently, a test of significance seems appropriate to decide, objectively, whether or not the ob-

TABLE 16-2-1 / Ten-year Records of Heart Patients

	Response		
Treatment	No Further Coronary	Further Coronary	Total
Therapy	56	32	88
No Therapy	52	28	80
Total	108	60	168

served difference in rates of subsequent attack is indicative of a real difference in population percentages or is simply attributable to random sampling variation with no real difference.

A number of comments must first be made about the data. The population studied is such that all admissible persons must have had an acute heart attack. In addition, they came from one part of the country, and, presumably, had some special reason for being included in the study. Perhaps they were referred from one of several clinics or a veteran's hospital. In any case, similar studies in other parts of the country and from other segments of the general heart-patient population would seem required before a broad generalization could be made.

Next, it seems important to ask where randomization fits into the picture, since this is the basis for application of the laws of chance. It was fairly clear from the newspaper description of the study that the patients were assigned to the treated and untreated groups by a randomization process. This assures us that no differential response, apart from a possible one due to the medical therapy or one due to chance, should have been introduced as the result of a conscious or unconscious bias-inducing assignment of the individuals to the two groups.

The random assignment is also assurance that we have independent samples and this is essential to our argument, quite apart from considerations of bias and the validity of applying probability theory.

Our problem is now to decide whether or not the two binomial samples can be considered to have come from populations with the same parameter. Each sample provides an independent estimate of a binomial parameter, namely the probability of a subsequent heart attack. The two samples are of fixed size, controlled by the investigator within the limits of patient availability.

To begin, let p_1 be the population probability of no further coronary for patients on therapy. Then $1 - p_1$ is the probability of a further coronary for a patient in this group. Let p_2 and $1 - p_2$ be the corresponding probabilities in the no-therapy group. Neither of the parameters p_1 and p_2 is known or hypothesized.

The proposed null hypothesis is simply $H_0: p_1 = p_2$, a hypothesis of equality or *homogeneity*. The alternative is $H_1: p_1 \neq p_2$. Let us fix the significance level of our test at $\alpha = .05$.

Assume that the two independent estimates of the possibly common parameter are approximately normally distributed and, consequently, that the same is true of their difference. This suggests Z as a statistic. It will be based on the difference between the two independent estimates as a random variable. These estimates are $\hat{p}_1 = 56/88$ and $\hat{p}_2 = 52/80$. The mean of the random variable, $\hat{p}_1 - \hat{p}_2$, will be $p_1 - p_2$, equal to 0 under the null hypothesis. The variance of the difference equals the sum of the variances when the two quantities are independent, as here. Equation 13-4-6 is applicable for the standard deviation, in general. Here, to estimate the variance, assume the null hypothesis to be true and pool the information concerning the common value of p to obtain $\hat{p} = 108/168$. Now, estimate the variance of the difference by Equation 16-2-1. The symbol $\hat{\sigma}^2$ or s^2 may be used, with appropriate subscript.

$$\begin{aligned}\hat{\sigma}_{\hat{p}_1-\hat{p}_2}^{\ 2} &= \hat{\sigma}_{\hat{p}_1}^{\ 2} + \hat{\sigma}_{\hat{p}_2}^{\ 2} \\ &= \frac{\hat{p}(1-\hat{p})}{n_1} + \frac{\hat{p}(1-\hat{p})}{n_2} \qquad \text{under } H_0 \\ &= \hat{p}(1-\hat{p})\left(\frac{1}{n_1} + \frac{1}{n_2}\right) \qquad (16\text{-}2\text{-}1)\end{aligned}$$

The sample statistic is the Z-value of Equation 16-2-2.

$$Z = \frac{\hat{p}_1 - \hat{p}_2}{\sqrt{\hat{p}(1-\hat{p})(1/n_1 + 1/n_2)}} \qquad (16\text{-}2\text{-}2)$$

Illustration 16-2-1

For the data of Table 16-2-1, test the null hypothesis that the probability of no subsequent heart attack, for single-heart-attack patients, is the same for the two therapy groups.

Let $p_1 = P(\text{no subsequent attack} \mid \text{therapy})$

and $p_2 = P(\text{no subsequent attack} \mid \text{no therapy})$

$$H_0{:}p_1 = p_2 \text{ versus } H_1{:}p_1 \neq p_2$$

$$\hat{p}_1 = 56/88 \qquad \hat{p}_2 = 52/80$$

$$\hat{p}(\text{under } H_0) = \frac{56+52}{88+80} = \frac{108}{168}$$

$$\hat{\sigma}_{\hat{p}_1-\hat{p}_2}^{\ 2}(\text{under } H_0) = (108/168)(60/168)(1/88 + 1/80)$$

Hence,

$$\begin{aligned} z &= \frac{56/88 - 52/80}{\sqrt{(108/168)(60/168)(1/88 + 1/80)}} \\ &= .184 \end{aligned}$$

Because of the alternative hypothesis, this is a two-tailed test. The critical region will include values of Z numerically larger than 1.96.

The observed value of the test criterion falls in the acceptance region, so the null hypothesis is accepted. This means that this hypothesis and chance offer an adequate explanation of the data.

The χ^2 test criterion of Equation 15-5-3 also leads to the same conclusion. Its value will be the square of the z-value just obtained. Expected values corresponding to the observed values are required.

Expected values are computed on the assumption that the null hypothesis is true, implying that the numbers in the two coronary classes are in the same proportions for each sample. This calls for our best estimate of the true p under H_0, namely, $\hat{p} = 108/168$. For the other category, $1 - \hat{p} = 60/168$. Expected values are now found by multiplying $\hat{p}$ and $1 - \hat{p}$ by the appropriate sample sizes, 88 and 80.

Illustration 16-2-2

Compute expected values and deviations-from-expected for the data of Table 16-2-1 under the null hypothesis that the two binomial parameters are equal.

$$\hat{p} = \frac{56 + 52}{88 + 80} \qquad \text{under } H_0$$

$$1 - \hat{p} = \frac{32 + 28}{88 + 80} \qquad \text{under } H_0$$

Expectations for the two samples are computed in Table 16-2-2. Deviations are −.57 and +.57, respectively, in the first sample and +.57 and −.57 in the second.

TABLE 16-2-2 / Expected Values for the Data of Table 16-2-1

	Response		
Treatment	No Further Coronary	Further Coronary	Total
Therapy	(108/168)(88) = 56.57	(60/168)(88) = 31.43	88
No Therapy	(108/168)(80) = 51.43	(60/168)(80) = 28.57	80
Total	108.00	60.00	168

Note that row and column sums are the same as in Table 16-2-1, that the deviations sum to 0 in each row and column, and that, necessarily, there is only one numerically distinct deviation.

To determine the value of the degrees of freedom for the χ^2 test criterion, the following sort of argument is generally satisfactory. We begin with the number of cells and subtract the number of restrictions imposed by the sampling scheme and the number of independent parameters to be estimated under the null hypothesis.

Illustration 16-2-3

Find the degrees of freedom for the χ^2 test criterion as applied to the null hypothesis for the heart patients. Complete the χ^2 test.

For this four-cell table, two sampling restrictions were imposed, the sample sizes of 88 and 80. In addition, it was necessary to estimate the parameter p, called for by the model under the null hypothesis. Hence,

$$\text{df} = 4 - 2 - 1 = 1$$

This is, of course, suggested by the set of deviations since there was a single, numerically distinct one.

From Equation 15-5-2, repeated here as Equation 16-2-3, we find

$$\chi^2 = \sum \frac{(O - E)^2}{E} \tag{16-2-3}$$

$$= \frac{(-.57)^2}{56.57} + \frac{(.57)^2}{31.43} + \frac{(.57)^2}{51.43} + \frac{(-.57)^2}{28.57}$$

$$= .0338$$

Observe that this is approximately the square of $z = .184$, the difference being attributable to rounding in the computations. Since it is distributed approximately as χ^2 with 1 degree of freedom, we compare it with $\chi^2_{.05} = 3.84$. There is no evidence to refute the null hypothesis.

Suppose we present data for a 2×2 table in the symbolic form of Table 16-2-3. Let the first subscript indicate the row or sample or treatment and the second indicate the column or category or response. We will use a dot in a subscript to show that summation has taken place over an index of summation formerly in this position. In this notation, a possibly more convenient computational form of Equation 16-2-3, for 2×2 tables, is given by Equation 16-2-4.

$$\chi^2 = \frac{(n_{11}n_{22} - n_{12}n_{21})^2 n}{n_{1.}n_{2.}n_{.1}n_{.2}} \qquad 1 \text{ df} \tag{16-2-4}$$

Note that the numerator is the square of the difference between diagonal products times the grand total and that the denominator is the product of the four marginal totals.

TABLE 16-2-3 / Symbolic Data for a 2 × 2 Table

	Category		
Sample	1	2	Total
1	n_{11}	n_{12}	$n_{1.}$
2	n_{21}	n_{22}	$n_{2.}$
Total	$n_{.1}$	$n_{.2}$	$n_{..} = n$

For the coronary data,

$$\chi^2 = \frac{[56(28) - 52(32)]^2 168}{88(80)108(60)} \quad 1 \text{ df}$$

$$= .0338$$

This is the same result as when Equation 16-2-3 is used.

EXERCISES

16-2-1 Smith and Nielson (1951) studied clonal isolations of Kentucky bluegrass from a variety of pastures. Clones have identical genetic constitutions. For two good, moderately grazed lowland fields, they observed the numbers of clones with and without rust, obtaining the accompanying data. Is there evidence that the proportion of clones showing rust differs for the two fields? What can one say about randomization in this experiment?

Field	Rust	No Rust
1	372	24
5	330	48

16-2-2 Two groups of asthmatic children were available for a behavioral study. One group had received treatment in a hospital, the other in a residential open air school (ROAS). For each group, the children were diagnosed as being sociopathic, that is, of antisocial behavior, or nonsociopathic (see accompanying table). Test the null hypothesis that the proportion of children diagnosed as sociopathic is the same for the two treatment groups.

	Diagnosis	
Treatment	Sociopathic	Nonsociopathic
Hospital	24	137
ROAS	30	25

Source: Pinkerton (1970).

16-2-3 The number of fatalities for bobwhite or quail, in confinement, due to quail disease, ulcerative enteritis, was becoming serious. A treatment group was selected at random and an antibiotic added to their food. Over a 7-week period, the two groups were compared for number of deaths (see accompanying table). Was treatment effective in reducing the number of deaths? What alternative is suggested by this question?

	Response	
Treatment	Alive	Dead
Treated	796	17
Untreated	1,136	230

Source: Kirkpatrick and Moses (1953).

16-2-4 For a study of the relationship of experimental pain tolerance to pain threshold, 110 individuals might be available. Say 40 of these are selected at random to constitute an experimental group while the remaining 70 constitute a control group. Suppose measurements are made, the median is determined for all individuals, each individual is recorded as above or below this median, and the results are presented as in the accompanying table. Test the null hypothesis that there is no difference in the proportion of responses above the median for the two groups. This can be considered as a test of whether or not the experimental and control populations have the same location.

	Location	
Treatment	Above	Below
Experimental	15	25
Control	40	30

16-2-5 Gastrointestinal parasites cause diarrhea in lambs when they first begin to graze in pastures. In a study on permanent pastures, one group of lambs might be treated while the other is not. The frequency of diarrhea might then be observed 10 to 30 days after grazing begins. Suppose the results are as in the accompanying table. Test the null hypothesis that the probability of having diarrhea is unrelated to the treatment. What do you think makes a reasonable alternative hypothesis?

	Response	
Group	Diarrhea	None
Untreated	15	35
Treated	15	25

Other observations might be made on lambs 10 to 30 days after grazing begins on permanent and on new pastures. Suppose the results are as in the accompanying table.

	Response	
Pasture	Diarrhea	None
Permanent	25	50
New	5	45

Test the null hypothesis that the probability of having diarrhea is the same for the two types of pasture. Are you using the same alternatives as before?

16-2-6 A university admissions committee considered a number of suspended students applying for readmission in the semester following their suspension. Of this number, 279 chose to make written application only; 127 were admitted and 152 were denied admission. The remaining 46 persons chose to make a personal appearance; 11 were admitted and 35 denied admission.

Test the null hypothesis that the probability of being admitted is the same for the two groups.

What can you say about randomization in justification of the use of a statistical procedure in this problem?

16-2-7 The susceptibility of grafts of white pine to blister rust was considered, using scions or shoots from parent trees that were either less than or more than 10 years old. Susceptibility was recorded as healthy or diseased (see accompanying table). Test the hypothesis that the proportion of healthy grafts is the same, regardless of the age class of the parent tree. (This age classification is for illustration purposes only in this exercise. Ages were given in more detail in the original data, and the information was used in the analysis.)

	Health	
Age	Healthy	Diseased
Younger than 10 Years	13	25
Older than 10 Years	26	13

Source: Courtesy of R. F. Patton, Department of Plant Pathology, University of Wisconsin.

16-2-8 In Exercise 15-4-4, two samples were used in testing the null hypothesis of a 1:1 ratio. However, it may be more valid simply to test whether or not they can be considered to have come from a population with the same binomial parameter. The data are as in the accompanying table. Test this hypothesis of homogeneity.

	Child	
Mother	Boy	Girl
Nonpill	24	38
Pill	7	23

16-3 THE COMPLETELY RANDOM 2 × 2 TABLE

Suppose we draw a random sample of 1,000 eligible voters and classify each as male or female and as participant or nonparticipant with respect to some election. We have not determined the number of males and females, or participants and nonparticipants, prior to sampling; this has been left to chance. The resulting data might well be those presented in Table 16-1-1.

These data differ from the heart-patient data of Table 16-2-1 in spite of the similarities. There, we were not concerned with estimating a parameter such as the population proportion receiving therapy since those 88 individuals were a sample of predetermined size. Here, we may estimate the proportions of males and females in the population. We may also estimate the proportions of participants and nonparticipants. We may even estimate the proportion of participating males, and so on. In Table 16-2-1, we had two potentially different populations, whereas the data of Table 16-1-1 came from a single population. Consequently, the same model cannot apply to both situations.

For the completely random sample only the total sample size is fixed. There is, then, a probability of a randomly selected individual falling in any one of the cells of the table. Table 16-3-1 describes the situation. Again, we resort to a dot subscript notation. For data from such a model, one might be interested in the question of whether or not the probability of participation depended on whether the individual was male or female. Thus, we want to know if the following equality holds. Does

$$P(\text{participant} \mid \text{male}) = P(\text{participant} \mid \text{female})$$

or

$$\frac{p_{11}}{p_{11} + p_{21}} = \frac{p_{12}}{p_{12} + p_{22}} \; ?$$

In turn, this implies $p_{11}p_{22} = p_{12}p_{21}$. This equality recalls the numerator of Equation 16-2-4. As a matter of fact, Equation 16-2-4 provides a test criterion for the implied hypothesis.

Because of the form of the equality first used in terms of p_{ij}'s to state the null hypothesis, it might be termed a hypothesis of *independence*. This follows because

TABLE 16-3-1 / Probabilities for a Model for the Data of Table 16-1-1

	Sex		
Action	Male	Female	Sum
Participant	p_{11}	p_{12}	$p_{1.}$
Nonparticipant	p_{21}	p_{22}	$p_{2.}$
Sum	$p_{.1}$	$p_{.2}$	$\sum_{i,j} p_{ij} = 1$

we are asking whether or not voting participation depends upon the sex of the voter. Or we may rewrite the equation in its second form as $p_{11}/p_{21} = p_{12}/p_{22}$ and have a hypothesis of proportionality. In the case of samples of fixed size, the hypothesis is one of equality.

The problem of the completely random sample can be looked at from a slightly different viewpoint. Suppose we stress the dependency-independency aspect of the original question. If there is no dependency, then

$$P(\text{participant} \mid \text{male}) = P(\text{participant} \mid \text{female}) = P(\text{participant})$$

Also $\quad P(\text{male} \mid \text{participant}) = P(\text{male} \mid \text{nonparticipant}) = P(\text{male})$

Now, we can multiply the probability of being a participant by the probability of being a male and have the probability of being a male participant. In particular, Equation 6-2-1 applies, while Equations 6-3-3 give a formal definition of independence. In other words, a null hypothesis of independence requires that $p_{11} = p_{1.}p_{.1}$, $p_{12} = p_{1.}p_{.2}$, $p_{21} = p_{2.}p_{.1}$, and $p_{22} = p_{2.}p_{.2}$. Now

$$p_{11}p_{22} = p_{1.}p_{.1}p_{2.}p_{.2} = p_{12}p_{21}$$

as in the previous development. If the null hypothesis is denied, we say that we have *dependence* of or *interaction* between the two systems of classifying individuals. In other words, the probability of falling in a particular category of a characteristic depends upon the category of the other characteristic. Use of the terms homogeneity and independence will not generally be found to depend on the sampling scheme; instead, they are used interchangeably.

The model and hypothesis for the completely random sample are different from those for independent samples. For the random sample, there is one population and a symmetry relative to the two systems of classification, whereas in the case of independent samples, parameters for two populations are being compared. For the completely random sample, the null hypothesis of independence is

$$H_0\colon p_{ij} = p_{i.}p_{.j} \qquad i, j = 1, 2$$

where $\sum_{i,j} p_{ij} = 1$. It is tested by the same test criterion as is the test of equality of probabilities in independent samples. For independent samples, the null hypothesis is $H_0\colon p_1 = p_2$ or, in the notation of this section,

$$p_{1j} = p_{2j} \qquad j = 1, 2$$

where $\sum_j p_{1j} = 1 = \sum_j p_{2j}$ and the first subscript denotes the sample. This common test criterion is χ^2 as computed in Equations 16-2-3 and 16-2-4, with 1 degree of freedom.

To determine the degrees of freedom for χ^2 in this 2×2 table, we proceed as follows. We have imposed one sample size restriction. However, under the null hypothesis it was necessary to provide two independent estimates of parameters, for example, $p_{1.}$ and $p_{.1}$. From these and the null hypothesis, all other probabilities, and the cell probabilities in particular, could be computed. Hence degrees of freedom $= 4 - 1 - 2 = 1$, as in the case of independent samples but for different reasons.

Illustration 16-3-1

Compute expected values for the data of Table 16-1-1 for use in Equation 16-2-3.

First estimate $p_{1.}$, $p_{2.}$, $p_{.1}$, and $p_{.2}$ as $\hat{p}_{1.} = 830/1{,}000$, $\hat{p}_{2.} = 170/1{,}000$, $\hat{p}_{.1} = 580/1{,}000$, and $\hat{p}_{.2} = 420/1{,}000$. Then

$$E_{11} = (\hat{p}_{1.}\hat{p}_{.1})n = \frac{830}{1{,}000} \times \frac{580}{1{,}000} \times 1{,}000 = \frac{830(580)}{1{,}000}$$

and so on. This computation is seen to be the same as for independent samples but the argument is different.

We will not illustrate the case of the completely random sample since the computations leading to the numerical value of the test criterion are the same as for independent samples.

EXERCISES

16-3-1 In studying the relationship between age and coffee consumed, a researcher obtained the data given in Table A on 235 individuals. Test the null hypothesis of no relationship.

TABLE A

	Cups per Day	
Age	Less than 3	3 or More
Under 30	82	94
Over 30	18	31

Source: Goldstein and Kaizer, 1969.

If the individuals had been classified somewhat differently in terms of their coffee drinking habits, then the data would have been as in Table B. Test the null hypothesis of no relationship. Are your conclusions the same as when the lower cutoff point on coffee drinking was used?

TABLE B

	Cups per Day	
Age	Less than 5	5 or More
Under 30	134	42
Over 30	30	19

16-3-2 At Yale University during a flu epidemic, 276 freshmen were classified as to blood group and whether or not they were infected with A_2 Hong Kong influenza. There were 224 individuals in the more common A and O groups and they were classified as in the accompanying table. Test the hypothesis of independence of blood classification and infection.

	Infection	
Group	Infected	Not
A	54	56
O	44	70

Source: Evans et al. (1972).

16-3-3 We grew some plants and classified them to produce the accompanying table. Test the null hypothesis that height and color are independent classification systems.

	Flower Color	
Plant Height	Red	White
Tall	44	9
Short	18	3

16-3-4 Suppose we are breeding fruit flies and observing whether they are normal or dark in color and whether they have normal or vestigial wings. The results might be as in the accompanying table. Test the null hypothesis that the two characteristics, color and wing type, are independently inherited.

	Wings	
Color	Normal	Vestigial
Normal	260	98
Dark	75	28

16-3-5 At a flea market, we take a random sample of 200 individuals and classify them according to sex and whether or not they bought anything. We find the accompanying

	Habit	
Sex	Lookers	Buyers
Male	23	64
Female	47	66

data. Is there evidence that buying habit depends upon sex? What was your alternative hypothesis?

16-3-6 It was stated in this section that $p_{11}/(p_{11} + p_{21}) = p_{12}/(p_{12} + p_{22})$ was equivalent to saying $p_{11}p_{22} = p_{12}p_{21}$. Show that this is so.

16-4 TESTING THE EQUALITY OF BINOMIAL PARAMETERS: RELATED SAMPLES

Suppose we plan to conduct an experiment to compare two headache remedies. One way is that of Section 16-2 where individuals are assigned by a random procedure, n_1 to remedy A and n_2 to remedy B. The response to each remedy is measured as a cure or a not-cure. We then compute the proportion of cures for each remedy and compare them by the z-test or the χ^2 test.

An alternative plan is to use both remedies on the same person for different headaches. Here, randomness would determine which remedy to use first. This plan seems better in that both remedies are applied to the same individuals, thereby eliminating group differences unrelated to remedies.

More generally, this latter plan involves meaningful pairing of individuals or matched samples. In other words, pairing is on the basis that when the treatments are the same, the responses of members of a pair are expected to be more alike than without pairing. Treatments are assigned at random within each pair, and any observed difference should be attributable largely to differences between treatments, if any. The samples are not independent.

For our headache experiment, we might have data such as in Table 16-4-1, presented as in Table 16-2-1. However, Table 16-2-1 dealt with two distinct and, consequently, independent samples, whereas Table 16-4-1 deals with one sample where there are two observations on each individual. The independent-samples requirement for the technique of Section 16-2 to be valid is not satisfied. Meaningful pairing always results in related samples.

TABLE 16-4-1 / Possible Responses of 40 Headache Sufferers to Two Remedies

Remedy	Cure	Not-Cure	Total
A	29	11	40
B	23	17	40

For related samples, we collect the data so that the degree of dependence can be measured or otherwise taken account of in the computational technique. This is done by observing the responses of each individual as an ordered pair, that is, as (1, 1), (1, 0), (0, 1), or (0, 0), and recording in the appropriate cell of Table 16-4-2. Note that marginal totals for this table are the entries in Table 16-4-1. Table 16-4-2 shows a true total of 40 individuals with their bivariate responses, whereas Table 16-4-1 shows the 80 univariate responses, 40 on each remedy, but sacrifices the information relevant to the degree of association or dependence.

TABLE 16-4-2 / Bivariate Responses of 40 Headache Sufferers to Two Remedies

Response to B Is	Response to A Is: Cure	Not-Cure	Total
Cure	17	6	23
Not Cure	12	5	17
Total	29	11	40

Let us use A and B to designate cures with remedies A and B, respectively, and $\not{A}$ and $\not{B}$, read as "not-A" and "not-B," to designate failure to cure. The probabilities for the four cells of Table 16-4-2 may be symbolized as $P(AB)$, $P(\not{A}B)$, $P(A\not{B})$, and $P(\not{A}\not{B})$, where the sum of the four is unity. The null hypothesis to be tested is that $P(A) = P(B)$ where clearly $P(A) = P(AB) + P(A\not{B})$ and $P(B) = P(AB) + P(\not{A}B)$. Hence, we have

$$H_0: P(AB) + P(A\not{B}) = P(AB) + P(\not{A}B)$$

Since $P(AB)$ is common to the two probabilities stated to be equal under the null hypothesis, the inference to be drawn is seen to depend upon $P(A\not{B})$ and $P(\not{A}B)$, and consequently on the experimental results in the upper right and lower left corners of Table 16-4-2. In particular, we need to know if these numbers can be considered to have been assigned by chance and according to a binomial probability of .5 to the two cells involved. For this test, the apparent sample size is seen to be determined by the number of observations actually in these two categories.

Illustration 16-4-1

Test the null hypothesis of no difference in response to treatments A and B, using the data of Table 16-4-2.

Equation 16-2-3 provides an appropriate test criterion, namely χ^2 with 1 degree of freedom. We find

$$\chi^2 = \frac{(6-9)^2}{9} + \frac{(12-9)^2}{9} = 2 \qquad \text{with 1 df}$$

This is compared with $\chi^2_{.05}$, 1 degree of freedom, and we conclude that there is insufficient evidence to lead us to reject the null hypothesis. Exercise 15-5-3 provides an alternative computational procedure for computing χ^2.

It is also possible to argue as follows. The AB and $\not{A}\not{B}$ cells contain no information to suggest one remedy is better than another since both either cured or failed to cure. The information that suggests one remedy is better than another is obtained only from those cells where one cured and the other failed to. Clearly one remedy is better than another when the numbers in these cells are markedly different. The information in the AB and $\not{A}\not{B}$ cells is not truly thrown away. The variance of a difference between proportions is given by the equation

$$V(\hat{p}_1 - \hat{p}_2) = V(\hat{p}_1) + V(\hat{p}_2) - 2\,\mathrm{cov}(\hat{p}_1, \hat{p}_2)$$

where cov stands for *covariance*, a measure of the relationship between the observations within the bivariate observations. We are using the information but not in an obvious manner.

EXERCISES

16-4-1 Responses to radiation treatments, nausea and vomiting, of 28 individuals follow. This was a paired test using a placebo, that is, a sugar pill containing no drug, and 4 milligrams of perphenazine. A + indicates a favorable response (see accompanying table).

Treatment with Placebo	Treatment with 4 mg Perphenazine	
	+	−
+	16	0
−	9	3

Source: Chu et al. (1969).

Did treatment with perphenazine reduce the incidence of nausea and vomiting? What have you used as null and alternative hypotheses?

16-4-2 The drug of Exercise 16-4-1 was also used at another dosage and a second drug was used at two dosages. The resulting data are given in the accompanying tables. Are the treatments responsible for reducing the incidence of nausea and vomiting? Specify your null and alternative hypotheses in each case.

Treatment with Placebo	Treatment with 8 mg Perphenazine	
	+	–
+	15	1
–	11	3

Treatment with Placebo	Treatment with 2.5 mg Metopimazine	
	+	–
+	14	0
–	7	4

Treatment with Placebo	Treatment with 5.0 mg Metopimazine	
	+	–
+	14	2
–	9	5

Source: Chu et al. (1969).

16-4-3 The drugs of the preceding exercise were used in a single experiment with the results as given in the accompanying table. Is one treatment better than the other in controlling the incidence of nausea and vomiting? Specify your null and alternative hypotheses.

Treatment with 2.5 mg Metopimazine	Treatment with 4 mg Perphenazine	
	+	–
+	19	2
–	4	1

Treatment with 5.0 mg Metopimazine	Treatment with 8 mg Perphenazine	
	+	–
+	24	4
–	5	2

Source: Chu et al. (1969).

16-4-4 Suppose that 1 month prior to voting on a public issue a random sample of 100 individuals is questioned as to their expected vote. During the following month, much debate occurs and many statements are made about the issue, all intended to affect voting behavior. When the voting is completed, the same 100 individuals are asked how they voted. Let the resulting data be as in the accompanying table. Estimate the proportion of individuals in the original population who intend to vote favorably. Estimate the proportion of individuals who eventually voted "yes" on the issue. Test the null hypothesis that the two population proportions are the same.

	First Preference	
Vote	Yes	No
Yes	33	27
No	19	21

16-5 THE $r \times c$ CONTINGENCY TABLE

The 2×2 table for classification of discrete data is a special case of the $r \times c$ table with r rows and c columns. Such tables are called *contingency tables*, and the ideas of Sections 16-2 and 16-3 are easily generalized to cover them.

Contingency tables arise from completely random and stratified random sampling schemes. Models depend upon the sampling scheme, with a single set of probabilities summing to unity for completely random sampling and a set summing to unity for each sample when such are independent.

The hypotheses to be tested are ones of independence of row and column probabilities or of equality of probabilities over independent samples. The terms homogeneity, heterogeneity, independence, and interaction are also used without much attempt being made to associate them with a particular sampling scheme.

The test criterion in general use is χ^2 as defined by Equation 16-2-3. It is distributed approximately as χ^2 with $(r - 1)(c - 1)$ degrees of freedom, and is often called interaction χ^2. The random variable χ^2 is defined as the sum of squares of normally and independently distributed variables with zero means and unit variances. Means and variances are taken care of in the test criterion. Since we sum over all $r \times c$ cells, the more obvious problem is that of deciding upon the degrees of freedom.

Expected values for the test criterion are computed from row and column totals. The expected value for the entry in the ith row at the jth column is the product of the ith row total by the jth column total, divided by the grand total. Justification for this computation and for the degrees of freedom depends upon the sampling scheme and, in turn, the model and null hypothesis. While the arguments differ, the results are the same. The arguments are generalizations of those used in Sections 16-2 and 16-3.

At one time, there was concern over the possibility that small expected frequencies, even as large as 5 and 10, resulted in too large contributions to our test criterion and so spoiled its approximation to χ^2. Pooling rows and/or columns containing small expectations was the suggested alternative. At the same time, this appeared to be counterproductive in that the best source of evidence to distinguish among distributions is in the tails, where expected values are necessarily small. More recently, Cochran (1952) has shown that the 5 percent χ^2 value, on which an "accept" or "reject" decision is often made, is not much disturbed by an expectation as low as 1 if only about one in five expectations is less than 5.

Section 16-4 has its extension also. Here we are concerned with more than two treatments so that meaningful pairing has to be extended to a *block* of matched samples, one sample for each treatment. Again, we have gotten away from independent sampling, and, consequently, now require an ordered sequence of 1s and 0s, that is, a multivariate observation. The analysis of these data requires us to make use of the pairwise dependencies among the responses within the multivariate observation. This analysis goes beyond the scope of this text.

EXERCISES

16-5-1 The data first introduced in Exercise 16-3-2 are given in their entirety in the accompanying table. Is this a case of completely random or stratified random sampling? Test the null hypothesis of independence of blood group and Hong Kong flu infection. How do you write this null hypothesis symbolically? What is your alternative hypothesis?

	Infection	
Group	Infected	Not
A	54	56
O	44	70
B	14	15
AB	6	17

Source: Evans et al. (1972).

16-5-2 In a study concerned with the relationship between residence as a youth and frequency of participation in hunting during youth, the data given in the accompanying table were collected. Is there a significant association between the two variables? What null hypothesis are you testing?

	Residence	
Participation	Rural	Nonrural
Frequent	293	74
Occasional	115	56
Never	43	68

Source: Sofranko and Nolan (1971).

16-5-3 Data similar to those above are also given for fishing. Is there a significant association between the variables? Are they independent?

	Residence	
Participation	Rural	Nonrural
Frequent	187	76
Occasional	91	60
Never	13	24

Source: Sofranko and Nolan (1971).

16-5-4 The data of Exercise 16-3-1 are given in more detail in the accompanying table. Suppose we are interested in the question of independence of age and cups of coffee consumed. What do you think should be done with the "unknown" row? Would it be unreasonable to leave it out? If you leave out the unknowns, should the over 40s get special treatment? Test the null hypothesis that age and number of cups of coffee consumed are independent.

	Cups per Day			
Age	0	1–2	3–4	5+
Under 20	2	2	2	1
20–29	37	41	50	41
30–39	13	2	12	17
40+	2	1	0	2
Unknown	2	1	0	1

Source: Goldstein and Kaizer (1969).

16-5-5 In a controlled experiment on the use of court probation for drunk arrests, three groups of chronic drunks received different treatments and were later classified as to the number of arrests (see accompanying table). Are the three groups homogeneous? How did you state your null hypothesis symbolically?

	Rearrests		
Treatment	None	1	2+
1 None	32	14	27
2 Alcoholism clinic	26	23	33
3 Alcoholics Anonymous	27	19	40

Source: Ditman et al. (1967).

16-5-6 In the study of Exercise 16-5-5, data were also collected on time to first rearrest. What sort of inferences can you draw from the accompanying data?

	Time		
Treatment	Within First Month of Treatment	After First Month of Treatment	No Rearrests
1	16	25	32
2	17	39	26
3	19	40	27

Source: Ditman et al. (1967).

Chapter 17

TESTS OF HYPOTHESES III: NORMAL-DISTRIBUTION PARAMETERS

17-1 INTRODUCTION

In Chapters 15 and 16, testing hypotheses concerning proportions was considered. The same basic principles apply when tests are made using continuous data. Such tests assume a null hypothesis, that is, one for which we can compute the probability of making a Type I error and, so, require a test criterion with a known sampling distribution. Critical values of the test statistic will depend upon the alternative hypothesis.

In this chapter, tests of hypotheses for one or two means and for two variances are treated, where the parent distributions are assumed to be normal or approximately so.

17-2 TESTS OF A SINGLE MEAN

We first deal with tests which enable us to decide from the evidence of a random sample whether or not a specific hypothesized value of a population location parameter offers an adequate explanation of the data. Thus, we may wish to ascertain whether the true mean weight of male students attending a university could be 165 pounds, a value possibly known for male students of a previous year of interest, or whether the true mean income of employed residents of a community could be $9,000, perhaps last year's census figure for the state.

To illustrate the test, we use the data of Exercise 3-6-3, the heart rates of 13 cats, 30 seconds after receiving an intravenous injection of 10 milligrams of procaine in the femoral vein. Procaine is a local anesthetic which blocks the sensory nerve and tends to increase the heart rate. The ordered values are 105, 120, 123, 126, 135, 138, 140, 150, 160, 168, 170, 186, and 198.

Let us assume that in the population of which the 13 observations are a random sample, each observation consists of the sum of the true sought-after population mean, that is, a parameter, and a random deviation. At this point, we have assumed *linearity* with respect to the mean and *additivity* of the two components describing an observation. Linearity is a simple assumption in comparison with a more complex one where the mean might be raised to a power dependent on the observation. Similarly, an additive model is simpler than a multiplicative one. We now have a *linear additive model.*

We assume the random components to be *independent* of each other as a consequence of the random sampling procedure, and to be from a normal distribution with mean 0 and a single variance.

To be formal and concise, we describe the ith observation by Equation 17-2-1, where μ represents the population mean and ε_i, read as "epsilon-sub-i," the random component.

$$Y_i = \mu + \varepsilon_i \qquad (17\text{-}2\text{-}1)$$

The requirements on the random components are a part of the model and may be written as $\varepsilon_i \sim \text{nid}(0, \sigma^2)$. We read this as "the ε's are *n*ormally and *i*ndependently *d*istributed with zero mean and variance σ^2," the lack of a subscript on σ^2 saying that all ε's are from the same population or are *homogeneous.* This model was first introduced in Section 13-5.

We have now made rather strong assumptions, namely, a linear additive model requiring a number of characteristics for the ε's, including a normal distribution. Random sampling makes the rules of probability apply, and the assumption of normality says that that specific distribution, or one derived from it, is to be used in computing probabilities. In particular, the derived population of $\bar{Y}$'s is normally distributed, approximately so even if the Y's aren't, with mean μ and variance σ^2/n. This leads us to use the Z-distribution discussed at length in Chapter 10 and the t-distribution mentioned in Chapter 12. Here, the standard deviation, rather than the variance, will be needed. Both these distributions were used in Chapter 13.

If σ^2 is known, the test criterion for the null hypothesis about the location of the mean is given by Equation 17-2-2.

$$Z = \frac{\bar{Y} - \mu_0}{\sigma/\sqrt{n}} \tag{17-2-2}$$

where $\bar{Y}$ = sample mean
n = number of observations in sample
μ_0 = hypothesized population mean
σ = parent population standard deviation

If, however, σ^2 is estimated by a sample s^2, then our test criterion is given by Equation 17-2-3.

$$t = \frac{\bar{Y} - \mu_0}{s/\sqrt{n}} \tag{17-2-3}$$

Here, s is the sample standard deviation.

The t-test was proposed by W. S. Gossett, 1867–1937, as an alternative to and a more satisfactory criterion than Z, particularly for small samples. Gossett was a student of Karl Pearson, 1857–1936, and a mathematician at the Guinness brewery. Interesting and entertaining material on Gossett and Pearson has been written by McMullen (1947), E. S. Pearson (1938, 1966), and Walker (1958).

Notice that the numerator of the test criterion measures how far $\bar{Y}$ is from μ_0. If they are close together, then the data are evidence in support of the hypothesized mean μ_0. However, we would not consider it unusual to find $\bar{Y} - \mu_0$ larger for more-variable data than for less-variable data. Consequently, we measure the difference in units of the appropriate standard deviation, that is, by dividing $\bar{Y} - \mu_0$ by its own standard deviation, namely $\sigma/\sqrt{n}$ or $s/\sqrt{n}$.

If the alternative is that the true mean μ can only be less than μ_0, then large negative values of z or t would be evidence in support of this alternative; positive values or small negative values would say that μ_0 and chance offer an adequate explanation of the data. Here, we have a one-sided alternative calling for a one-tailed test.

On the other hand, if the alternative is simply that μ is different from μ_0, then large values of Z or t, either positive or negative, are evidence in support of the alternative hypothesis, a two-tailed one, whereas numerically small values say that μ_0 and chance offer an adequate explanation of the data.

Now we need only decide what values of Z or t are to be considered as large and what ones as small; we need the separation points between acceptance and rejection regions, the critical values. This is an arbitrary decision, and it is customary to say that an event which, when the null hypothesis is true, can occur by chance no more than 1 in 20 times on the average is really evidence in favor of the alternative hypothesis. That is, we commonly set $\alpha = .05$, where α is the probability of a Type I error as defined in Chapter 15; other choices of α are clearly possible. The acceptance region, then, will have an associated probability of $1 - \alpha = .95$, computed under the condition that the null hypothesis is true.

To carry out the test, we compare the sample Z or t with the critical values or compute the probability of a more extreme result than that observed and compare it with the chosen α.

Illustration 17-2-1

Using the heart-rate data on cats, test the null hypothesis that the population parameter μ is $\mu_0 = 160$ heartbeats per minute.

For the cat data, $\bar{y} = 147$, $s = 27.5$, and $n = 13$. Since σ^2 is unknown, it was necessary to estimate it from the data. Using Equation 17-2-3, we find

$$t = \frac{147 - 160}{27.5/\sqrt{13}} = -1.71 \qquad 12 \text{ df}$$

In other words, the sample $\bar{y}$ is 1.71 standard deviations to the left of μ_0.

Let us suppose that the only acceptable alternative to $\mu = 160$ is for it to be less than 160, that is, $\mu < 160$. This is a one-sided alternative so that values of t that support it must be numerically large and negative.

Table B-2 is arranged to make selection of a critical value easy, regardless of the alternatives. The top of the table reads, "probability of a larger value of t, sign ignored," meaning that the probability covers both positive and negative values of t and so is to be used as the value of α for two-tailed tests. At the bottom, the table reads, "probability of a larger value of t, sign considered," and this tells us to take the probability as the value of the significance level α for one-tailed tests, but that we must also consider the sign of t, observing whether or not it is in the direction called for by the alternative.

From Table B-2, opposite $n - 1 = 12$ degrees of freedom, we find $P(t > 1.782) = .05$. Hence, any sample t to the left of -1.782 supports the alternative, whereas one to the right supports the null hypothesis. The sample $t = 1.71$ is to the right so does not deny the null hypothesis; that is, $\bar{y} = 147$ is close enough to $\mu_0 = 160$ for us to say that the parameter $\mu_0 = 160$ and random sampling offer an adequate explanation of the data. Hence, we accept the null hypothesis that $\mu_0 = 160$ because there is insufficient evidence to deny it.

Alternatively, we look for $t = 1.71$ opposite 12 degrees of freedom and find the two tabulated t-values that bracket the sample t, namely 1.356 and 1.782. Symmetry of the t-distribution allows us to drop the negative sign temporarily. We are then able to present the accompanying inequality concerning the probability of a more extreme t than that observed. Since this is a test against one-sided alternatives, we read the probabilities from the bottom of the table.

$$P(t < -1.782) = \underbrace{.05 < P(t < -1.71) < .10} = P(t < -1.356)$$

On the other hand, if our alternative is simply that $\mu \neq 160$, then we have a two-sided alternative hypothesis and the probabilities at the top of Table B-2 serve as values of the significance level α. Opposite $n - 1 = 12$ degrees of freedom we find $P(|t| > 2.179) = .05$. Since our sample $t = -1.71$ is numerically nowhere near this large, it is to be considered as one that can reasonably be expected under the null hypothesis when sampling is random. There is no evidence to deny that $\mu = 160$.

Alternatively, we may write the following, giving special attention to the inequality concerning $P(|t| > 1.71)$.

$$P(|t| > 1.356) = \underbrace{.2 > P(|t| > 1.71) > .1} = P(|t| > 1.782)$$

In other words, the probability that a random value of t will be numerically larger than that observed is somewhat larger than .1.

In practice, we would use either the two-tailed or the one-tailed test, this decision being made by the experimenter during the planning stage of the experiment. Making a decision concerning alternatives after observing the data means that the test will be result-guided. Valid probability statements cannot be made using the t-table under these circumstances.

EXERCISES

17-2-1 Exercise 3-6-5 gives the percent copper in 10 bronze castings as follows: 85.54, 85.72, 84.48, 84.98, 84.54, 84.72, 84.72, 86.12, 86.47, 84.98 (Wernimont, 1947). Test the null hypothesis $\mu = 86.20$ against the two-sided alternative $\mu \neq 86.20$ using a t-test with $\alpha = .05$. Find the probability of a larger value of t, sign ignored, than that observed.

17-2-2 For the grade data of Exercise 3-6-1, test $H_0\!:\mu = 70$ against $H_1\!:\mu > 0$. Use a t-test with $\alpha = .01$. Find the probability of a larger value of t than that observed.

17-2-3 A random sample of 200 trucks were driven, on the average, 16,300 miles a year with a sample standard deviation of 3,100 miles. Test the null hypothesis that the average truck mileage in the parent population is 17,000 miles a year against the alternative hypothesis that the average is less. Use the 5 percent level of significance. What is your conception of the parent population?

17-2-4 A survey of 30 families ("Survey of Current Business," 1965) in a medium-sized city during 1968 indicated a weekly expenditure for food of \$39.75 with a sample standard deviation of \$6.12. Evaluate the hypothesis that the average weekly expenditure for food for all families in this city in 1968 was \$42.00 per week. Let the alternative hypothesis be that it was not \$42.00 and use the 1 percent level of significance. Comment on the fact that the city consists of a finite number of families and its possible effect on your testing procedure.

17-2-5 A manufacturer claims that the nicotine content of his cigarettes is less than 18 milligrams. A sample of 20 analyses of the cigarettes shows a mean nicotine content of 20.2 milligrams and a standard deviation of 1.7 milligrams. Can we conclude, using a t-test and $\alpha = .01$, that the product meets the manufacturer's claim? Test against the hypothesis that the nicotine content is higher than 18 milligrams, using the .05 significance level.

17-2-6 Exercise 3-6-2 gave lengths in millimeters of a lower molar from 16 fossils of the mammal *Hyopsodus* from the Eocene Age. The lengths were: 4.87, 4.77, 4.93, 4.71, 4.68, 4.52, 4.77, 4.85, 4.80, 4.65, 4.45, 4.95, 4.76, 4.90, 4.72, and 4.49 millimeters. Test the null hypothesis that $\mu = 4.60$ against the alternative $\mu \neq 4.60$, using $\alpha = .05$ and a t-test.

17-3 COMPARISON OF TWO SAMPLE MEANS: GENERAL

An important test of significance deals with the problem of whether the observed difference between two sample means may be attributable to chance or whether it is evidence that the samples came from populations having different means. For example, we may wish to decide if the true average weekly grocery bills in two communities differ, based on a random sample of 250 families with an average expenditure of \$48.45 in community A and a random sample of 300 families similarly defined with an average expenditure of \$46.65 in community B. Or we may wish to decide if the true average length of trout in stream A is greater than in stream B, based on a random sample of 75 trout averaging 15.5 inches in length from stream A and a random sample of 60 trout averaging 14.1 inches in length from stream B.

Such data consist of two samples, and it is necessary to use two subscripts to identify an observed value. Any random value can be represented by Y_{ij}, where the first subscript, i, denotes the sample or population sampled, and the second subscript, j, denotes the individual observation within the sample. For the two-sample case, $i = 1$ or 2 and $j = 1, \ldots, n_1$ for $i = 1$ and $j = 1, \ldots, n_2$ for $i = 2$. The subscript on n is dropped when $n_1 = n_2$. The test procedure depends upon:

1 Whether the two samples are independent or dependent
2 Assumptions concerning the parent populations, more particularly, the random component of the model

For independent sampling, the selection of the observations in one sample must in no way affect the selection of the observations in the other. Observation 1 of the first sample, Y_{11}, is to be no more related to observation 1 of the second, Y_{21}, than to any other observation of the second sample.

When sampling is independent, we refer to the sample design, or layout, as a *nested, hierarchal,* or *completely random design*. This is not limited to two samples. Since fixing the sample sizes is not a restriction on randomization, no restrictions have been imposed on the random sampling.

Independent sampling actually applies to two different situations. First, consider an experiment or sampling plan to determine if any difference exists between the average weights of the fish of a species inhabiting two different lakes. A random sample should be taken from each lake. There are, of course, many practical problems involved in such a wildlife project. Second, consider an experiment to compare two rations using rats as test subjects. The rats are not randomly selected from two, or even one, population. Instead, those that are available are used. They will ordinarily be animals bred and raised under homogeneous conditions. Here, the experimental units, the animals, are assigned to the two rations using a random process based on a table of random numbers. All possible randomizations are to be equally likely. Here, we have an *experimental* design rather than a sampling design. The arithmetic for analyzing the data from these two different situations is the same.

In contrast to independent sampling or the completely random design, we may have the situation where experimental units are intentionally and nonrandomly paired. Pairing is according to expected response, with the members of a pair being expected to respond more similarly if treated alike than would those in a pair where the individuals were randomly selected. Consequently, any observed difference must be due largely to treatments, certainly more so than if pairing were random with no attempt being made to minimize the part of the observed difference not attributable to treatments. Now Y_{11} is more like Y_{21} than like any other $Y_{2j}, j \neq 1$.

Here, we refer to the design or layout as a *two-way* or *cross classification*, since the data are conveniently presented in this way, or a *randomized* (*complete*) *block design*. It is not limited to two-treatment experiments but may be applied when there are sufficient sets of multiple experimental units, the multiple being equal to the number of treatments, and all members of a multiple being expected to respond similarly if treated similarly. Now we have a restriction on the random sampling or the assignment of treatments, the former requiring sampling from a population of pairs, or larger multiples, and the latter requiring a separate randomization within each pair or block.

As an illustration, consider an experiment to compare the effects, based on gains in weight, of two different rations using twin lambs. For each pair of twins, the members are assigned randomly, one to each ration. For a before-and-after experiment, the two samples could be the blood pressures of the individuals in a single group before and after receiving a drug intended to lower pressures. Obviously, the observations are in pairs. Here we must assume a randomness in the responses since we cannot provide it in the assignment of treatments.

The assumptions referred to for determining what test is appropriate are concerned with the random components in the equation for the model. These are discussed in the following sections.

17-4 COMPARISON OF TWO SAMPLE MEANS: INDEPENDENT SAMPLES WITH EQUAL VARIANCE

To illustrate the test procedures for two independent samples, we compare the mean length in millimeters of the lower molar for two species of *Hyopsodus* but, first, we discuss the problem. The data and computations are given in Table 17-4-1.

Hyopsodus is found as a fossil in beds of the Eocene Age. It was a small hoofed mammal with many primitive characteristics. The measurements were made on specimens of the *Hyopsodus* collection at the American Museum of Natural History, New York, by E. C. Olson and R. L. Miller, paleozoologists of the University of Chicago.

The null hypothesis is that population means for lower molar length are the same for the two species. We consider both two- and one-sided alternative hypotheses, using a rejection level of $\alpha = .05$. In real life, the choice of an alternative hypothesis is made when the experiment is planned.

For independent samples from two populations, we begin by assuming a linear additive model where each population has its own mean so that an observation consists of the appropriate mean plus a random component. The model of Section 17-2 has simply been extended to two populations. We also assume that the two populations are normally distributed with the only possible difference being their locations. In turn, this says that the random components came from the same normal distribution with mean assumed to be 0 and an unknown variance.

We may write the equation model concisely as Equation 17-4-1.

$$Y_{ij} = \mu_i + \varepsilon_{ij} \qquad i = 1, 2; j = 1, \ldots, n_i \tag{17-4-1}$$

The population means are the parameters μ_1 and μ_2; the assumptions about the ε's may also be written concisely as $\varepsilon_{ij} \sim \text{nid}(0, \sigma^2)$.

Our interest is in inferring whether or not $\mu_1 = \mu_2$. We, then, propose the null hypothesis $H_0: \mu_1 = \mu_2$. The alternative is likely to be $H_1: \mu_1 \neq \mu_2$, although it could be one-sided if we had external evidence to suggest this. If we accept the null hypothesis, we claim that the evidence indicates $\mu_1 = \mu_2 = \mu$, say; whereas if we reject the null hypothesis, we conclude that the evidence indicates that $\mu_1 \neq \mu_2$.

The test criterion is constructed to be similar to that for the one-sample problem. We hypothesize a value, usually 0, for the difference between population means. We then measure the difference between sample means. The difference between these two quantities, one a parameter and the other a random variable, becomes the numerator of our test criterion. It must, in turn, be measured in appropriate units, namely, the standard deviation applicable to a difference between sample means. Thus, our test criterion is given by Equation 17-4-2.

$$t = \frac{(\bar{Y}_1 - \bar{Y}_2) - (\mu_1 - \mu_2)}{s_{\bar{Y}_1 - \bar{Y}_2}} \tag{17-4-2}$$

Under our null hypothesis, $\mu_1 = \mu_2$ or $\mu_1 - \mu_2 = 0$. However, we are free to specify any difference we wish.

Once again, we are dealing with a derived distribution, this time that of a population of $(\bar{Y}_1 - \bar{Y}_2)$s. The mean of this distribution is $\mu_1 - \mu_2$. For independent random samples, the variance of the difference is the sum of the variances; that is, $V(\bar{Y}_1 - \bar{Y}_2) = \sigma_1^2/n_1 + \sigma_2^2/n_2$. However, the t of Equation

17-4-2 is distributed as Student's t only if $\sigma_1^2 = \sigma_2^2$. Hence, let us first assume a common variance, σ^2. Now

$$V(\bar{Y}_1 - \bar{Y}_2) = \sigma^2 \left(\frac{1}{n_1} + \frac{1}{n_2} \right)$$

Customarily, we must estimate σ^2.

To estimate σ^2, we could use either sample 1 or sample 2 and, in either case, have an unbiased estimate. However, in practice, we use all the available information about σ^2 by using a weighted average, the weights being the degrees of freedom of the sample variances. The appropriate weighted average is given by Equation 17-4-3.

$$s^2 = \frac{(n_1 - 1)s_1^2 + (n_2 - 1)s_2^2}{(n_1 - 1) + (n_2 - 1)} \qquad (17\text{-}4\text{-}3)$$

Recall that $(n - 1)s^2$ is a sum of squares, so that the numerator is the pooled sum of squares while the denominator is the pooled degrees of freedom. When $n_1 = n_2$, the sample variances have equal weights. The weighted average is an unbiased estimate of σ^2 when the ε's are random and from a common population with mean 0 and variance σ^2.

For Equation 17-4-2, $s_{\bar{Y}_1 - \bar{Y}_2}$ is given by Equation 17-4-4.

$$s_{\bar{Y}_1 - \bar{Y}_2} = \sqrt{s^2 \left(\frac{1}{n_1} + \frac{1}{n_2} \right)} = \sqrt{s^2 \frac{n_1 + n_2}{n_1 n_2}} \qquad n_1 + n_2 - 2 \text{ df} \qquad (17\text{-}4\text{-}4)$$

When both samples are the same size, this reduces to Equation 17-4-5.

$$s_{\bar{Y}_1 - \bar{Y}_2} = \sqrt{\frac{2s^2}{n}} \qquad 2(n - 1) \text{ df} \qquad (17\text{-}4\text{-}5)$$

The test criterion in the general form of Equation 17-4-2 for testing $H_0\!:\mu_1 - \mu_2 = 0$, using independent random samples, may now be written conveniently as Equation 17-4-6.

$$t = \frac{\bar{Y}_1 - \bar{Y}_2}{\sqrt{s^2[(n_1 + n_2)/n_1 n_2]}} \qquad n_1 + n_2 - 2 \text{ df} \qquad (17\text{-}4\text{-}6)$$

This is distributed as Student's t when the assumptions are valid. In fact, considerable departures from normality may not seriously affect the distribution, especially near the commonly used 5 and 1 percent values.

When σ^2 is known, the test criterion becomes that of Equation 17-4-7.

$$Z = \frac{\bar{Y}_1 - \bar{Y}_2}{\sqrt{\sigma^2(n_1 + n_2)/n_1 n_2}} \qquad (17\text{-}4\text{-}7)$$

Illustration 17-4-1

We now illustrate the t-test for comparing independent samples using the *Hyopsodus* data of Table 17-4-1, where the computations are also shown. Sample sizes, means, and sums of squares are given. Sums of squares are pooled to estimate a common σ^2 and, in turn, $s_{\bar{Y}_1 - \bar{Y}_2}$. Finally, the value of the test criterion is computed.

If our alternative hypothesis is to be two-sided, that is $H_1: \mu_1 \neq \mu_2$ or $\mu_1 - \mu_2 \neq 0$, our rejection value will be found in Table B-2 under the required α probability for $n_1 - 1 + n_2 - 1 = 20$ degrees of freedom.

TABLE 17-4-1 / Length in Millimeters of the Lower Molar of Two Species of *Hyopsodus*

Species A y_1	Species B y_2	Species A y_1	Species B y_2
3.81	5.22	3.29	4.87
4.18	5.08	4.13	4.29
3.78	4.74	3.38	4.42
4.07	4.01	4.22	4.85
3.88	4.69	3.64	
4.33	4.39	3.74	

$\sum y_1 = 46.45$ $\qquad$ $\sum y_2 = 46.56$

$\sum y_1^2 = 180.9961$ $\qquad$ $\sum y_2^2 = 218.0506$

$\bar{y}_1 = 3.87$ $\qquad$ $\bar{y}_2 = 4.67$

$$\sum y_1^2 - (\sum y_1)^2/n_1 = 180.9961 - 179.8002 = 1.1959 = (n_1 - 1)s_1^2$$

$$\sum y_2^2 - (\sum y_2)^2/n_2 = 218.0506 - 216.7834 = 1.2672 = (n_2 - 1)s_2^2$$

$$s^2 = \frac{[\sum y_1^2 - (\sum y_1)^2/n_1] + [\sum y_2^2 - (\sum y_2)^2/n_2]}{(n_1 - 1) + (n_2 - 1)}$$

$$= \frac{1.1959 + 1.2672}{11 + 9} = 1.232$$

$$\text{df} = (n_1 - 1) + (n_2 - 1) = 20$$

$$s_{\bar{y}_1 - \bar{y}_2} = \sqrt{s^2 \frac{n_1 + n_2}{n_1 n_2}} = \sqrt{.1232 \frac{12 + 10}{12(10)}} = 0.15$$

$$t = \frac{\bar{y}_1 - \bar{y}_2}{s_{\bar{y}_1 - \bar{y}_2}} = \frac{3.87 - 4.67}{0.15} = \frac{-0.80}{0.15} = -5.33$$

Source: Data courtesy of Olson, E. C., and P. L. Miller: "Morphological Integration." University of Chicago Press, Chicago, 1958.

For our problem and $\alpha = .05$, the tabulated t value for 20 degrees of freedom is 2.086 and for $\alpha = .01$ is 2.845. Since $|-5.33| > 2.845$, we say that the difference is highly significant and conclude that the alternative hypothesis is required as an explanation of the data.

If our alternative hypothesis is one-sided and states that $\mu_1 > \mu_2$ or $\mu_1 - \mu_2 > 0$, no calculations are required since $\bar{y}_1 < \bar{y}_2$ is an event which must be attributed to chance and $H_0: \mu_1 = \mu_2$. If, however, our alternative is one-sided and states that $\mu_1 < \mu_2$ or $\mu_1 - \mu_2 < 0$, then the rejection values for probability levels of 5 and 1 percent are found under the 10 and 2 percent headings or above the 5 and 1 percent designations at the foot of Table B-2 which, for 20 degrees of freedom, are 1.725 and 2.528, respectively. The null hypothesis is rejected and the alternative accepted.

EXERCISES

17-4-1 Suppose the life in thousands of miles for 10 tires of each of two brands is as follows:

A	26.3	31.7	20.9	25.2	27.6	28.1	26.9	25.8	23.0	32.2
B	29.2	28.1	33.2	32.1	26.3	30.1	28.2	33.4	25.6	30.7

Test the null hypothesis that the means of the two populations of tire lives of which these observations are random samples are the same. Use a 5 percent level of significance and a two-sided alternative. What is the probability of a numerically larger t than that observed?

17-4-2 For the preceding data, assume that it is reasonable to hold an alternative hypothesis that life-span for brand *B* is the greater. Now test the null hypothesis using a 1 percent significance level.

17-4-3 A laboratory reported burning times in hours for two types of batteries as follows:

A	32	25	29	31	30	31	33	28	29	
B	29	32	31	28	27	29	30	25	26	27

Determine if the sample evidence indicates a difference in burning times for population means for the two types of batteries. Use a 5 percent rejection level.

17-4-4 A random sample of 20 families in one city spent, on the average, \$46.24 per week for food with a sample standard deviation of \$8.25; and a random sample of 25 families in another city spent, on the average, \$41.18 with a sample standard deviation of \$6.92. Assume $\sigma_1^2 = \sigma_2^2$. Test the null hypothesis that the mean weekly expenditure for food is the same for the two cities. As an alternative, use $H_1: \mu_1 \neq \mu_2$ and $\alpha = .01$. Suppose the alternative is $H_1: \mu_1 > \mu_2$. Test against this alternative using $\alpha = .05$.

17-4-5 To test two methods of teaching arithmetic, 50 students were assigned at random to two classes, 25 students per class, and a different method was used in each class. After a suitable period of instruction, a test was given to evaluate the two methods. The average scores made by the students in the two classes were 65 and 72. The corresponding sample standard deviations were 16 and 18. Evaluate the hypothesis that the two methods are equal in effectiveness for teaching arithmetic. Consider $\sigma_1^2 = \sigma_2^2$ and use $\alpha = .05$. Criticize the experiment.

The question that follows is algebraic.

17-4-6 Prove the following when $n_1 = n_2$.

(a) $\dfrac{s_1^2 + s_2^2}{2} = \dfrac{(n_1 - 1)s_1^2 + (n_2 - 1)s_2^2}{(n_1 - 1) + (n_2 - 1)}$

(b) $(\bar{y}_1 - \bar{y}_2)^2$ is an estimate of $2\sigma_Y^2$.

17-5 COMPARISON OF TWO SAMPLE MEANS: PAIRED OBSERVATIONS

In two-treatment experiments, experimental units may be meaningfully paired in advance, as when twins are to be used, in an attempt to minimize the random component. This amounts to an effort to assign some of the random variation among all individuals to pair differences, by use of matched individuals, leaving a minimum associated with differences between members of the same pair. The observed difference between the responses of the members of any pair will, then, consist of any real treatment difference that there may be plus a small random component. If pairing has been successful, the ability of the experiment to detect a real treatment difference has been improved relative to an experiment without such pairing. The information that led to pairing has allowed us to make a part of the natural variability among all the individuals extraneous as far as the difference between treatment means is concerned.

Now we may describe an observation, formally and concisely, by Equation 17-5-1.

$$Y_{ij} = \mu_i + \beta_j + \varepsilon_{ij} \tag{17-5-1}$$

The observation Y_{ij} is that on the ith treatment, $i = 1, 2$, for the jth pair, $j = 1, \ldots, n$. It is composed of a treatment mean μ_i, a pair effect β_j, and a random element ε_{ij}. The treatment mean is often written as a general mean plus a treatment effect. The β's may be considered as part of some original random component, say, $\beta_j + \varepsilon_{ij}$ for $i = 1$ and 2, but it can be seen that this part need not be considered random. In other words, the pairs need not be chosen randomly if we can be assured that the ε_{ij}'s are random. Finally, we assume that the ε_{ij}'s are nid$(0, \sigma_i^2)$, where $i = 1$ or 2 according to the treatment involved. Notice that the variance can depend upon the treatment. These assumptions complete the model.

To test the null hypothesis that $\mu_1 = \mu_2$, the test criterion is again that of Equation 17-4-2 but with $s_{\bar{Y}_1 - \bar{Y}_2}$ computed by Equation 17-5-2.

$$s_{\bar{Y}_1 - \bar{Y}_2} = \sqrt{\frac{\sum (Y_{1j} - Y_{2j})^2 - [\sum (Y_{1j} - Y_{2j})]^2/n}{n(n-1)}} \tag{17-5-2}$$

Now we have a derived distribution of $(\bar{Y}_1 - \bar{Y}_2)$'s with mean $\mu_1 - \mu_2$ but a variance that is not the sum of the variances because pairing has not been random. We will not see its relation to σ_1^2 and σ_2^2 at this time, but Equation 17-5-2 does give an appropriate estimate of the variance of the difference between the sample means.

Notice that the computations are carried out using differences as defined by Equation 17-5-3.

$$\begin{aligned} D_j &= Y_{1j} - Y_{2j} \\ &= (\mu_1 - \mu_2) + (\varepsilon_{1j} - \varepsilon_{2j}) \end{aligned} \tag{17-5-3}$$

The mean, $\bar{D} = \bar{Y}_{1.} - \bar{Y}_{2.}$, contains no contribution from the β's but is an estimate of $\mu_1 - \mu_2$, in which we are interested. The numerator of Equation 17-5-2 is a sum of squares of n D_j's and, hence, contains no contribution from the β's. The divisor $n - 1$ takes care of the degrees of freedom while n is the number of differences in $\bar{D}$. Thus, Equation 17-5-2 gives a sample variance appropriate to $\bar{D}$. In other words, the problem has been reduced to that of a single sample, and Equation 17-4-2 may be written as Equation 17-5-4.

$$t = \frac{\bar{D}}{s_{\bar{D}}} \qquad n - 1 \text{ df} \tag{17-5-4}$$

It is clear that meaningful pairing of individuals makes for a better estimate of the true difference between population means. However, it would be pointless to do this if there were not some way to obtain a standard deviation appropriate to the situation, that of meaningful pairing. Pooling variances does not eliminate a source of variation such as that contributed by the β's. Only the use of differences between the responses of paired individuals is seen to be a suitable computational device.

It is also of interest that the validity of t is not based on the assumption that $\sigma_1^2 = \sigma_2^2$. By taking differences, we obtain a valid t-test even if σ_i^2 depends upon the treatment.

To illustrate the procedure, we use the data of Table 17-5-1, showing the effect of training on the triglyceride content of blood lipids in diabetic adolescent girls. This study was made by the Pediatric Department of Karolinska Institute, Crown Princess Lovisa's Children's Hospital, Stockholm, Sweden. The girls spent a week at a winter camp in the Swedish mountains. Daily skiing excursions of 7 hours duration were made, and the triglyceride content of the blood lipids was measured 2 days before and 5 days after the training program for each girl. Triglyceride is an ester of glycerol resulting from decomposition of body fats and oils. An increase in the triglyceride content of the blood lipids would indicate a decomposition of body fats.

TABLE 17-5-1 / Triglycerides in the Blood Lipids of Diabetic Adolescent Girls before and after Physical-fitness Training (Milligrams per 100 millilitres)

	Before y_1	After y_2	Difference $d = y_1 - y_2$
	72	107	−35
	60	86	−26
	102	123	−21
	54	102	−48
	57	70	−13
	42	70	−28
	45	73	−28
	56	61	− 5
	61	100	−39
	64	41	+23
	51	68	−17
$\sum y$	664	901	−237
$\sum_j (y_{1j} - y_{2j})^2$			8,747
$\bar{y}$	60.4	81.9	−21.5

$$s_{\bar{d}} = s_{\bar{y}_1 - \bar{y}_2} = \sqrt{\frac{\sum (y_{1j} - y_{2j})^2 - [\sum (y_{1j} - y_{2j})]^2/n}{n(n-1)}}$$

$$= \sqrt{\frac{8{,}747 - (-237)^2/11}{11(11-1)}} = 5.75$$

$$t = \frac{\bar{d}}{s_{\bar{d}}} = \frac{\bar{y}_1 - \bar{y}_2}{s_{\bar{y}_1 - \bar{y}_2}} = \frac{60.4 - 81.9}{5.75}$$

$$= -3.75 \qquad 10 \text{ df}$$

Source: Larson, Y. A. A., et al., 1962.

Illustration 17-5-1

The null hypothesis for the blood-lipid data is that the difference between the means of the two populations is 0 or that the mean of the population of differences is 0, that is, that $\mu_1 = \mu_2$ or $\mu_1 - \mu_2 = 0$. The two-sided alternative is that the difference between the population means is not 0.

The computations are based on differences. There is a minor check in that $\bar{D} = \bar{Y}_{1}. - \bar{Y}_{2}.$. The sample t is −3.75, with 10 degrees of freedom. This is numerically larger than tabulated $t = 3.169$ for

$\alpha = .01$ and 10 degrees of freedom. Thus, the evidence supports rejection of the null hypothesis and acceptance of the alternative, namely, that physical-fitness training affects the triglyceride content of the blood lipids of diabetic adolescent girls. The data also indicate that there is an increase in the population mean after training.

To test the same null hypothesis against the alternative that $\mu_1 < \mu_2$ or $\mu_1 - \mu_2 < 0$, we have the same sample t. We note also that the sample data are in a direction to support the alternative hypothesis. Presumably, the investigator had some prior knowledge to propose this alternative. Now the probability of a random sample t being to the left of -3.169 is .005 under the null hypothesis. The sample t is more extreme, so we are led to reject $H_0{:}\mu_1 = \mu_2$ and conclude that $\mu_1 < \mu_2$. There is a real treatment effect as proposed in the alternative.

EXERCISES

17-5-1 Thompson and Podmore (1953) determined the average crushing strengths, in pounds per square inch, for two zones in eight cottonwood logs (see accompanying table). Test the hypothesis of no difference between population means, using a 5 percent rejection level. Let the alternative be that there is a difference. Compute the 95 percent confidence interval for the mean difference. Compare this interval for length with those obtained by using the sign-test and rank-sum test procedures.

Log	1	2	3	4	5	6	7	8
Zone *A*	1,916	2,057	1,956	1,817	2,014	1,884	1,900	2,005
Zone *B*	2,198	2,107	1,969	1,963	2,094	1,933	2,191	2,171

Source: These data are reproduced by permission of the National Research Council of Canada, *Can. J. Bot.*, **31**: 675–692 (1953).

17-5-2 Petersen (1954) determined the number of seeds per pollination for top and bottom flowers of 10 lucerne plants (see accompanying data). Test the hypothesis that the top and bottom means are equal. Set $\alpha = .01$ and use the alternative that top flowers set more seeds than bottom flowers.

Plant	1	2	3	4	5	6	7	8	9	10
Top	1.4	3.3	2.0	0.4	2.1	1.9	1.1	0.1	0.9	3.0
Bottom	1.1	1.7	1.8	0.3	0.8	1.4	1.0	0.4	0.7	0.9

17-5-3 The average acre catch of muskrats at 12 locations in the Horicon Marsh, for 2 years, is given in the accompanying table (data are paired vertically). Determine at the 10 percent level whether the muskrat catch was different in size for the two years. Compute the 90 percent confidence interval for the mean difference. Compare the length of this interval with that given by the sign-test procedure.

Year 1	.53	.46	1.25	1.33	2.85	.71
Year 2	.98	2.50	.53	.84	3.18	.81
Year 1	2.85	3.11	3.29	5.60	8.72	3.24
Year 2	2.49	3.28	3.80	4.36	4.35	3.36

Source: James B. Hale.

17-5-4 Dr. R. E. Nichols, Department of Veterinary Science, University of Wisconsin, measured the apparent surface tension, in dynes, of rumen fluid taken on different days before and after feeding. The paired data are given in the accompanying table. Use a one-tailed test at the 5 percent level to evaluate the alternative that apparent rumen fluid tension is greater after than before feeding. Compute the 95 percent confidence interval for the mean difference. Compare this interval in length with that given by the rank-sum test procedure.

Before	52.9	49.1	50.9	51.2	49.1	48.5	51.7	53.8
After	56.1	51.4	54.1	50.9	56.7	55.8	54.4	53.5

17-5-5 The times required by 10 persons to perform a task in seconds, before and after receiving a mild stimulant, are given in the accompanying table. Test the null hypothesis that there is no difference between the mean times in the "before" and "after" populations. As an alternative, assume that the after population will have a lower mean. Find the probability of getting a more extreme difference between sample means than that observed.

Before	34	45	31	43	40	41	33	29	41	37
After	29	42	32	29	36	42	26	28	38	33

17-6 THE PROBLEM OF INDEPENDENT SAMPLES WITH UNEQUAL VARIANCES

Student's t is, basically, the ratio of a normally distributed random variable with zero mean and unit variance, that is, a Z-variable, to the square root of the sum of squares of such variables, that is, a $\sqrt{\chi^2}$ variable. Together, Z and χ^2 involve one or more σ's. However, the single, common σ cancels out of t in the independent-samples problem when there is equality of variance. When there is inequality of the variances, that is, when $\sigma_1{}^2 \neq \sigma_2{}^2$, the ratio of these variances remains as a parameter in the test criterion, and the criterion, in turn, is not distributed as Student's t. Consequently, one resorts to an approximate solution. Cochran and Cox (1957) give a sufficiently accurate one.

17-7 EQUALITY OF TWO VARIANCES

Section 17-4 dealt with testing the equality of two population means using independent samples with equal variances, while Section 17-6 pointed out that we resort to an approximate solution to this problem of testing when the variances are unequal. It is clear that a test of the equality of two variances could often be required.

To test the equality of two variances, a ratio, rather than a difference, is used. Two sample variances, say s_1^2 and s_2^2, are computed and the test is made.

In Section 13-8, we saw that $(n_1 - 1)s_1^2/\sigma_1^2$ is distributed as a χ^2 variable with $n_1 - 1$ degrees of freedom. To test the null hypothesis $H_0: \sigma_1^2 = \sigma_2^2$, using a ratio of s^2's, is to use a ratio of two χ^2 variables, each divided by its own degrees of freedom provided the σ^2's cancel as they must when $\sigma_1^2 = \sigma_2^2$. The resulting test statistic is known as the *F*-statistic, and the *F*-distribution is known and tabulated (Table B-7).

If $\sigma_1^2 \neq \sigma_2^2$, the quantity computed as F involves the ratio of these two unknown variances, some number other than 1, so that the computed ratio can be expected to be larger or smaller, on the average, than a random *F*-value ought to be. In other words, large or small values of F would tend to support an alternative hypothesis, say $H_1: \sigma_1^2 \neq \sigma_2^2$.

Table B-7 is seen to consist of a few points from each of many *F*-distributions, a distribution being specified by both the numerator and denominator degrees of freedom, in that order. Also, note that only large values of F are given. This turns out to be acceptable and a great saving of space since, if F follows an *F*-distribution with $n_1 - 1$ and $n_2 - 1$ degrees of freedom, then $1/F$ follows an *F*-distribution with $n_2 - 1$ and $n_1 - 1$ degrees of freedom.

An *F*-test is made, then, by computing the ratio given in Equation 17-7-1.

$$F = \frac{\text{larger estimate of } \sigma^2}{\text{smaller estimate of } \sigma^2} \tag{17-7-1}$$

This value is then referred to Table B-7 where tabulated probabilities are those for larger values of F than shown, probabilities associated with only one tail of the distribution. Consequently, a test against two-sided alternatives with a Type I error of size .05 requires us to find a critical value of F opposite a probability of .025.

The *F*-test is so-named after Sir Ronald A. Fisher, 1890–1962. Fisher, an English statistician, made numerous important contributions to statistics in this century. The *F*-statistic was named for him by George W. Snedecor, 1881–1974, who gave us F as a simpler form of a statistic being used by Fisher. Snedecor was an American statistician who pioneered in the valid application of statistics in cooperative research. He was also responsible for the promotion of statistics in America by bringing many outstanding foreign statisticians, as visiting lecturers, to the Statistical Laboratory at Iowa State University.

Illustration 17-7-1

Test the equality of the variances computed for the two samples in Table 17-4-1.

We have $H_0: \sigma_1^2 = \sigma_2^2$. Sample variances are:

For species *A*:

$$s_1^2 = \frac{1.1959}{11} = .108718 \qquad 11 \text{ df}$$

For species B:

$$s_2{}^2 = \frac{1.2672}{9} = .140800 \qquad 9 \text{ df}$$

Using Equation 17-7-1, compute

$$F = \frac{.140800}{.108718} = 1.295 \qquad 9 \text{ and } 11 \text{ df}$$

For a two-sided alternative hypothesis and $\alpha = .05$, we look in the F-table under 9 and opposite 11 and .025. The critical value is found to be 3.59.

Since $1.295 < 3.59$, there is insufficient evidence to deny the null hypothesis.

EXERCISE

17-7-1 Test the null hypothesis of equal variances for the two samples given in Exercises 17-4-1, 17-4-3, 17-4-4, and 17-4-5. In all cases, assume $H_1: \sigma_1{}^2 \neq \sigma_2{}^2$ is the appropriate alternative hypothesis.

Chapter 18

NONPARAMETRIC TEST PROCEDURES

18-1 INTRODUCTION

Tests of hypotheses for discrete data, primarily when from the binomial distribution, were considered in Chapters 15 and 16; tests involving continuous data, but just the normal distribution, were the subject of Chapter 17. To assume a distribution is to make strong assumptions concerning the parent population. In the present chapter, we consider tests of hypotheses involving one and two samples where only weak assumptions are made because a specific parent distribution is not specified. Such tests need *distribution-free* statistics, that is, procedures that are not dependent upon a specific parent distribution. Such procedures are also called *nonparametric.* There are statisticians who maintain a distinction between the two terms. Assumptions under such circumstances were discussed in Chapter 14.

Procedures of this chapter, like those of Chapter 14, are not the traditional ones that depend upon the validity of the assumption of a binomial or normal distribution. They may be omitted by those whose interests do not extend this far. However, they are direct and simple, and study of them should help in one's overall understanding of statistics.

18-2 THE SIGN TEST WITH A SINGLE SAMPLE

In Chapter 14, the sign-test estimation procedure was used to establish confidence limits for a population median. The procedure, which requires minimum assumptions concerning the sampled population, namely, independent observations and a common unknown median, is based on a test procedure which can be used to evaluate the null hypothesis that the population median is a specified value.

The test procedure requires a count of the number of observations which are greater than, or less than, the hypothesized median. Observations equal to the hypothesized median may be ignored. The count is then used to test the null hypothesis that pluses and minuses occur with equal frequency in the parent population.

The test criteria given in Chapter 15 are applicable. In particular, the χ^2 test criterion applies. Here, it is presented in a new form as Equation 18-2-1 (see Exercise 15-5-11). This form is applicable to test $H_0:p = 0.5$, the equal-frequency null hypothesis, customarily against two-sided alternatives, that is, $H_1:p \neq 0.5$.

$$\chi^2 = \frac{(n_1 - n_2)^2}{n_1 + n_2} \quad 1 \text{ df} \qquad (18\text{-}2\text{-}1)$$

The values n_1 and n_2 are the counts of pluses and minuses.

Note that $n_1 - n_2$ may be positive or negative, the test statistic making no distinction since it squares this quantity. If a test against one-sided alternatives is desired, then we must observe this sign. Thus, suppose that a test against a one-sided alternative calls for minuses to be present with probability greater than .5; that is, $H_1:p(-) > 0.5$ and $p(+) < 0.5$. Now if n_1 is the number of positives and $n_1 - n_2 > 0$, there is an excess of positives and we simply conclude that the null hypothesis is not denied. On the other hand, if $n_1 - n_2 < 0$, we are observing a result in the direction that tends to deny the null hypothesis and we compute χ^2. However, on entering the χ^2 table, we look for the critical value under 2α for a test declared to be at the α-level of significance.

Illustration 18-2-1

To illustrate the sign test, we again use the data on the heart rates of 13 cats, data recently used in Section 17-2. The ordered values are: 105, 120, 123, 126, 135, 138, 140, 150, 160, 168, 170, 186, and 198 heartbeats per minute.

Let the hypothesized median be 160; in other words, $H_0:Q_2 = 160$ or $P(X < 160) = .5 = P(X > 160)$. Recall that Q_2, the symbol for the second quartile, is appropriate for the median. The two-sided alternative is to be $H_1:Q_2 \neq 160$ or $P(X < 160) \neq 0.5$. Set $\alpha = .05$ as the level of significance.

To test the null hypothesis, observe that the number of observations greater than 160 is $n_1 = 4$ and the number less than 160 is $n_2 = 8$. Ignore 160, since it equals the hypothesized median. Substitution of n_1 and n_2 in Equation 18-2-1 gives

$$\chi_2 = \frac{(n_1 - n_2)^2}{n_1 + n_2} = \frac{(4 - 8)^2}{4 + 8} = 1.33 \qquad 1 \text{ df}$$

Since 1.33 is less than 3.84, the tabulated χ^2 value for 1 degree of freedom and $\alpha = .05$, we have no basis for rejection of our null hypothesis that the population median is 160 in favor of an alternative calling for $Q_2 \neq 160$. In fact, the probability of observing a larger random value of χ_2 than that observed is seen to be very close to 0.25. We would not describe this sample as an unusual one if the population median is, in fact, $Q_2 = 160$.

If the hypothesized median had been 180, that is, $H_0:Q_2 = 180$, and the two-sided alternative had been $H_1:Q_2 \neq 180$, then $n_1 = 2$, $n_2 = 11$, and $\chi^2 = 6.23$ so that we would have rejected this null hypothesis, since the observed $\chi^2 = 6.23$ is greater than $\chi^2_{.05} = 3.84$. We can observe from the χ^2 table that the probability of a more extreme random value than that observed is larger than but fairly close to 0.01; compare tabulated $\chi^2 = 6.63$. We conclude that our sample could hardly have been drawn from a population where the true median is 180.

Again, suppose we have the same null hypothesis, namely, $H_0:Q_2 = 180$, but that the alternative is one-sided. In particular, suppose $H_1:Q_2 < 180$. Since 11 of the 13 observations are less than 180, this alternative must be considered seriously. On the other hand, if 11 observations had been greater than 180, we would have proceeded no further since such a result is not at all compatible with the alternative hypothesis. As before, the sample $\chi^2 = 6.23$. This value is compared with the tabulated $\chi^2_{.10}$ value for a test with $\alpha = 0.05$.

The determination of critical values of χ^2 when used to test against one- and two-sided alternatives is, perhaps, more apparent from an examination of Figure 18-2-1. It is seen that for one-sided alternatives we must think in terms of Z to establish the probability that must be used with tabulated χ^2 to assure the desired α. Perhaps Z is a more obvious test criterion when testing against one-sided alternatives. In any case, the density function on the left is not really that of χ^2 and should, perhaps, begin with a probability of .5 at $\chi^2 = 0$ to account for all z-values to the left of $z = 0$ for the one-sided alternative.

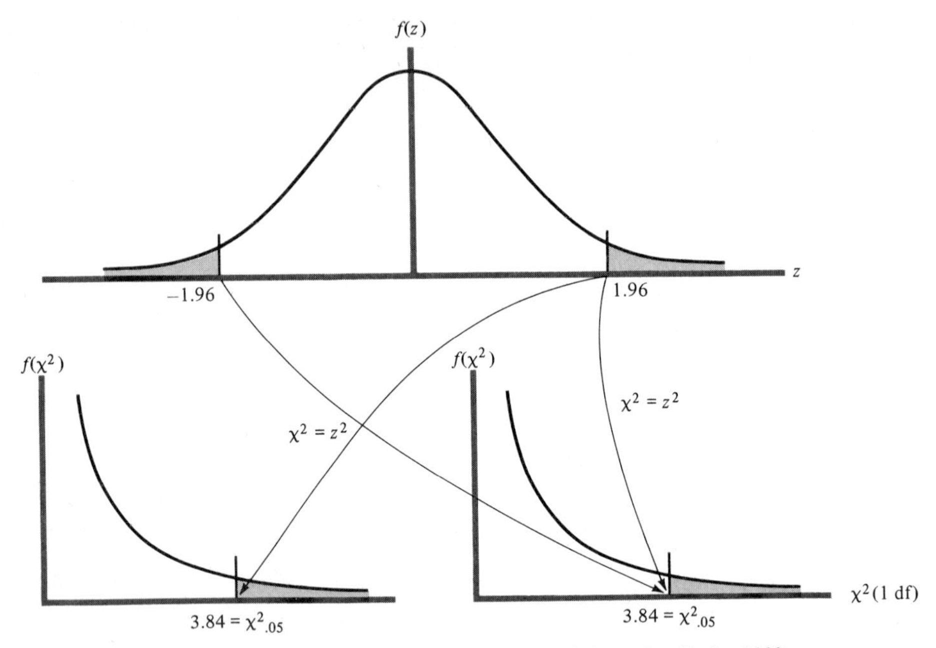

FIGURE 18-2-1 Relation Between z and χ^2 When χ^2 Is Used to Test Against One- and Two-Sided Alternatives in the Case of a Median.

EXERCISES

18-2-1 Exercise 17-2-1 gave the percent copper for 10 bronze castings. It was assumed that these observations were approximately normally distributed, and a t-test was made of the null hypothesis that $\mu = 86.20$ against two-sided alternatives, using $\alpha = .05$.

Use the sign test to test the null hypothesis that $Q_2 = 86.20$ against two-sided alternatives, using $\alpha = .05$. Remember that $\mu = Q_2$ for any symmetric distribution. What is the probability of finding a larger value of χ^2 than that observed?

18-2-2 For the grade data of Exercise 3-6-1, used with a t-test in Exercise 17-2-2, test the null hypothesis that $Q_2 = 70$ against $H_1: Q_2 > 70$, using $\alpha = .01$.

18-2-3 Exercise 17-2-6 repeats the *Hyopsodus* data of Exercise 3-6-2 and calls for the use of a t-test to test $H_0: \mu = 4.60$.

Use the sign test to test the null hypothesis that $Q_2 = 4.60$ against a two-sided alternative with $\alpha = .05$. What is the probability of observing a χ^2 value larger than that observed? What is the probability of observing not more than as many minus signs as appear? Is this probability appropriate to one- or two-sided alternatives?

18-2-4 Exercise 3-6-7 gives commodity price indices for 1964 based on 1957–1959 as 100.
(a) Meat, poultry, and fish: 98.3, 98.3, 97.2, 97.0, 96.6, 98.9, 99.2, 101.4, 100.6, 99.5, 99.0.

Test the null hypothesis that the population median is 100 against a two-tailed alternative. Use $\alpha = .05$ since the sample is small. What do you consider to be the sampled population?

(b) Dairy products: 105.0, 104.8, 104.5, 104.1, 103.9, 104.0, 104.3, 104.4, 104.6, 105.3, 105.3, 105.6.

Test the null hypothesis that the population median is still no higher than 105.0.

(c) Fruits and vegetables: 112.4, 113.7, 115.1, 115.7, 115.7, 120.2, 122.3, 117.3, 112.2, 111.7, 113.0, 114.5.

Test the null hypothesis that the population median is still 110.0 or lower.

18-3 THE WILCOXON SIGNED-RANK TEST WITH A SINGLE SAMPLE

On the very simple assumption of independent observations from distributions with a common median, the sign test was developed. Let us strengthen our assumptions somewhat and assume that we have a single sampled distribution which is symmetric. Now, positive and negative deviations from the median Q_2 will have the same probability if of the same numerical value. This is stated in Equation 18-3-1.

$$P(X - Q_2 \leq C) = P(X - Q_2 \geq C) \tag{18-3-1}$$

Consider that we are dealing with a real sample, and that we compute the deviations from the hypothesized median and rank them with respect to their magnitudes, ignoring signs. If we are sampling where the null hypothesis is true, then the sum of the ranks for deviations on the left side of the median should be about the same as the sum of the ranks for those on the right. This is, of course, introducing the signs of the deviations. The difference between these values, or either one of them since the sum of the ranks is fixed, should provide a test criterion that is more powerful than the sign test since it makes use of both the sign and the relative magnitude of each deviation. Figure 14-3-1 illustrates what might happen if an alternative hypothesis is true.

The test procedure suggested above and based on the assumption stated in Equation 18-3-1 has been formalized as the Wilcoxon signed-rank test or signed–rank-sum test. It was referred to in Section 14-3 in conjunction with the estimation of medians. The procedure for the Wilcoxon signed-rank test is as follows.

1 Compute the signed differences between the observed values and the hypothesized median, that is, the $(X_i - Q_2)$'s.
2 Rank the differences from smallest to largest without regard to sign. Ignore 0s; they are assumed to occur with zero probability. Tied ranks are given the average of the ranks which would have been assigned if there had been no ties.
3 Assign to each rank the sign of the corresponding difference.
4 Compute the sum of the ranks for the positive and for the negative differences. These are equivalent criteria since the sum of the ranks is a constant.

5 For a test against two-sided alternatives, compare the smaller rank sum with the critical value given in Table B-5, where the tabulated probability is that of a more extreme value regardless of sign.

5′ If the alternative hypothesis is one-sided, then we must first look at the sample value of the signed-rank sum and assure ourselves that the sum is in the appropriate direction to support the alternative hypothesis before using the table of critical values. If support is indicated, we enter the table with the smaller signed-rank sum; however, tabulated probability levels to be used in entering the table must be twice the desired α-level since we have had to consider the sign of the test criterion.

Table B-5 consists of critical values based on the smaller of the two rank-sum values. Consequently, values of the test criterion that are termed significant are those which are smaller than the tabulated values.

Illustration 18-3-1

We use the heart-rate data, as in Illustration 18-2-1 for the sign test, to illustrate Wilcoxon's signed-rank test. Let the hypothesized median be 160; that is, the null hypothesis is $H_0\!:Q_2 = 160$. The signed deviations of the observations from 160, with the ranks of their absolute values in parentheses, are (arranged sequentially for convenience):

−57(12) −40(11) −37(9) −34(8) −25(7) −22(6) −20(5)
−10(2.5) 0(ignore) +8(1) +10(2.5) +16(4) +38(10)

The sum of the negative ranks is 60.5 and that of the positive ranks is 17.5. Now $n = 12$ since one value equaled the hypothesized median and was ignored.

Suppose the alternative hypothesis is two-sided; that is, $H_1\!:Q_2 \neq 160$. Since the smaller observed rank sum, 17.5, is greater than 13.8, the critical value given in Table B-5 for a two-sided $\alpha = .05$ rejection level when $n = 12$, there is insufficient evidence to reject the null hypothesis.

On the other hand, if the alternative hypothesis is $H_1\!:Q_2 < 160$, a one-sided test is in order and the critical value for a rejection level of α is found under the probability 2α. For a 5 percent α-level, the critical value is 22, found under 10 percent for $n = 12$. If Q_2 is, in fact, less than 160, we have computed a signed-rank sum with a false Q_2 equal to 160 that is to the right of the true value. We will expect an excess of negative ranks if this alternative is true. Since the sum of the negative ranks is 60.5, that is, in a direction tending to support the alternative hypothesis, we proceed with the test. Since $17.5 < 22$, rejection of the null hypothesis and acceptance of the alternative hypothesis that Q_2 is less than 160 are indicated.

EXERCISES

18-3-1 In Exercise 17-2-1, data were presented for 10 bronze castings and the null hypothesis $H_0: Q_2 = 86.20$ was tested using the t-test. The data were used again in Exercise 18-2-1 where a sign test was carried out. Use Wilcoxon's signed-rank test to test the same hypothesis. Compare your result with those obtained in Exercises 17-2-1 and 18-2-1.
18-3-2 For the grade data of Exercise 3-6-1, use Wilcoxon's signed-rank test to test the null hypothesis $H_0: Q_2 = 70$ against the alternative $H_1: Q_2 > 70$, using $\alpha = .01$. Compare the result of this test with those given for the t-test and sign test obtained in Exercises 17-2-2 and 18-2-2.
18-3-3 For the *Hyopsodus* data given in Exercise 17-2-6, use Wilcoxon's signed-rank test to test the null hypothesis $H_0: Q_2 = 4.60$. Use a two-sided alternative and $\alpha = .05$. Compare your result with those obtained for the t-test and sign test used in Exercises 17-2-6 and 18-2-3.
18-3-4 Exercise 18-2-4 gives commodity price indices for three commodities. Use Wilcoxon's signed-rank test to test the null hypotheses proposed in that exercise. Use the same alternative hypotheses and significance levels as for sign test and compare your results.

The final exercise is concerned directly with the test criterion for a particular sample size rather than with its application to a particular set of data.

18-3-5 Consider a problem where the sample size is to be $n = 4$. It is planned to use a signed-rank test for the hypothesis to be tested. Will it be possible to get a test-criterion value equal to 0? A value of -1? Of -2? Is it possible for any test-criterion value to be generated by more than one sample outcome? What will be the negative value farthest from 0? Prepare a table showing all possible combinations of ranks which may be negative. What will be the sum of each set of negative ranks? Find the distribution of the negative rank sum. Will a sample of size $n = 4$ ever allow us to declare a result to be significant for $\alpha = .05$?

18-4 THE SIGN TEST: COMPARISON OF TWO INDEPENDENT SAMPLES

In Section 17-4, the use of the t-test to compare two sample means to see whether or not they might have come from the same population was discussed. Strong assumptions concerning the sampled population were made, assumptions of normality and homogeneity of variance, so that the only remaining possible difference would be the population means. Such strong assumptions are not always justified. However, a two-sample sign test can be used with the minimum assumptions that the random components of the observations are independent and each is from a continuous distribution with median 0. The observations of any sample are, of course, assumed to come from distributions with a common median.

For the two-sample case where the null hypothesis is that they have a common population median, the test procedure is as follows.

1 Rank the two samples together from smallest to largest. Tied observations are given the average of the ranks which would have been assigned if there had been no ties.

2 Find the median of all the observations.
3 For each sample, count the number of observations above and below the median and arrange in a 2 × 2 contingency table. If $n_{1.}$ and $n_{2.}$ are the two sample sizes and $n_{1.} + n_{2.}$ is odd, the number above the median is $(n_{1.} + n_{2.} + 1)/2$; and if $n_1 + n_2$ is even, the number above the median is $(n_{1.} + n_{2.})/2$.
4 Use χ^2 as discussed in Section 16-2 for the test criterion if samples are of size 10 or above. Equation 18-4-1, originally Equation 16-2-4, is convenient. Otherwise, use the procedure based on exact probabilities, given later in this section.

$$\chi^2 = \frac{(n_{11}n_{22} - n_{12}n_{21})^2 n}{n_{1.}n_{2.}n_{.1}n_{.2}} \qquad 1 \text{ df} \qquad (18\text{-}4\text{-}1)$$

Recall that n is the total number of observations.

Illustration 18-4-1

Table 17-4-1 consists of data from two independent samples for two species of *Hyopsodus*. The ranked data follow with species A data underlined and the median indicated by a vertical line.

3.29 3.38 3.64 3.74 3.78 3.81 3.88 4.01 4.07 4.13 4.18 |
4.22 4.29 4.33 4.39 4.42 4.69 4.74 4.85 4.87 5.08 5.22

The contingency table called for by the test is given below.

Notation

Species	Above	Below	Total
A	n_{11}	n_{12}	$n_{1.}$
B	n_{21}	n_{22}	$n_{2.}$
	$n_{.1}$	$n_{.2}$	$n_{..} = n$

Data

Species	Above	Below	Total
A	2	10	12
B	9	1	10
	11	11	22

The sample χ^2-value, using Equation 18-4-1, is then

$$\frac{[(2 \times 1) - (10 \times 9)]^2 22}{12(10)(11)(11)} = 11.46 \qquad 1 \text{ df}$$

Since χ^2 is greater than tabulated $\chi^2 = 3.84$ for 1 degree of freedom, and $\alpha = .05$ (Table B-3), we have sufficient evidence to reject our null hypothesis of a common median and to accept the two-sided alternative that these species do not have a common median. Observation of the data indicates that the population median for species B must lie to the right of that for species A.

If our alternative hypothesis had been that the median for length of the lower molar in species A was less than that in species B, a rejection level of $\alpha = .05$ would call for a critical value of $\chi^2 = 2.71$ with 1 degree of freedom, read under $P = .10$ in Table B-3. For these data, species B observations tend to be to the right of those for species A, as expected under the alternative hypothesis, so we compare the observed $\chi^2 = 11.46$ with the tabulated $\chi^2 = 2.71$ and again reject the null hypothesis.

When the numbers in a 2 × 2 table are small, for example, if the sample sizes are no greater than 10, it may be best to use *the exact probability method.* For this method, we compute the probability of a result or table as extreme or more so than that observed when all marginal totals are fixed. For the *Hyopsodus* data, the only tables that qualify are those below.

Observed

Species	Above	Below	Total
A	2	10	12
B	9	1	10
	11	11	22

More Extreme

Species	Above	Below	Total
A	1	11	12
B	10	0	10
	11	11	22

The probability associated with any 2 × 2 table with fixed marginal totals can be computed by Equation 18-4-2.

$$P = \frac{n_{1.}!\,n_{2.}!\,n_{.1}!\,n_{.2}!}{n_{11}!\,n_{12}!\,n_{21}!\,n_{22}!\,n!} \qquad (18\text{-}4\text{-}2)$$

Illustration 18-4-2

For the *Hyopsodus* data just used, the probabilities follow:

$$P(\text{observed table}) = \frac{12!\;10!\;11!\;11!}{2!\;10!\;9!\;1!\;22!} = .000936$$

$$P(\text{more extreme table}) = \frac{12!\;10!\;11!\;11!}{1!\;11!\;10!\;0!\;22!} = .000017$$

Hence, the probability of a result at least as extreme as that observed is .000953.

Notice that extremeness has been defined as being in the direction observed. Clearly this test is against a one-sided alternative in that direction. If this direction had not been foreseen, then the test would be result-guided and the computed probability unreliable.

For a test against a two-sided alternative hypothesis, we must include in our considerations those tables that deviate similarly but in the opposite direction to those already involved in our computations. Since the same numbers will be involved, the appropriate probability of a result at least as extreme as that observed, regardless of direction, is simply 2(.000953) = .001906. In the case of either alternative, we must reject the null hypothesis as an unacceptable explanation of the data. In other words, the evidence is that the two populations are not identically located.

EXERCISES

18-4-1 Using the tire-life data of Exercise 17-4-1, test the null hypothesis that the sampled populations have a common median against a two-sided alternative. As test criteria, use χ^2 and the exact probability procedure. In both cases, try to evaluate the probability of a more extreme value than obtained. Compare all results, including that obtained in Exercise 17-4-1.

18-4-2 Using the battery data of Exercise 17-4-3, test the null hypothesis that the sampled populations have a common median against a two-sided alternative. As test criteria, use χ^2 and the exact probability method. In both cases, try to evaluate the probability of a more extreme value than obtained. Compare all results, including that obtained in Exercise 17-4-3.

18-4-3 For the *Hyopsodus* data used in Illustration 18-4-1, compute probabilities for each of the possible 2 × 2 tables with marginal totals fixed as indicated. Show that the sum of these probabilities is 1.

18-5 THE WILCOXON RANK-SUM TEST FOR COMPARING TWO POPULATIONS

Suppose we have two populations with the median of the first lying to the left of that of the second. If we draw a random sample of the same size from each and rank the two together, then the sample from the first population will tend to have the smaller ranks and the smaller rank sum. Wilcoxon (1964) used this idea to propose a rank-sum test.

It is also clear that it would be valid to pair observations randomly when from independent random samples, observe the signs of the differences, and use a sign test. Clearly this would be less efficient than if we had meaningful pairing. On the other hand, one could suggest that each observation in one sample be compared with every one in the other and that only the sign of the difference be considered. The signs would not, of course, be independent, and this would have to be taken into account. Mann and Whitney (1947) proposed such a test with assumptions the same as those proposed by Wilcoxon and, in addition, allowed for unequal sample sizes.

Wilcoxon's test uses ranks, whereas the Mann-Whitney test makes use of a count. The test statistics are different but the tests are equivalent. We present Wilcoxon's test using ranks.

The Wilcoxon rank-sum test may be considered as a test of the null hypothesis that two populations are identical, or, if we are prepared to assume that the populations are the same except for location, a test that they have the same median or other location parameter. The assumptions are that the observations are independent and from a continuous distribution. Under these assumptions, all permutations of the observations are equally likely when the null hypothesis is true. Consequently, it is easy to construct tables of critical values.

The Wilcoxon rank-sum test is carried out as follows:

1 Rank the observations of both samples together, from smallest to largest. Tied observations are given the average of the ranks which would have been assigned if there had been no ties.
2 When $n_1 \neq n_2$, add the ranks of the sample with the smaller n, say n_1. Call this T. Compute $T' = n_1(n_1 + n_2 + 1) - T$, the value that would have been obtained if the observations had been ranked from largest to smallest.
3 Compare T or T', whichever is smaller, with the critical values given in Table B-6. Notice that small values of the test criterion are evidence to reject the null hypothesis.

Tabulated critical values are for several probability levels and for two-sided tests. For one-sided tests at an α-level of significance, read critical values under a tabulated probability of 2α.

When $n_1 = n_2$, compare the first T computed with the critical values in Table B-6. If it is readily apparent that this sample sum is the smaller one, some computation may be saved.

Illustration 18-5-1

Again we use the two independent samples of *Hyopsodus* data given in Table 17-4-1 to illustrate the rank-sum test. We order and rank the observations as follows:

3.29(A), 1	3.38(A), 2	3.64(A), 3	3.74(A), 4
3.78(A), 5	3.81(A), 6	3.88(A), 7	4.01(B), 8
4.07(A), 9	4.13(A), 10	4.18(A), 11	4.22(A), 12
4.29(B), 13	4.33(A), 14	4.39(B), 15	4.42(B), 16
4.69(B), 17	4.74(B), 18	4.85(B), 19	4.87(B), 20
5.08(B), 21	5.22(B), 22		

The sum of the ranks for the smaller sample, sample B with 10 observations, is $T = 169$. $T' = 10(10 + 12 + 1) - 169 = 61$.

Since T' is less than 76, the tabulated 1 percent critical value for the two-sided test, we reject the null hypothesis and accept the alternative that the true difference in mean first molar length for these species of *Hyopsodus* is not 0.

If the alternative hypothesis had been one-sided, for example, if the length of the first molar in species B lay to the right of that for species A, the probability levels given in Table B-6 would be double the significance level.

The sum of n integers, $1[1]n$, is $n(n + 1)/2$, a fact used in developing the formula for T'. It also shows that the sum of the rank sums for the two samples equals $(n_1 + n_2)(n_1 + n_2 + 1)/2$. For our data, the rank sum for A is 84 and the rank sum for B is 169. Also $(n_1 + n_2)(n_1 + n_2 + 1)/2 = 22(23)/2 = 253 = 84 + 169$. This gives a partial check on our arithmetic.

If the tables of critical values are inadequate, we make use of the fact that we are sampling a known population of ranks and so can compute μ_T and σ_T, the mean and standard deviation of Wilcoxon's T. These are given in Equations 18-5-1 and 18-5-2.

$$\mu_T = \frac{n_1(n_1 + n_2 + 1)}{2} \tag{18-5-1}$$

$$\sigma_T = \sqrt{\frac{n_1 n_2(n_1 + n_2 + 1)}{12}} \tag{18-5-2}$$

For a sufficiently large sample size, we now assume T to be approximately normally distributed with the above mean and standard deviation, and compute a value of the normal random variable Z by Equation 18-5-3.

$$Z = \frac{T - \mu_T}{\sigma_T} \tag{18-5-3}$$

Table B-1 may now be used to judge significance.

EXERCISES

18-5-1 Using the tire-life data of Exercise 17-4-1, test the null hypothesis that the sampled populations have a common median. Let the alternative hypothesis be two-sided, set $\alpha = .05$, and use Wilcoxon's rank-sum test. Compare your conclusions with those obtained in Exercise 18-4-1.

18-5-2 Using the battery data of Exercise 17-4-3, test the null hypothesis that the sampled populations have a common median. Let the alternative hypothesis be two-sided, set $\alpha = .05$, and use Wilcoxon's rank-sum test. Compare your conclusions with those obtained in Exercise 18-4-2.

The next exercise is concerned directly, rather than through data, with the test criterion. However, it makes use of material covered in Chapter 7.

18-5-3 Suppose we are considering a problem where $n_1 = 2$ and $n_2 = 3$. We will be using independent samples and Wilcoxon's rank-sum test. The assumptions imply that all permutations of the ranks are equally likely when the null hypothesis is true. Consequently, we are able to generate our own frequency table for T.

In how many ways can we draw $n_1 = 2$ ranks for the criterion T? What is the least possible value that T may take? The greatest? Prepare a frequency table for T. Is it a symmetric distribution?

Wilcoxon's rank-sum test also calls for us to compute T'. Do this for all values of T.

Since the usual table of critical values supposes a two-tailed alternative hypothesis, only the lesser of T and T' is used as a test criterion. Prepare a frequency table of values of the test criterion. Will it be possible to carry out any test at a reasonable value of α?

This final exercise considers the relationship between the Wilcoxon and Mann-Whitney tests, by example.

18-5-4 In Exercise 18-5-3, you needed all possible permutations of the ranks to obtain values of T and, consequently, a frequency table for T.

Each permutation of ranks comes from a permutation of the observations which we can write as $X_{(1)}, \ldots, X_{(5)}$. By underlining all possible pairs of observations as the X's of the first sample with $n_1 = 2$, it is easy to visualize all the permutations and to count, for each, the number of times the smaller of the two observations exceeds an observation of the second sample and to count the number of times the larger exceeds an observation of the second sample. The sum of these two values is the Mann-Whitney test criterion as described in this section. Show the 1:1 relationship between the Wilcoxon and the Mann-Whitney criteria for this case.

18-6 THE SIGN TEST: MEANINGFULLY PAIRED OBSERVATIONS

Section 17-5 contains procedures for testing a null hypothesis concerning the location of the population mean of differences between meaningfully paired observations. Now consider similar data but make only weak assumptions. In particular, for each difference, $Y_{1j} - Y_{2j}$, we assume only that it has a probability of .5 of being less than the median and the same probability of being greater. Nothing is said about the distribution of differences as they move away from the median, and the distribution may change from trial to trial.

In this case of meaningfully paired observations, we have what appears to be a two-sample problem. However, by taking differences, we have reduced it to a one-sample problem. When the assumptions of the preceding paragraph are valid, we may test a hypothesis that the median of the differences is any particular value, using the sign test of Section 18-2.

The sign-test procedure is as follows:

1 Determine the sign of the differences between members of each pair, $Y_{1j} - Y_{2j}$. Ignore 0s. The minimum assumptions really include $P[(Y_{1j} - Y_{2j}) = Q_2] = 0$, because we are assuming that a continuous distribution is involved and so do not associate probabilities with points but only with sets of points. In practice such events do occur.

2 Count the number of pluses and of minuses, say n_1 and n_2, and use the χ^2 test criterion to test the null hypothesis that the differences have a median of 0, that is, that pluses and minuses occur with equal frequency in the population of signed differences.

3 Compare the sample χ^2 with the tabulated values of χ^2 for 1 degree of freedom given in Table B-3. The test may be two-sided or one-sided depending upon the alternative hypothesis. Equation 18-6-1 may be used for computing the sample χ^2.

$$\chi^2 = \frac{(n_1 - n_2)^2}{n_1 + n_2} \quad 1 \text{ df} \tag{18-6-1}$$

Illustration 18-6-1

Table 17-5-1 gives the signed differences of the change in triglyceride content of the blood lipids of 11 diabetic adolescent girls. The null hypothesis is that the population of such differences has a median $Q_2 = 0$. That is, we have a binomial distribution of signs and are hypothesizing that $P(+) = .5 = P(-)$.

The test criterion is χ^2 with 1 degree of freedom, computed by Equation 18-6-1 to test the null hypothesis that the parameter p equals .5. We find $n_1 = 1$ and $n_2 = 10$.

$$\chi^2 = \frac{(n_1 - n_2)^2}{n_1 + n_2} = \frac{(1 - 10)^2}{1 + 10} = 7.36 \quad 1 \text{ df}$$

The tabulated χ^2 with 1 degree of freedom for $\alpha = .05$ when testing against two-sided alternatives is 3.84; for $\alpha = .05$ when testing against one-sided alternatives, tabulated χ^2 is read under $P = .10$ as 2.71. The sample χ^2 is compared with this value only if the response is in the direction called for by the alternative.

In testing the null hypothesis of a common median against the two-sided alternative, we observe that $5.82 > 3.84$ and reject the null hypothesis that physical-fitness training has no effect on the triglyceride content of diabetic adolescent girls and accept the alternative hypothesis that there was an effect. If our alternative hypothesis had been one-sided, specifying that physical-fitness training increased the triglyceride content of diabetic adolescent girls and so moved that median to the right, then, since $5.82 > 2.71$, the evidence would support acceptance of this alternative.

EXERCISES

18-6-1 Exercise 17-5-1 contains paired observations on crushing strengths for two zones in eight cottonwood logs. Test the null hypothesis of no difference between population medians for the two log zones against the alternative that there is a difference. Use $\alpha = .05$ and the test criterion χ^2. Since $n = 8$, use the binomial distribution to find the probability of a result as extreme as or more so than that observed. Compare the two results.

18-6-2 Numbers of seeds set for top and bottom flowers of 10 lucerne or alfalfa plants are given in Exercise 17-5-2. Test the null hypothesis that population medians for top and bottom flowers are the same. Use $\alpha = .01$ and the alternative that top and bottom flowers set are not the same. Use $\alpha = .01$ and the alternative that top flowers set more seed. As test criterion, use χ^2. Also compute the probability of a result as extreme or more so than that obtained, using the binomial distribution.
18-6-3 Average acre catch of muskrats at 12 locations is given for two different years in Exercise 17-5-3. Test the null hypothesis of no difference between population medians for the two years against a two-sided alternative, using $\alpha = .10$. As test criterion, use χ^2. Compute the probability of a result as extreme as or more so than that obtained, using the binomial distribution.
18-6-4 For the data of Exercise 17-5-4, carry out a χ^2 test for the same null hypothesis. Use the same alternatives and α-values. Use the binomial distribution to compute just how extreme the observed result was.
18-6-5 For the data of Exercise 17-5-5, carry out a χ^2 test of the same null hypothesis. Use the same alternatives and α-value. Use the binomial distribution to compute just how extreme the observed result was.

18-7 THE WILCOXON SIGNED-RANK TEST: PAIRED OBSERVATIONS

The assumptions here are those discussed in Section 18-3, primarily that the distribution of differences is symmetric. This says more about the distribution than is required for the sign test but much less than is required for a t-test. The resulting test is generally considered to be a test of location.

The Wilcoxon signed-rank test procedure is as follows:

1 Determine the signed difference between the members of each pair, $Y_{1j} - Y_{2j}$.
2 Rank the differences from smallest to largest without regard to sign. Tied differences are given the average of the ranks which would have been assigned if there had been no ties. Ignore 0s.
3 Assign to each rank the sign of the original difference.
4 Compute the sum of the positive or negative ranks, whichever is the smaller.
5 Compare the sum from step 4 with the critical value given in Table B-5.

Illustration 18-7-1

For the adolescent girl data in Table 17-5-1, the sum of the positive ranks, the smaller of the two sums, is +5. This is compared with the critical values 11 and 5 for $\alpha = .05$ and $\alpha = .01$, respectively, from Table B-5. These values are for testing against two-sided alternatives. The corresponding values for testing against one-sided alternatives are 14 and 7. Since small values of the test criterion have a lower probability of occurring due to chance when the null hypothesis is true, we accept the alternative hypothesis whether it be two-sided or one-sided.

EXERCISES

18-7-1 The data of Exercise 17-5-1, used again in Exercise 18-6-1, are appropriate for illustrating the test procedure of this section. Test the null hypothesis proposed and compare all results obtained to date.

18-7-2 See Exercise 18-6-2 for data and repeat Exercise 18-7-1.

18-7-3 See Exercise 18-6-3 for data and repeat Exercise 18-7-1.

18-7-4 See Exercise 18-6-4 for data and repeat Exercise 18-7-1.

Chapter 19

THE ANALYSIS OF VARIANCE

19-1 INTRODUCTION

In Section 17-3, we considered the problem of designing a sampling scheme or assigning treatments in an experiment intended to help determine whether the difference between two sample means, measures of location, could be attributed to chance. For this consideration, we made strong assumptions concerning a parent population. In Chapter 19, we generalize the procedures of Chapter 17 to differences among two or more sample means.

The procedure used is called the *analysis of variance* or, simply, the *anova.* It was developed by Sir Ronald A. Fisher, and the mechanics are essentially an arithmetic process for partitioning the total sum of squares for a set of data into components associated with nominal sources of variation.

In this chapter, we discuss the analysis of variance and models for experiments where the experimental design of sampling plan is analogous to the unpaired and to the paired sample situations of Chapter 17. Following these sections, there is a discussion of error-rate definitions, multiple-comparisons procedures, and generalizations or modifications of the two experimental designs. These latter sections should be read, at least, for their general information about statistical techniques.

19-2 THE COMPLETELY RANDOM DESIGN OR ONE-WAY CLASSIFICATION

Suppose we wish to determine if four different methods of teaching a subject produce real differences in test performance. Many classes are available, and methods are assigned at random to classes, probably with equal numbers of classes for each method. Grades from appropriate tests are to be the basis of our conclusions. Alternatively, we may want to determine if any real differences exist in the life-spans of three makes of television tubes, based on the average life in hours of a random sample of 10 tubes of each make.

It is seen, then, that we intend to discuss larger and more comprehensive experiments than in the earlier chapters. Some terms, for example, "treatment" and "experimental unit," were used there without formal definition when there seemed little chance of misunderstanding. We now define these terms. The term "random experiment" was defined in Section 5-2 and this definition should be recalled.

In experimentation, it is customary to talk about treatments, regardless of whether or not anything was done to any experimental material by the experimenter. Thus, the teaching methods referred to above would be four treatments where the experimenter did something to the experimental material, the students, or had something done. The makes of television tubes would be three treatments where the experimenter did nothing but where something was clearly done. Again, if we randomly select 100 men and 100 women and ask them their opinions about flying saucers, we have obviously done nothing that could be called a treatment; but, we say we have two treatments, namely, male and female.

The term treatment, then, is intended to be used in a broad sense, as is the term experiment. In all cases, different recognizable procedures have been followed to provide nominal groups, either with or without our help, and the experiment is intended to evaluate any differential effects among these treatments. Briefly, we might say the following:

A *treatment* is a procedure whose effect is to be measured and compared with others.

A *response* to a treatment is measured on a unit of experimental material. These experimental units have been assigned at random to treatments, or the treatments to the units, and the measurements are used in the calculations by which comparisons of treatment effects are made. Deciding upon what constitutes an experimental unit requires thoughtful consideration.

In the case of a class of students taught by one method calling for an instructor, the treatment was probably assigned to the class. This means that each student in the class has been subject to more than simply a teaching method. Like everyone else in the class, he has also had the same instructor at the same time as everyone else. All have been treated identically in too many ways for each to be considered as an experimental unit with its own individually assigned application of a treatment that is simply a method. Each true experimental unit must receive a different application of the same treatment by a randomization that deals with it as an individual. We will, of course, measure the performance of each student, but we must obtain a class average or total as a single response to the treatment.

If we get a carton of television tubes, then the carton is the experimental unit because this is the unit selected at random. Very possibly these tubes came off the production line in sequence and, consequently, may vary less than if chosen at random. To say that each is an experimental unit is to say that differences among them satisfactorily measure the true variability that exists in the population, a variability that we have agreed can be measured only by means of a random sample. The carton must provide a single measurement even though we make a measurement on each tube in the process of getting it. The randomly selected carton will not provide an appropriate sample of 10 randomly selected tubes.

Finally, each individual who responds to our question is an experimental unit because each was subject to his or her own individual influences before he or she provided us with a response relative to flying saucers.

An *experimental unit* is an amount of *material* to which a treatment is applied at random in a single trial.

The term material is intended to be a very general one, as was the term treatment.

Suppose that for the experiment in teaching or in manufacturing, all the experimental units are expected to perform pretty much alike if treated alike. In the case of the student classes, all have attained the same scholastic level. We are unable to say that some are in private while others are in public schools or that there are any other recognizable sources that might make some units perform differently from others. For the television tubes, all are intended to serve the same function in the television set and, so, should be interchangeable. The manufacturing process or components have not been changed at any stage within any company. If there are such differences, they are from company to company but not within. In other words, no existing sources of variation over and above the natural variation can be recognized in the experimental units of either experiment.

In the conduct of the experiment, treatments are assigned at random to the experimental units or the units to the treatments. This was made clear for the teaching experiment. For the industrial experiment, a random sample is drawn from each treatment population, that is, from the tubes produced by each company. We may say then that there are essentially no restrictions on this randomization procedure and that we have a completely random design. Each observation can be labeled only by the treatment received by the experimental unit, and the data can be presented only in a one-way classification.

Prior to the application of the treatments, potential observations could be considered to be coming from a single population and described as consisting of a mean and a random component as was done in Section 17-2. The mean is a fixed parameter and we need to make some assumptions about the random components. However, there have been treatments. Any one of these may raise or lower the mean of the observations concerned but has no effect on the random component.

For example, it may be that in the population of classes to which our teaching methods might be applied, the true class average is 75. Individual class averages differ from this value. Suppose the three teaching methods have real effects of -5 for method A, $+5$ for method B, and $+9$ for method C. The resulting true means are 70, 80, and 84. We cannot know these because of random sampling fluctuations but can only try to estimate them.

Our experiment is appropriate to estimate the final population means, but not the contributions due to methods. To begin with, we have no way to estimate the original mean from data with method effects added; some external information would be required. We are, however, able to estimate the method means and differences among them, and we do so. Finally, our experiment can also provide a measure of variability in the standard deviation.

Notice that variability can be measured only if we have several observations on a treatment. This repetition of a treatment is called *replication* and is a most important aspect of statistics.

Much of the above discussion of a completely random design is stated concisely by Equation 19-2-1, a generalization of Equation 17-4-1.

$$Y_{ij} = \mu + \tau_i + \varepsilon_{ij} \qquad i = 1, \ldots, t; j = 1, \ldots, n_i \tag{19-2-1}$$

where Y_{ij} = observed value
μ = general mean
τ_i = a contribution due to the ith of t treatments or associated with the ith population sampled and called tau-sub-i
ε_{ij} = random component associated with the jth of n_i observations on that treatment.

This is a linear additive model. Since we cannot estimate μ from the experiment alone, we also cannot estimate the τ's. Consequently, we agree to consider the τ's as deviations from some value such that they sum to 0. In a sense, this defines or redefines μ and allows us to proceed.

The assumptions to complete the model are the usual strong ones for the random components and are necessary for tests, customarily based on an underlying normal distribution, to be valid. Assumptions for the treatment effects vary.

The following assumptions are made in conjunction with Equation 19-2-1 for data from a completely random design and for a *fixed effects linear additive model*.

1. μ is an unknown, fixed parameter.
2. The τ_i's are unknown parameters with $\sum \tau_i = 0$.
3. The ε_{ij}'s are independent and from a normal distribution with mean equal to zero and unknown variance equal to σ^2.

When the fixed-effects-model assumptions apply, the treatments or populations sampled constitute the whole set of interest. Inferences apply only to this specific set, and if the experiment or sampling were repeated, the new experiment would involve precisely the same set of treatments. Its purpose would be to estimate true treatment means or differences among them. If we had proposed μ_i as a symbol for the mean of the ith population, as in Equation 17-4-1, it would now be convenient to define μ by Equation 19-2-2, which is essentially what has been done.

$$\mu = \frac{1}{t} \sum \mu_i \qquad (19\text{-}2\text{-}2)$$

Equations 19-2-3 now hold.

$$\mu_i = \mu + \tau_i \qquad \tau_i = \mu_i - \mu \qquad \sum \tau_i = 0 \qquad (19\text{-}2\text{-}3)$$

Sometimes a random-effects model is appropriate. Here, treatment effects are a random sample from a population of treatment effects with zero mean and unknown variance. Repetition of the experiment would involve a new set of treatments. The τ's are no longer parameters; now the parameter of interest is their unknown variance, say σ_τ^2. We will not be concerned with this model.

Finally, the random component measures the failure of the data to conform to a model that calls for a general mean and a treatment effect only. It is measured as a deviation by Equations 19-2-4.

$$\begin{aligned} \varepsilon_{ij} &= Y_{ij} - \mu_i \\ &= Y_{ij} - \mu - \tau_i \end{aligned} \qquad (19\text{-}2\text{-}4)$$

The variance in the population of ε's is generally symbolized by σ^2 or, if necessary, by σ_ε^2.

A comparison of the features of the two models is presented in concise form in Table 19-2-1.

TABLE 19-2-1 / A Comparison of the Fixed and Random Models

	Model	
Contribution	Fixed	Random
μ	A fixed value	A fixed value
τ's	Parameters; fixed in repeated experiments	Parameter is σ_τ^2; vary in repeated experiments but are from one population
ε's	Sample from a single population with $\mu = 0$ and variance σ^2	Sample from a single population with $\mu = 0$ and variance σ^2
Parameters involved	μ, $\{\tau_i\}$, σ^2	μ, σ_τ^2, σ^2
Inferences to be made	About the treatments used	Generalization to a treatment population

19-3 ANALYSIS OF VARIANCE: COMPLETELY RANDOM DESIGN

Consider t independent sets of n observations each, either samples from nominally different populations or samples that have received different treatments. These are represented symbolically at the top of Table 19-3-1, a convenient form for the necessary computations. Capital Y's would be used if we were thinking in terms of random variables rather than observations at hand.

A realistic set of such data is given in Table 19-3-2 and will be used to illustrate the anova. In this experiment, three groups, each of 10 university students, received a different type of training. The response measured was the reaction time to perform a standardized task. The three treatments will be compared.

To review the notation, recall that the first subscript of y_{ij} denotes the treatment or population while the second gives the relatively unimportant information as to the position of the observation in the random selection. When a subscript is replaced by a dot, summation has taken place over that subscript; thus, $y_{1.}$ is the total of the observations in the first treatment while $y_{..}$ is the grand total. If squaring an observation is required, we write y_{ij}^2; if these are to be summed, a $\sum$ is required and we write $\sum_j y_{ij}^2$ for the sum of squares of the observations on the ith treatment and $\sum_{i,j} y_{ij}^2$ for the sum of squares of all the observations. If a total is to be squared, we may write $y_{i.}^2$ or $(\sum_j y_{ij})^2$ for the square of the sum of the observations on the ith treatment. In particular, if a dot and a square are both present, then the dot or summation operation is the first to be carried out. Thus, if we want a sum of squared totals, we use a dot for the total and a $\sum$ for the sum; for example, $\sum_i y_{i.}^2$ is a sum of squared treatment totals. If we want a sum of squared numbers, we must use a $\sum$-notation. Finally, we use SS for a sum of

TABLE 19-3-1 / Symbolic Presentation of n Observations on Each of t Treatments or of t Samples of n Observations

	Treatment = Sample						
	1	2	$\cdots$	i	$\cdots$	t	Pooled Data
	y_{11}	y_{21}		y_{i1}		y_{t1}	
	y_{12}	y_{22}		y_{i2}		y_{t2}	
	$\cdots$	$\cdots$	$\cdots$	$\cdots$	$\cdots$	$\cdots$	
	y_{1j}	y_{2j}		y_{ij}		y_{tj}	
	$\cdots$	$\cdots$	$\cdots$	$\cdots$	$\cdots$	$\cdots$	
	y_{1n}	y_{2n}		y_{in}		y_{tn}	
Sum	$T_1 = y_{1.}$	$T_2 = y_{2.}$		$T_i = y_{i.}$		$T_t = y_{t.}$	$T = y_{..}$
Mean	$\overline{T}_1 = \bar{y}_{1.}$	$\overline{T}_2 = \bar{y}_{2.}$		$\overline{T}_i = \bar{y}_{i.}$		$\overline{T}_t = \bar{y}_{t.}$	$\bar{y} = \bar{y}_{..}$
Unadjusted sums of squares	$\sum_j y_{1j}^2$	$\sum_j y_{2j}^2$		$\sum_j y_{ij}^2$		$\sum_j y_{tj}^2$	$\sum_i \sum_j y_{ij}^2$
Correction terms	$\frac{T_1^2}{n} = \frac{y_{1.}^2}{n}$	$\frac{T_2^2}{n} = \frac{y_{2.}^2}{n}$		$\frac{T_i^2}{n} = \frac{y_{i.}^2}{n}$		$\frac{T_t^2}{n} = \frac{y_{t.}^2}{n}$	$\frac{T^2}{tn} = \frac{y_{..}^2}{tn}$
Adjusted within-treatment or error sums of squares	$SS_1 = \sum_j y_{1j}^2 - \frac{y_{1.}^2}{n}$	$SS_2 = \sum_j y_{2j}^2 - \frac{y_{2.}^2}{n}$		$SS_i = \sum_j y_{ij}^2 - \frac{y_{i.}^2}{n}$		$SS_t = \sum_j y_{tj}^2 - \frac{y_{t.}^2}{n}$	Error SS $= \sum_{i,j} y_{ij}^2 - \frac{\sum_i y_{i.}^2}{n}$

TABLE 19-3-2 / Reaction Time in Minutes of Three Random Groups of Male University Students Required to Perform a Standardized Task after Receiving Different Types of Training

	Type 1	Type 2	Type 3	Pooled Data
	21	15	17	
	19	18	13	
	23	21	18	
	20	23	21	
	17	17	17	
	22	14	15	
	22	24	16	
	27	14	20	
	20	16	20	
	19	18	15	
Sum = $y_{i.} = T_i$	210	180	174	564
Mean = $\bar{y}_{i.} = \bar{T}_i$	21.0	18.0	17.4	18.8
Unadjusted sums of squares = $\sum_j y_{ij}$	4,478	3,356	3,102	10,936
$C = \frac{y_{i.}^2}{n} = \frac{T_i^2}{n}$	4,410.0	3,240.0	3,027.6	10,603.2
Within-treatments or error SS	68.0	116.0	74.4	258.4

squares of deviations from a mean. The last lines of Tables 19-3-1 and 19-3-2, then, contain within-samples SS or within-treatments SS.

We also use capital letters for totals, with T implying treatment; thus, T_i represents the total for the ith treatment, so that $T_i = y_{i.}$. This alternative notation may be helpful for some students but cannot be used to indicate all the necessary computations.

From Table 19-3-1 and Equation 19-2-1, we see that the observations on any treatment vary only in their random components. Consequently, such observations can provide us with an estimate of σ^2. These estimates are pooled over treatments or samples, and so we use all the direct information about σ^2. The resulting within-sample variance is a "pure error" because it is among observations treated alike and provides us with a standard or yardstick, namely, a standard deviation, for deciding whether observed treatment differences are attributable to chance or not.

Also from Table 19-3-1 and Equation 19-2-1, we see that sample treatment means consist of a population treatment mean plus a random component; for example, $\bar{y}_{i.} = \mu + \tau_i + \bar{\varepsilon}_{i.}$. Now, if the null hypothesis of no treatment effects is true, then variation among these treatment means can also be used to estimate σ^2 because, again, the only variation is associated with their random components, the $\bar{\varepsilon}_{i.}$'s.. When the alternative is true, there are real treatment effects and variation among treatment means consists of two parts, the one associated with the random components and one associated with the treatment effects. In this case, variation among treatment means will tend to be larger than the pure error estimate of σ^2.

Finally, we can construct a test criterion, from the estimate of pure error and the variance among treatment means, to test the null hypothesis of no real treatment effects against an alternative that there are such effects.

We now proceed to illustrate the arithmetic of the anova for this design of an experiment; pertinent explanations are also provided. Because of the latter, we will not label this presentation as an illustration. Formulas and computations are given side by side and Tables 19-3-1 and 19-3-2 should be referred to continuously.

For each sample, obtain the treatment total, $T_i = y_{i.}$, and the unadjusted sum of squares $\sum_j y_{ij}^2$. In Table 19-3-2, these are shown for each value of i. These are totaled to give the grand total,

$$T = \sum_i y_{i.} = y_{..} = 564$$

and the unadjusted total sum of squares,

$$\sum_i \left(\sum_j y_{ij}^2\right) = \sum_i \sum_j y_{ij}^2 = 10{,}936$$

All these quantities are needed in computing entries for the anova.

Next compute the correction term C by Equation 19-3-1.

$$C = \frac{y_{..}^2}{tn} = \frac{T^2}{tn} \tag{19-3-1}$$

$$= \frac{564^2}{30} = 10{,}603.2$$

The correction term is the squared sum of all the observations divided by the total number of observations. Also, compute the total sum of squares by Equation 19-3-2, a computing formula.

$$\text{Total SS} = \sum_i \sum_j y_{ij}^2 - C \tag{19-3-2}$$

$$= 10{,}936 - 10{,}603.2 = 332.8$$

This may be shown to equal the sum of squares of the deviations from the sample mean, Equation 19-3-3, a readily interpretable definition formula.

$$\text{Total SS} = \sum_{i,j} (y_{ij} - \bar{y}_{..})^2 \tag{19-3-3}$$

The total sum of squares or sum of squares adjusted for the mean is the usual starting point in an anova. Here, we have fitted a model calling only for an overall mean and a random component, that is, the model $Y_{ij} = \mu + \varepsilon_{ij}$. In the total sum of squares, we have an estimate of the sum of squares of the residuals for this model; that is, $\sum (y_{ij} - \bar{y}_{..})^2$ estimates $\sum (y_{ij} - \mu)^2 = \sum \varepsilon_{ij}^2$. But the true model may include nonzero treatment effects, so we partition this sum of squares and assign the resulting components to at least nominal sources of variation, namely treatments and pure error. The former will be subjected to a test of significance to see whether or not the evidence supports the hypothesis of true treatment differences.

The pure error component is a sum of squares among individuals within treatments, also called the *within-treatments* sum of squares, *residual* sum of squares, *error* sum of squares, or *discrepancy*, and may be found by pooling the within-treatments sums of squares as in Equation 19-3-4 (see under "pooled data" in Tables 19-3-1 and 19-3-2).

$$\begin{aligned}
\text{Error SS} &= \sum \text{SS}_i \\
&= \sum_i \left(\sum_j y_{ij}^2 - \frac{y_{i.}^2}{n} \right) = \sum_{i,j} y_{ij}^2 - \frac{\sum y_{i.}^2}{n} \qquad (19\text{-}3\text{-}4) \\
&= \left(4{,}478 - \frac{210^2}{10}\right) + \left(3{,}356 - \frac{180^2}{10}\right) + \left(3{,}102 - \frac{174^2}{10}\right) \\
&= 258.4
\end{aligned}$$

The arithmetic may be rearranged for equally convenient computation by use of the second algebraic equality of Equation 19-3-4.

The treatment component is based on treatment means. Variation among treatment means can be measured by computing the variance $\sum (\bar{y}_{i.} - \bar{y}_{..})^2/(t - 1)$. Notice that $\bar{y}_{..}$ is the mean of the means as well as the grand mean. Since $\bar{y}$'s from a single population, as when there are no treatment effects, have variance σ^2/n, the variance based on sample means will have to be multiplied by n if it is to provide an estimate of σ^2. If there are real treatment effects, then this variance estimates σ^2 plus a treatment component so should, on the average, be larger than the estimate of σ^2 based on pure error. The definition formula, with multiplier n, has a convenient computing formula; Equation 19-3-5 shows the relationship in terms of sums of squares.

$$\begin{aligned}
\text{Treatment SS} &= n \sum (\bar{y}_{i.} - \bar{y}_{..})^2 \\
&= \frac{\sum y_{i.}^2}{n} - \frac{y_{..}^2}{tn} \qquad (19\text{-}3\text{-}5) \\
&= \frac{\sum T_i^2}{n} - \frac{T^2}{tn} \\
&= \frac{210^2 + 180^2 + 174^2}{10} - 10{,}603.2 \\
&= 74.4
\end{aligned}$$

The treatment sum of squares is also called the *among-treatments* or *among-samples* sum of squares and even the *between-treatments* sum of squares.

Examination of Equations 19-3-4 and 19-3-5 shows that the treatment and error sums of squares add to the total sum of squares, Equation 19-3-2, a fact that provides a computational check or an alternative means of finding the error sum of squares, namely, Equation 19-3-6.

$$\text{Error SS} = \text{total SS} - \text{treatment SS} \qquad (19\text{-}3\text{-}6)$$
$$= 332.8 - 74.4 = 258.4$$

In most analyses of variance, the error sum of squares is found by subtraction, a fact which adds meaning to the term residual sum of squares.

Equation 19-3-7, an identity in that all terms on the right cancel except for Y_{ij}, and Figure 19-3-1 provide additional understanding of the analysis of variance. We use random-variable notation.

$$Y_{ij} = \text{general mean} + \text{treatment effect} + \text{random component}$$
$$= \bar{Y}_{..} + (\bar{Y}_{i.} - \bar{Y}_{..}) + (Y_{ij} - \bar{Y}_{i.}) \qquad (19\text{-}3\text{-}7)$$

Both equation and figure show an observation to be composed of the overall sample mean $\bar{Y}_{..}$, a treatment contribution $\bar{Y}_{i.} - \bar{Y}_{..}$, and the residual or discrepancy $Y_{ij} - \bar{Y}_{i.}$, the last being needed to complete the identity. Each component is a sample quantity that serves as an estimate for a corresponding quantity in Equation 19-2-1, the description of an observation according to the model. Thus $\bar{Y}_{..}$ estimates μ, $\bar{Y}_{i.} - \bar{Y}_{..}$ estimates τ_i, and $Y_{ij} - \bar{Y}_{i.}$ estimates ε_{ij}. We see $Y_{ij} - \bar{Y}_{i.}$ as the part of the observation unexplained by the model since it is needed to complete the identity and as a discrepancy between the jth observation on the ith population and the estimate of the mean of the population. These last components are used to estimate the true error variance, σ^2.

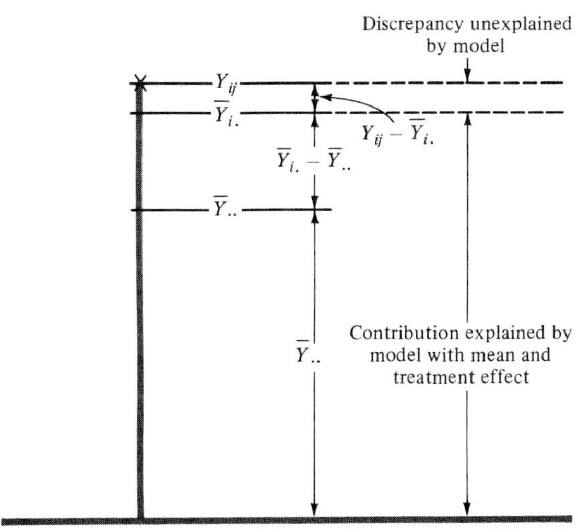

FIGURE 19-3-1 Makeup of an Observation in a One-Way Classification.

If we square both sides of Equation 19-3-7 and sum over all observations, we arrive at Equation 19-3-8.

$$\text{Unadjusted SS} = C + \text{treatment SS} + \text{error SS} \qquad (19\text{-}3\text{-}8)$$

This leads to Equation 19-3-9, which shows the additivity of sums of squares mentioned in connection with Equation 19-3-6 for finding the error sum of squares.

$$\text{Total SS} = \text{treatment SS} + \text{error SS} \qquad (19\text{-}3\text{-}9)$$

Finally, the anova table, Table 19-3-3, is completed by dividing sums of squares by degrees of freedom to give *mean squares*. The degrees of freedom should be clear. For the total, use $tn - 1$; for treatments, use $t - 1$; and for within-treatments, each sample provides $n - 1$ degrees of freedom for a total of $t(n - 1)$.

TABLE 19-3-3 / Analysis of Variance, One-way Classification

Source of Variation	Degrees of Freedom	Sums of Squares: Definition	Sums of Squares: Computing	Mean Square† F‡
Among treatments	$t - 1$	$n \sum_i (\bar{y}_{i.} - \bar{y}_{..})^2$	$\frac{\sum_i y_{i.}^2}{n} - \frac{y_{..}^2}{tn}$	MST
Within treatments	$t(n - 1)$	$\sum_i \sum_j (y_{ij} - \bar{y}_{i.})^2$	by subtraction	MSE
Total	$tn - 1$	$\sum_i \sum_j (y_{ij} - \bar{y}_{..})^2$	$\sum_i \sum_j y_{ij} - \frac{y_{..}^2}{tn}$	

† A mean square is a sum of squares divided by the corresponding degrees of freedom.
‡ The F-value is the among-groups mean square divided by the within-groups mean square.

The *error mean square* or *mean square error* is denoted by s^2 or MSE and is an unbiased estimate of σ^2 provided the assumption of a common variance is true but regardless of whether or not there are real treatment effects. Individual sample s^2's provide unbiased estimates of σ^2. Normally, we pool these, as shown, to give a *generalized error*. Thus, the generalized error is the pooled error sum of squares divided by the pooled degrees of freedom.

The treatment mean square, often denoted by MST, is also an estimate of σ^2 provided the null hypothesis is true. Mean squares in an anova have this property, and the computations are said to have been on a *per-observation* basis. In addition, these mean squares are independent. Consequently, if the null hypothesis is true, the two estimates differ by chance alone, and the F-statistic is an appropriate test criterion for testing the null hypothesis of no treatment differences.

The F-statistic was defined in Section 17-7 as the ratio of two independent estimates of the same σ^2, to be computed as in Equation 17-7-1 when the alternative hypothesis is simply that there can be two σ's. In the present case, a one-sided alternative is appropriate since it calls for real treatment effects with a consequent, on the average, increase in the treatment mean square. Hence, we compute F by Equation 19-3-10.

$$F = \frac{\text{treatment mean square}}{\text{error mean square}} \tag{19-3-10}$$

For a one-sided alternative, the tabulated probability level of F is also the significance level.

For the data of Table 19-3-2 and corresponding anova Table 19-3-4, $F = 37.2/9.57 = 3.89$, with 2 and 27 degrees of freedom. This value is usually included in the anova table and is to be compared with 3.35 for $\alpha = .05$ or 5.49 for $\alpha = .01$. Since $3.89 > 3.55$, that is, since calculated F exceeds the .05 rejection value, we conclude that the null hypothesis is not supported by the data and that the experiment provides evidence of real differences among treatment means. It is customary to mark such an F with an asterisk; if it exceeds tabulated F for $\alpha = .01$, then two asterisks are used.

TABLE 19-3-4 / Analysis of Variance for Data of Table 19-3-2

Source of Variation	Degrees of Freedom	Sums of Squares	Mean Square	F
Among treatments	2	74.4	37.2	3.89*
Within treatments	27	258.4	9.57	
Total	29	332.8		

In repeated sampling, that is, repetition of the whole experiment under either the null or the alternative hypothesis, the average value or expectation of the error mean square is always σ^2. This is unbiasedness. In repeated sampling under the alternative hypothesis, the expectation of the among-treatments mean square is given by Equation 19-3-11 for the fixed-effects model.

$$E\,(\text{treatment mean square, fixed model}) = \sigma^2 + n\,\frac{\sum_i \tau_i^2}{t - 1} \tag{19-3-11}$$

This expected value is sometimes shown in anova tables. The F-test is intended to detect the component other than σ^2, should it be present. Note the role played by n, the number of observations per treatment, in improving the ability of the test to detect this component.

Inferences in the form of confidence intervals for population parameters, as discussed in Chapter 13, are often important to the research worker. The within-treatments mean square is used to compute standard deviations for means, differences between means, and, in turn, confidence intervals, and to compute coefficients of variation.

EXERCISES

19-3-1 The accompanying table gives possible weight gains, in grams per 100 days, of rats fed different rations. Observations for a ration are shown horizontally for typing convenience. Compute the analysis of variance and evaluate the null hypothesis of no treatment effects. Compute the error sum of squares directly and demonstrate the additivity of sums of squares. What is the coefficient of variation? Find the standard deviation of a ration mean. Of a difference between two ration means. What assumptions have you made in performing the analysis of variance? What possible aspect of the physical conduct of the experiment would cause you to suspect conclusions based on the analysis of variance?

Ration	Weight Gain							
1	230	225	223	216	229	201	205	193
2	187	223	214	192	222	208	198	217
3	177	201	178	179	207	182	178	195
4	129	139	160	149	122	146	166	169

19-3-2 Twelve dogs were assigned at random to four groups. Each group received a different amount of IPA. The accompanying table gives the percentage absorption of IPA for the four groups of animals. Evaluate the null hypothesis of no treatment effects using the analysis of variance. What is the coefficient of variation? Find the standard deviation of a group mean, and of the difference between two group means. Is it all right if the three animals of a group are housed together?

Group = Treatment			
1	2	3	4
39.2	42.2	38.6	61.0
48.5	44.0	51.1	61.1
41.1	37.1	39.5	51.8

Source: Wax et al. (1949).

19-3-3 Twenty-eight male students were assigned at random to four groups. Each group received a different training program. The training programs were evaluated by determining how long it took in seconds for each student to perform a task. Observations could be those given in the accompanying table, shown horizontally for typing convenience. Evaluate the hypothesis of no treatment effects for the training programs. Compute the coefficient of variation, the standard deviation of a group mean, and the standard deviation of the difference between two groups means. What aspect of the conduct of the training programs could make any conclusions suspect?

Groups	Observations						
1	36	45	31	38	40	41	32
2	31	38	37	40	35	28	30
3	34	30	43	42	38	46	32
4	34	26	39	33	27	28	35

19-4 THE RANDOMIZED BLOCK DESIGN OR TWO-WAY CLASSIFICATION

In Section 17-5 we examined the problem of comparing the effects of two treatments using meaningfully paired experimental units. This is a special case of the randomized complete block design or two-way classification, which we now generalize to $t \geq 2$.

The randomized complete block design is used when the experimental units can be meaningfully grouped according to expected response when treated alike, the number of units in a group being equal to the number of treatments. Such a group is called a *complete block* or *replicate.* In animal experiments, the animals are often placed in outcome groups or blocks on the basis of such characteristics as weight, age, condition, breed, and sex. Since each treatment is present in every block, the variation among treatment means is free of any contribution from blocks. In turn, the arithmetic of the analysis of variance must be such that experimental error, our yardstick for evaluating observed treatment differences, is also free of such a contribution.

Data on the coefficients of digestibility of total carbohydrates of linseed oil meal, a component of some cattle diets, are used to illustrate this design. Five rations, consisting of 3 kilograms of hay per animal per day with increasing amounts of linseed oil meal, approximately 1, 2, 3, 4, or 5 kilograms per animal per day, were compared. These treatments were assigned at random to the animals in each block. Coefficients of digestibility, in percent, were determined for different nutrients, including total carbohydrates. A high coefficient of digestibility is desirable if the particular source of the nutrient is to be kept in the diet.

Illustration 19-4-1

Instructions for analyzing the nutrition data follow. The notation used is discussed at greater length in Section 19-5, along with the model. The data and results of the computations are given in Table 19-4-1.

TABLE 19-4-1 / Coefficients of Digestibility for Five Rations (Treatments) for Total Carbohydrates from Linseed Oil Meal Fed to Steers

Treatment = Ration*	Block = Steer Group						$y_{i.} = T_i$	$\sum_j y_{ij}^2$	$\bar{y}_{i.} = \bar{T}_i$
	1	2	3	4	5	6			
1	86.5	74.5	68.8	79.9	78.2	86.8	474.7	37,799.43	79.1
2	78.2	76.9	67.8	74.2	72.5	76.5	446.1	33,239.83	74.4
3	74.7	72.3	72.7	76.3	75.8	76.1	447.9	33,451.21	74.7
4	72.9	76.9	64.7	73.2	73.2	73.2	434.1	31,488.83	72.4
5	70.8	73.5	67.2	74.5	71.5	70.4	427.9	30,541.39	71.3
$B_j = y_{.j}$	383.1	374.1	341.2	378.1	371.2	383.0	$2{,}230.7 = y_{..} = 1$		
$\sum_i y_{ij}^2$	29,504.63	28,007.01	23,317.50	28,619.83	27,587.62	29,492.10	. . .	166,528.69	
$\bar{B}_j = \bar{y}_{.j}$	76.6	74.8	68.2	75.6	74.2	76.6	. . .	. . .	$74.4 = \bar{y}_{..} = \bar{T}$

* Number of kilograms of linseed oil meal fed per animal per day.

Source: Watson et al. (1943). The data are from the Division of Chemistry, Science Service, Department of Agriculture, Ottawa, Canada.

1 Arrange the raw data as in Table 19-4-1, a two-way table since observations are classified by both block or group and treatment or ration. Obtain treatment totals $T_i = y_{i..}$, block totals $B_j = y_{.j}$, and the grand total $T = y_{...}$ Simultaneously obtain the sum of the squared observations for each treatment and block, that is, $\sum_j y_{ij}^2$ for each i and $\sum_i y_{ij}^2$ for each j, respectively. Obtain the grand total by summing treatment totals and block totals separately, an arithmetic check. Obtain the unadjusted total sum of squares by summing the unadjusted sums of squares for each treatment and for each block separately, another check. We find

$$y_{..} = \begin{cases} 383.1 + \cdots + 383.0 = 2{,}230.7 & \text{from block totals} \\ 474.7 + \cdots + 427.9 = 2{,}230.7 & \text{from ration totals} \end{cases}$$

Also,

$$\sum_{i,j} y_{ij}^2 = \begin{cases} 29{,}504.63 + \cdots + 29{,}492.10 = 166{,}528.69 \\ \qquad\qquad\qquad\qquad \text{from block totals} \\ 27{,}799.43 + \cdots + 30{,}541.39 = 166{,}528.69 \\ \qquad\qquad\qquad\qquad \text{from ration totals} \end{cases}$$

2 Compute the correction term and total sum of squares adjusted for the mean as shown. Block sum of squares and treatment sum of squares are computed as for treatments in the completely random design. Finally, compute error sum of squares by subtraction. Justification of the use of this term as an error sum of squares is provided with the discussion of the model.

$$C = \frac{2230.7^2}{6(5)} = 165{,}867.41$$

$$\text{Total SS} = 166{,}528.69 - 165{,}867.41 = 661.28$$

$$\text{Block SS} = \frac{383.1^2 + \cdots + 383.0^2}{5} - 165{,}867.41$$

$$= 246.97$$

$$\text{Treatment SS} = \frac{474.7^2 + \cdots + 427.9^2}{6} - 165{,}867.41$$

$$= 261.07$$

$$\text{Error SS} = 661.28 - 246.97 - 216.07 = 198.24$$

Degrees of freedom for the total, block, and treatment sums of squares are $6(5) - 1 = 29$, $6 - 1 = 5$, and $5 - 1 = 4$, respectively, all of which were to be expected. For error degrees of freedom, we simply subtract those for blocks and treatments from the total to get 29 − 5 − 4 = 20. This value is also the product of the block and treatment degrees of freedom, an easy way to remember the value.

3 The degrees of freedom and the adjusted sums of squares are entered in the anova table, Table 19-4-2, opposite the appropriate sources of variation.

TABLE 19-4-2 / Anova for the Data of Table 19-4-1

Source of Variation	Degrees of Freedom	Sum of Squares	Mean Square	F
Blocks	5	246.97	49.39	
Treatments	4	216.07	54.02	5.45**
Error	20	198.24	9.91	
Total	29	661.28		

4 Mean squares, except for the total, are computed and entered.

5 The sample F-value for testing the null hypothesis of no treatment differences is found by dividing the treatment mean square by the error mean square. Very rarely are comparisons made among block means.

$F = 54.02/7.91 = 5.45^{**}$ with 4 and 20 degrees of freedom. It is significant at the 1 percent level, as indicated by the double asterisk. This is evidence in support of the alternative hypothesis that there are real differences among the population treatment means or evidence of real treatment effects.

Variances and standard deviations for treatment means and differences between treatment means can be computed as in Section 19.3. The same is so for blocks if such computations are meaningful. Using the standard deviations, one can then find confidence intervals.

EXERCISES

19-4-1 Four crops were assigned at random to plots, that is, small areas of land, within each of six blocks. From the straw or stems for each crop, the percent sulfur was determined. Results for each of the 24 experimental units are given in the accompanying table. For these data, compute the anova. Test the null hypothesis that the four crop populations have the same mean. Rank the means and use t to test each of the three null hypotheses that neighboring crops in this ordering have equal population means against two-sided alternatives. Is this a result-guided test procedure? Should we regard the tabulated probability level as overestimating or underestimating the true α-level?

Crop = Treatment	Block					
	1	2	3	4	5	6
Oats	.17	.17	.19	.19	.17	.18
Barley	.23	.28	.25	.29	.30	.28
Wheat	.24	.31	.24	.34	.29	.26
Flax	.28	.27	.29	.26	.22	.21

Source: Clagget et al. (1952).

Compute the coefficient of variation for the experiment. Compute the standard deviation of a treatment mean and of the difference between two treatment means.

19-4-2 The optical score for spray coverage of grapefruit for five different sprayers using three replicates of a randomized block design is given in the accompanying table. For typing convenience, the blocks are shown vertically rather than horizontally. For these data, carry out the anova. Test the null hypothesis that the five sprayers are from populations with a common mean. Do you think the five sprayers constitute a population of sprayers or are only a random sample?

Block	Sprayer = Treatment				
	1	2	3	4	5
1	2.20	2.58	2.11	1.47	2.25
2	2.42	2.25	1.90	2.29	2.36
3	2.40	2.21	2.48	1.64	1.82

Source: Edwards et al. (1961).

Compute the coefficient of variation for the experiment. Compute the standard deviation of a treatment mean, and of the difference between two treatment means.

19-4-3 The plankton caught by five different nets, that is, blocks, were classified into four types. The accompanying table gives the logarithms of the number of each type in each net. These transformed numbers tend to conform more closely to the assumptions discussed than do the original numbers. Is this an experimental design or a sampling type of problem?

Type of Plankton = Treatment	Net = Block				
	1	2	3	4	5
1	2.95	2.73	3.01	2.67	2.63
2	3.18	3.21	3.28	3.13	2.99
3	4.64	4.52	4.46	4.54	4.44
4	4.04	3.93	3.92	3.99	3.88

Source: Winsor and Clarke (1940).

Perform an anova for these data. Test the null hypothesis that there are no differences among plankton population means. This is one case where comparisons among block means might be made. Test the null hypothesis that there are no differences among population means for the nets. Compute the standard deviation of a type mean, a net mean, the difference between type means, and the difference between net means.

19-4-4 The time required in minutes by five drivers to plow two rows 200 feet long on three different days (blocks) is given in the accompanying table. For typing convenience, blocks are presented in columns. Carry out the anova for these data. Test the null hypothesis that there are no differences among the means of the populations represented by each of the five drivers. Test the null hypothesis that there are no day-to-day differences.

	Driver = Treatment				
Block = Day	*A*	*B*	*C*	*D*	*E*
1	3.89	2.58	2.94	2.60	2.46
2	3.81	2.10	2.86	2.16	2.39
3	4.01	2.08	2.29	2.09	2.59

Source: Barker and VanRest (1962).

Compute the standard deviation of a driver mean, a day mean, the difference between two driver means, and the difference between two day means.

19-5 THE RANDOMIZED BLOCK MODEL

Equation 19-5-1 describes an observation for this experimental design, the observation receiving the ith treatment and being in the jth block.

$$Y_{ij} = \mu + \tau_i + \beta_j + \varepsilon_{ij} \tag{19-5-1}$$

The observation includes a general mean μ, a treatment component τ_i, a block component β_j, and a random component ε_{ij}. The model is clearly additive and is also linear in the parameters. Equation 19-5-1 can be used to replace Equation 17-5-1, used for paired observations, where $\mu + \tau_i$ was written simply as μ_i. Assumptions will be discussed later.

A symbolic representation of observations from an experiment with b blocks of t experimental units is given in Table 19-5-1, which is comparable to Table 19-4-1 with realistic data. Notice that the y_{ij}th observation appears at the intersection of the ith row and jth column. Additional symbols for quantities used in anova computations are also given.

Table 19-5-2 is a symbolic presentation of the working and definition formulas for the sums of squares and degrees of freedom in the analysis of the data from the randomized complete block design. In contrast to the completely random design, the subscript j is more than a tag describing the position in which the observation was recorded; it tells us the block from which the observation came. The block totals are new quantities that are not in a completely random design.

TABLE 19-5-1 / Symbolic Representation of Observations in b Blocks of t Treatments

Treatment or sample	Block 1	2	$\cdots$	j	$\cdots$	b	Treatment Sums	Treatment Means	Sums of Squares
1	y_{11}	y_{12}		y_{1j}		y_{1b}	$T_1 = y_{1.}$	$\bar{y}_{1.}$	$\sum_j y_{1j}^2$
2	y_{21}	y_{22}		y_{2j}		y_{2b}	$T_2 = y_{2.}$	$\bar{y}_{2.}$	$\sum_j y_{2j}^2$
i	y_{i1}	y_{i2}		y_{ij}		y_{ib}	$T_i = y_{i.}$	$\bar{y}_{i.}$	$\sum_j y_{ij}^2$
t	y_{t1}	y_{t2}		y_{tj}		y_{tb}	$T_t = y_{t.}$	$\bar{y}_{t.}$	$\sum_j y_{tj}^2$
Block sums	$B_1 = y_{.1}$	$B_2 = y_{.2}$		$B_j = y_{.j}$		$B_n = y_{.b}$	$T = y_{..}$		
Block means	$\bar{y}_{.1}$	$\bar{y}_{.2}$		$\bar{y}_{.j}$		$\bar{y}_{.b}$	$\cdots$	$\bar{y}_{..}$	
Sums of Squares	$\sum_i y_{i1}^2$	$\sum_i y_{i2}^2$		$\sum_i y_{ij}^2$		$\sum_i y_{ib}^2$	$\cdots$	$\cdots$	$\sum_i \sum_j y_{ij}^2$

TABLE 19-5-2 / Analysis of Variance: Randomized Blocks Design

Source of Variation	Degrees of Freedom	Sums of Squares	
		Definition	Computing†
Blocks	$b - 1$	$t \sum_j (\bar{y}_{.j} - \bar{y}_{..})^2$	$= \frac{\sum_j y_{.j}^2}{t} - \frac{y_{..}^2}{tb} = \frac{\sum B_j^2}{t} - C$
Treatments	$t - 1$	$b \sum_i (\bar{y}_{i.} - \bar{y}_{..})^2$	$= \frac{\sum_i y_{i.}^2}{b} - \frac{y_{..}^2}{tb} = \frac{\sum T_i^2}{b} - C$
Error	$(b - 1)(t - 1)$	$\sum_i \sum_j (y_{ij} - \bar{y}_{i.} - \bar{y}_{.j} + \bar{y}_{..})^2$	= total SS − block SS − treatment SS
Total	$bt - 1$	$\sum_i \sum_j (y_{ij} - \bar{y}_{..})^2$	$\sum_i \sum_j y_{ij}^2 - \frac{y_{..}^2}{tb} = \sum_{i,j} y_{ij}^2 - C$

† $\frac{y_{..}^2}{tb}$ is the correction term.

A better understanding of the basis underlying the computations is possible from the following considerations. We will use random-variable notation, that is, capital letters, since we will be thinking of the problems of estimation and testing.

Once again there is no real way to estimate μ without external information. For example, if every treatment had a positive effect, how could we learn this from the experiment alone? Consequently, we provide for μ to be defined so that it is estimated by $\bar{Y}_{..}$.

Also, $\sum (Y_{ij} - \bar{Y}_{..})^2$ is an estimate of the sum of squares of residuals when the model calls only for a mean and a random component. It is this quantity that we partition in the anova.

It should follow with reasonable ease from the discussion of nonrandom pairing in Section 17-3 that the treatments SS contains contributions from treatments and error but none from blocks, and that the blocks SS contains contributions from blocks and error but none from treatments. This is because of the balance built into the design by having each treatment appear once and only once in each block. There is a symmetry between blocks and treatments, and, under some circumstances, blocks might even be a second kind of treatment.

Treatment effects are estimated by deviations of treatment means from the overall mean, that is, by the $(\bar{Y}_{i.} - \bar{Y}_{..})$'s. Similarly, block effects are estimated by the $(\bar{Y}_{.j} - \bar{Y}_{..})$'s.

Since the variance of means of n observations is σ^2/n, multipliers of t and b shown in the definition–sums-of-squares column of Table 19-5-2 result in all mean squares being estimates of the same σ^2 when there are no block or treatment effects. The same reasoning applied to totals accounts for the divisors t, b, and tb in the computing sums of squares column.

Finally, we come to error. Equation 19-5-1 lets us write an ε as in Equation 19-5-2, the difference between an observation and the mean of the population from which it was drawn. Notice that we have a different mean for every cell in the table.

$$\varepsilon_{ij} = Y_{ij} - \mu - \tau_i - \beta_j \tag{19-5-2}$$

Since we have just seen how to estimate μ, τ_i, and β_j, we can, in turn, estimate a cell mean and an ε_{ij}. Equation 19-5-3 estimates an ε_{ij} as an e_{ij}.

$$\begin{aligned} e_{ij} &= Y_{ij} - \bar{Y}_{..} - (\bar{Y}_{i.} - \bar{Y}_{..}) - (\bar{Y}_{.j} - \bar{Y}_{..}) \\ &= Y_{ij} - \bar{Y}_{i.} - \bar{Y}_{.j} + \bar{Y}_{..} \end{aligned} \tag{19-5-3}$$

This is our measure of the difference between an observation and the estimate of its corresponding population mean.

The sum of the squares of the e_{ij}'s is the error sum of squares and is a measure of the failure of the data to fit a linear additive model calling for a population mean to consist of a general mean, a treatment effect, and a block effect. The error sum of squares is the least possible value obtainable, given that we are allowed to use any values as estimates of μ, the τ's, and the β's. In other words, we have *least-squares* estimates of these parameters, least squares being an estimation procedure that guarantees this particular property.

Error sum of squares can be shown to have an expected value of $(b - 1)(t - 1)\sigma^2$. This provides the divisor that allows us to obtain an unbiased estimate of σ^2. This value, $(b - 1)(t - 1)$, is the degrees of freedom. It is now apparent that, for the randomized block design, degrees of freedom as well as sums of squares are additive.

Replace all parameters in Equation 19-5-1 by estimates. Since we derived Equation 19-5-3 by doing this, we will have an identity; that is, the terms on the right will cancel one another until what remains is identical with what is on the left. However, the identity will help establish the validity of the arithmetic of the analysis of variance, including the fact that error sum of squares can be found by subtraction. First, we have Equation 19-5-4.

$$Y_{ij} = \bar{Y}_{..} + (\bar{Y}_{i.} - \bar{Y}_{..}) + (\bar{Y}_{.j} - \bar{Y}_{..}) + (Y_{ij} - \bar{Y}_{i.} - \bar{Y}_{.j} + \bar{Y}_{..}) \quad (19\text{-}5\text{-}4)$$

Squaring both sides of Equation 19-5-4 and summing over all observations will give Equation 19-5-5.

$$\sum_{i,j} Y_{ij}^2 = b\bar{Y}_{..}^{\ 2} + b\sum_i (\bar{Y}_{i.} - \bar{Y}_{..})^2 + t\sum_j (\bar{Y}_{.j} - \bar{Y}_{..})^2 + \sum_{i,j} (Y_{ij} - \bar{Y}_{i.} - \bar{Y}_{.j} + \bar{Y}_{..})^2 \quad (19\text{-}5\text{-}5)$$

All cross-product terms have summed to 0.

By comparing Equation 19-5-5 with Table 19-5-2, particularly the definition formulas, we can recognize that the equation is the basis of the arithmetic of the analysis of variance. Notice that we can compute error sum of squares directly, rather than by subtraction, though inconveniently and subject to the possibility of considerable rounding errors.

In summary, Equation 19-5-1 describes an observation. Beginning here, we discussed the estimation of the parameters in the equation and the arithmetic of the anova. This arithmetic is based on an algebraic identity and is not dependent on any assumptions. However, assumptions are necessary for a proper interpretation of the data, including estimation and testing. The usual assumptions are basically of the same sort as discussed in Section 19-2 for the completely random design.

For a randomized complete block design, the following assumptions complete a reasonable model.

1 μ is an unknown, fixed parameter.
2 The τ_i's are unknown parameters, treatment effects, with $\sum \tau_i = 0$. This implies that the only treatments of interest in drawing inferences are included in the experiment; if the experiment were to be repeated, the same treatments would be used.
3 The β_j's are independent values of a random variable which is normally distributed with mean 0 and variance σ_β^2. This means that we are trying out the treatments under a varying set of conditions which we hope are representative of a broader population. In studying teaching methods, we

would not use the same children every time; in testing cultivars of wheat, we want our recommendations to apply throughout a geographic or climatic region and not just to the fields in the experiment. In other words, repetition of the experiment would involve a different set of blocks. Here, σ_β^2 is the block parameter of interest.

4 The ε_{ij}'s are independent and from a normal distribution with zero mean and homogeneous variance, say σ^2.

This is a *mixed model*, fixed with respect to treatments but random with respect to blocks. Clearly, one could propose a fixed model or a random model, at least.

EXERCISES

19-5-1 For Exercise 19-4-1, should we regard the experiment as one calling for a fixed or a random model with respect to treatment effects? Will we want to estimate differences among the population means for treatments or a variance component, σ_τ^2, for crops?

Should we regard the model as fixed or random with respect to blocks? Do you think we would want to generalize our results to blocks rather than those in the experiment?

Do you think that sulfur content was measured for the whole experimental unit or for only a part of it?

19-5-2 For Exercise 19-4-2, do you think the assumption of normally and independently distributed error components is valid for this experiment? Can you suggest any questions that you might ask the experimenter to help make your decision? Are you proposing a fixed or random model with respect to sprayers?

19-5-3 For Exercise 19-4-3, do you think the type means should be considered as representing a fixed or random set of types? Do you thin the nets should be considered as a population of fixed effects or a random sample of effects? Did either of your decisions here depend upon the results of your earlier tests of significance? Would you propose a mixed model?

19-5-4 For Exercise 19-4-4, do you think that drivers should be considered as fixed or random in the model? What brought you to your decision? Did you consider days as constituting fixed or random effects?

19-6 ERROR-RATE DEFINITIONS AND MULTIPLE COMPARISONS

It is of special interest to note what has happened to the error rate as we have gone from the one- or two-treatment experiment with a single *t*-test to the larger experiment with an *F*-test. Its definition has changed. In the earlier experiment, we made only one inference, namely, that a mean or difference between two means was or was not equal to some hypothesized value. A Type I error was defined as the probability of declaring this null hypothesis to be false when, in fact, it was true.

For an experiment with more than two treatments, it is possible to make many paired comparisons and, indeed, to make comparisons involving more than two but possibly less than all of the treatments. For example, one might compare the average of several treatments expected to act similarly on the experimental material with a standard or check treatment mean. Such comparisons are seen to be pooled, in a sense, and to be considered simultaneously in an F-test. To make a Type I error now is to conclude that somewhere in the experiment, some comparison is being declared falsely significant. In other words, the unit to which the probability of a Type I error applies has changed from the individual comparison to the whole experiment with all its implied simultaneous comparisons.

In the two-treatment experiment, we clearly had a *comparisonwise* or *per-comparison* error rate. In the many-treatment experiment, we have an *experimentwise* error rate. In the latter experiment, it may be desired to make comparisons among and within groups of treatment means or, in general, to use units other than the single comparison or the experiment when defining error rate. In particular, there is a per-experiment error rate which is defined differently from the experimentwise one, indicating the need for being careful with our terms.

A significant F-value is evidence that true differences exist but gives no indication as to where they may be. This raises the possibilities of using some test after the F-test to help locate these differences or bypassing F altogether and going straight to some predetermined set of comparisons. At the same time, we are faced with the problem of deciding whether to have our error rate or confidence coefficient cover our decisions one at a time with a comparisonwise error rate or simultaneously with an experimentwise one.

For a comparisonwise error rate, a number of possible approaches are available. If, prior to the conduct of the experiment, it is possible to write down a specific set of meaningful, single-degree-of-freedom comparisons, then these may be tested by Student's t and the error rate will be the tabulated one. This actually allows us to make all possible comparisons of paired treatment means. The one thing to be avoided is the comparison suggested by the data, the result-guided comparison. For example, we cannot wait until we see the data and then compare only the largest treatment mean with the smallest, by means of a t-test. This also implies that we cannot decide to make t-tests only after finding a significant F-value. Such a procedure requires some modification of the tabulated probability levels for t or of the critical values.

For an experimentwise error rate, we may also choose a meaningful set of comparisons, again including comparison of all possible paired treatment means. A number of tests for comparisons of possible general interest have been proposed, for example, a test comparing each treatment with a control or check treatment, and comparison of all possible pairs of means. In addition, several procedures allow us to include all possible comparisons regardless of how many means may be involved in any one. Such a procedure clearly allows us to make tests suggested by the data. It is also possible to conclude that the null hypothesis, consisting of a set of simultaneous hypotheses, is false when this is the case, but to conclude so for the wrong reason.

The most suitable procedure for any particular experiment has to be decided upon by the experimenter. Pretty obviously, procedures that call for us to make simultaneous inferences with an experimentwise error rate will call for larger critical values if we use the same α-value as we would have for a comparisonwise error rate. Figure 19-6-1 gives an idea of what is going on. It shows us making a confidence-interval statement about a mean in a plane, that is, a statement about (μ_1, μ_2), the interval being centered on $(\bar{Y}_1, \bar{Y}_2)$. The probability P is that of covering (μ_1, μ_2). For the comparisonwise error rate, we assume independent comparisons, and so multiply the two probabilities to give a probability or area of $.95^2 = .9025$. For the experimentwise error rate, we start by setting this probability or area at the larger value of .95. For the figure, we have simplified the problem by using independent comparisons and assuming σ^2 to be known. In a real-life situation, we would certainly have to estimate σ^2 and use the same value for all comparisons, thus always introducing a dependency into the problem. The topic of multiple comparisons is covered in greater depth in Steel and Torrie (1960).

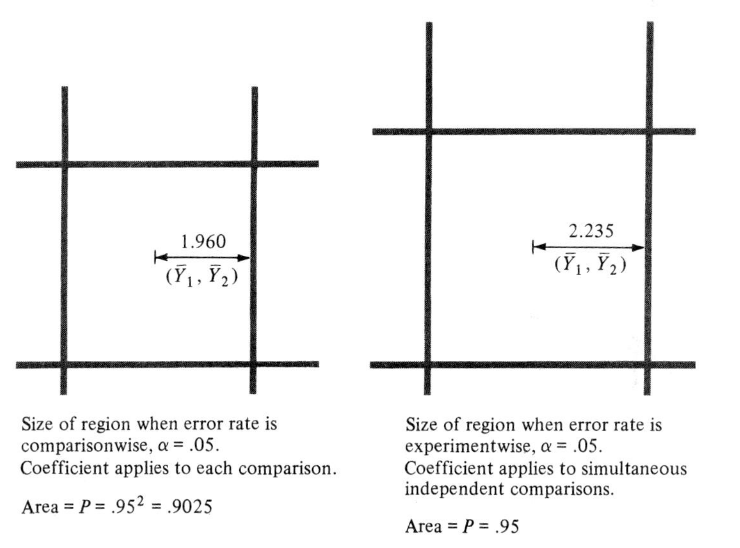

Size of region when error rate is comparisonwise, $\alpha = .05$.
Coefficient applies to each comparison.

Area = $P = .95^2 = .9025$

Size of region when error rate is experimentwise, $\alpha = .05$.
Coefficient applies to simultaneous independent comparisons.

Area = $P = .95$

FIGURE 19-6-1 Comparison of Two Error Rates; Standardized Normal Variables.

19-7 GENERALIZATIONS OF THE COMPLETELY RANDOM DESIGN

This experimental design is also available when there are unequal numbers of experimental units in each group or treatment. Modification of the arithmetic of the analysis of variance is, of course, required.

The correction term is computed precisely as before, although the algebraic definition, $y_{..}^2/\sum n_i$, may look a little different.

The among-treatments sum of squares now becomes a weighted sum of squares of the deviations of treatment means from the general mean. The weights are the numbers of observations in the corresponding treatment means. Definition and computing formulas on a per-observation basis are shown together in Equation 19-7-1.

$$\sum_i n_i(\bar{y}_{i.} - \bar{y}_{..})^2 = \sum_i \frac{y_{i.}^2}{n_i} - \frac{y_{..}^2}{\sum n_i} = \sum \frac{T_i^2}{n_i} - \frac{T^2}{\sum n_i} \tag{19-7-1}$$

Note that each squared total is divided by the number of observations in it. Recall that variances of means and totals are σ^2/n and $n\sigma^2$, respectively, and that we are working on a per-observation basis. When totals are based on unequal n_i's, division of a squared total must be made prior to any further summation and we must be careful how the $\sum$ is placed. See the first term on the right of either equality sign in Equation 19-7-1.

Finally, while the within-treatments sum of squares may be found by subtraction in an anova, it may also be found by application of the right side of Equation 19-7-2 which equates definition and computing formulas for this sum of squares.

$$\sum_{i,j} (y_{ij} - \bar{y}_{i.})^2 = \sum_{i,j} y_{ij}^2 - \sum_i \frac{y_{i.}^2}{n_i} \tag{19-7-2}$$

Expectation of the treatment mean square changes slightly but still consists of σ^2 plus a sum of squares, now weighted, for the fixed model.

The completely random design may also be used without measuring the whole experimental plot. For example, one may take a preharvest measurement of some crop and perform a chemical analysis, essentially a destructive technique, say for soluble solids in the case of grapes, using only small samples from the experimental plot so that no measurement is ever made on the total experimental unit. Either one or more sampling units within each experimental plot may be measured. Analysis and interpretation of such data, when from a completely random design, again need modification.

Additional *nestings* may be encountered in classifying data, as when observations are made on farms within counties or within soil types which are, in turn, within states or climatic regions. Again, the analysis needs to be extended. The interested reader is referred to Steel and Torrie (1960).

As already suggested, one may propose a random-effects model as an alternative to the fixed model. Still other possibilities exist.

EXERCISE

19-7-1 The cudding times in minutes per day for three groups of dairy cattle are shown in the accompanying table. *Observations are shown horizontally by groups for reasons of space.* Note the unequal numbers of observations per treatment. Evaluate the null hypothesis of equal population means using the anova. Demonstrate the additivity of the sums of squares. Compute the coefficient of variation. Find the standard deviation for the mean of group 2 and of group 3, and of the difference between the means for

Group	Cudding Times									
1	260	273	165	213	260	260	268	238		
2	165	253	228	143	195	240				
3	208	205	228	248	268	240	203	268	195	225

Source: Brumby (1959).

groups 2 and 3. Note that we now have the problem that no single standard deviation applies to all means and that there is no single standard deviation appropriate to the differences between all pairs of means.

19-8 GENERALIZATIONS OF THE RANDOMIZED BLOCK DESIGN

As in the case of the one-way classification, it is not always feasible or convenient to observe or measure an experimental unit in its entirety for a two-way classification; sampling must sometimes be relied upon with one or more samples per experimental unit. The design is still available, although modifications in the arithmetic and interpretation are required.

While we have discussed this design with all treatments present exactly once in each block, the more general requirement is that any treatment should be present the same number of times in each block. The use of extra replication for some treatments, such as a standard or check, can be desirable. Again, modifications in the arithmetic and interpretation are necessary.

Occasionally an experimenter will have limited amounts of some new treatment material or materials, which are not enough for an experimental unit in each block, yet which the experimenter would like to include in the testing. This is possible with modifications of the arithmetic. The situation is not unlike that where experimental units or data from them are lost through no fault of the treatment, that is, not because of some treatment effect. This poses a greater problem than do losses in a completely random design.

The fact that we have limited our discussion to only one model has already been stated.

Chapter 20

NONPARAMETRIC DATA ANALYSIS

20-1 INTRODUCTION

Chapter 19 presented the analysis of variance primarily as a procedure preliminary to testing the equality of several population means simultaneously, when strong assumptions concerning the parent populations, in particular that of an underlying normal distribution, were justified. In this chapter we present analogous procedures for testing hypotheses about the locations of two or more populations under weak assumptions. These are sign and rank tests. In the last section, randomization tests are discussed briefly. Their wide applicability is apparent.

20-2 THE SIGN TEST FOR THE ONE-WAY CLASSIFICATION

The sign test of Chapter 18 called for us to know a very minimum about the underlying distribution. In fact, reference is made to the median only, and nothing is said about the way observations are distributed as we move away from it.

Equation 14-2-1, repeated here as Equation 20-2-1, is all that is given about this distribution.

$$P(Y < Q_2) = .5 = P(Y > Q_2) \tag{20-2-1}$$

The sign test is readily extended to the one-way classification.

Suppose we have $t \geq 2$ treatments and wish to know whether their average responses, when defined as their medians, are the same.

First consider the overall median of the data. If the treatments do not give very different responses, then each treatment should have about half of its observations above and half below this median. If one treatment is very much better than the others, then its responses will tend to be mostly on one side of the common median, say above if a high response is a good one. Similarly, a poor treatment will result in most of its observations being below this common median.

We can, then, prepare a table showing the number of times each treatment exceeds the overall mean and how many times it does not. Such a table, a $2 \times t$ table of counts in two rows for t treatments, provides sufficient information for a test comparing the performance of the treatments. The test criterion could be χ^2 as discussed in Chapter 16.

Illustration 20-2-1

The data of Table 19-3-2, repeated here as Table 20-2-1, are now used to demonstrate the procedure. Thirty individuals were randomly assigned to 3 different groups, each of which underwent a different sort of training for a task. The 30 final times required for the task are ranked from 1 to 30, with ties being given the average of those ranks that would be assigned if the observations were different. Ranks are shown in parentheses.

The overall median lies between the 15th and 16th observations. Ranks 15 and 16 were not assigned, but the observation 18 appeared three times where ranks 13, 14, and 15 were next to be assigned. They were all assigned rank 14. The observation 19 was next, and there were two of them. The ranks 16 and 17 were to be assigned, so both observations received rank $16\frac{1}{2}$. Hence the observations qualified to receive ranks 15 and 16 are the numbers 18 and 19, respectively. It is readily seen that 15 observations are less than or equal to 18, while 15 observations are greater than or equal to 19. Consequently, the overall median lies between 18 and 19, say at 18.5.

Observe how the observations for each treatment fall with respect to the common median. For treatment A, 9 of 10 are above; for B, 3 are above; and for C, 3 are above. This information is presented in Table 20-2-2, a 2×3 contingency table. Notice that all marginal totals are fixed.

TABLE 20-2-1 / Time in Minutes Taken to Perform a Standardized Task by Three Groups of Male University Students after Different Types of Training (Ranks in parentheses)

Groups		
A	*B*	*C*
21 (22)	15 (5)	17 (10.5)
19 (16.5)	18 (14)	13 (1)
23 (27.5)	21 (22)	18 (14)
20 (19)	23 (27.5)	21 (22)
17 (10.5)	17 (10.5)	17 (10.5)
22 (25)	14 (2.5)	15 (5)
22 (25)	24 (29)	16 (7.5)
27 (30)	14 (2.5)	20 (19)
20 (19)	16 (7.5)	22 (25)
19 (16.5)	18 (14)	15 (5)
r_i = 211	134.5	119.5

Note: Since n is 30 and even, the value of the median is the average of the 15th and 16th observations when arranged in an array and is $(18 + 19)/2 = 18.5$.

While it appears that treatment A is superior to the others, a test of hypothesis is desirable to see if this is simply attributable to chance. If we assume only that the observations on any treatment have the same median, then our null hypothesis should be that all treatments have the same median, that is, $H_0: Q_2(A) = Q_2(B) = Q_2(C)$. Let this be our null hypothesis. Note that it was not suggested by the data. The alternative is that the three medians are not the same since there was no information available prior to the conduct of the experiment to suggest any other alternative.

TABLE 20-2-2 / Numbers of Observations above and below Overall Median for Three Treatments

	Treatment			
Observations	*A*	*B*	*C*	Totals
Above Median	9	3	3	15
Below Median	1	7	7	15
Totals	10	10	10	30

One would expect to test this null hypothesis by the usual χ^2 test criterion, Equation 15-5-3, with $t - 1$ degrees of freedom for a $2 \times t$ contingency table. However, a computing formula giving somewhat different results is generally used and is considered to be a slightly better approximation to the true χ^2 distribution. It is used here without being defined; it is defined and explained after the illustration.

For these data, compute χ^2 as follows.

$$\chi^2 = \frac{30(29)}{15(15)}\left[\frac{1}{10}(9 - 5)^2 + \frac{1}{10}(3 - 5)^2 + \frac{1}{10}(3 - 5)^2\right]$$
$$= 9.28^{**} \qquad 2 \text{ df}$$

Since the observed $\chi^2 = 9.28$ is greater than tabulated $\chi^2 = 9.21$ at the .01 level for 2 degrees of freedom, we conclude that chance and the null hypothesis cannot account for the data and so we accept the alternative. We conclude that the population median responses to the three treatments differ.

In summary, when we have observations on t treatments but can make only minimum assumptions about their distribution, a sign test of the null hypothesis of a common median is appropriate. The number of observations need not be the same for all treatments. The test criterion is distributed approximately as χ^2.

We carry out the sign test as follows.

1 Compute the median of all the data, ignoring treatments.
2 For each treatment, observe how many times, say m_i, $i = 1, \ldots, t$, it is above this median and how many times not above, $n_i - m_i$. Notice that $n_i - m_i$ is the number of observations less than or equal to the median and will be larger by 1 than m_i when n_i is odd.
3 Prepare a $2 \times t$ table, such as Table 20-2-3, of the counts just made.

TABLE 20-2-3 / Numbers of Observations above and Not above the Overall Median

Observations	Treatment				Total
	T_1	T_2	$\cdots$	T_t	
Above	m_1	m_2	$\cdots$	m_t	m
Not above	$n_1 - m_1$	$n_2 - m_2$	$\cdots$	$n_t - m_t$	$n - m$
Total	n_1	n_2	$\cdots$	n_t	n

$$n = \sum n_i \qquad m = \sum m_i = \begin{cases} n/2 & \text{for } n \text{ even} \\ (n-1)/2 & \text{for } n \text{ odd} \end{cases}$$

4 Compute a value of χ^2 by Equation 20-2-2. This is distributed approximately as χ^2 with $t - 1$ degrees of freedom.

$$\chi^2 = \frac{n(n-1)}{m(n-m)} \sum_{i=1}^{t} \frac{1}{n_i}\left(m_i - \frac{mn_i}{n}\right)^2 \qquad (20\text{-}2\text{-}2)$$

5 Compare the value of the test criterion with tabulated χ^2 for $t - 1$ degrees of freedom.

This test is against two-tailed alternatives, that is, that the medians are not identical.

Under the null hypothesis, $(m/n)n_i = mn_i/n$ is simply the expected value for one of the top row cells. Since we expect about half the observations to be above the median and half below, the expected values will be 5s for our illustration.

EXERCISES

20-2-1 Use the sign test of this section on the data of Exercise 19-3-1 to test the null hypothesis of no differences among population medians for the four rations. Compare your present findings with those obtained by the *F*-test in the anova.

20-2-2 Apply the sign test of this section to the data of Exercise 19-3-2 to test the null hypothesis of no differences among treatment medians. Compare your present conclusion with those obtained earlier. Does this χ^2 test criterion sound as though it might have any drawbacks as a test criterion for use with these data?

20-2-3 Apply the test of this section to the data of Exercise 19-3-3. The null hypothesis is that of no differences among teaching methods. Compare your conclusion with those obtained earlier.

20-2-4 Apply the test of this section to the data of Exercise 19-7-1. Compare your present conclusion with those obtained in the earlier exercise.

20-3 KRUSKAL-WALLIS *t*-SAMPLE TEST: ONE-WAY CLASSIFICATION

Kruskal and Wallis (1952) developed a test criterion based on ranks which is appropriate for the completely random design. The procedure comes close to being an anova of ranks. However, the population variance is computed from the finite population of ranks, and this is used, in preference to an estimate, in computing the test criterion. The test criterion uses the treatment sum of squares in a χ^2 value rather than a treatment mean square in an *F*-value. As for other rank tests, we require that all populations sampled be continuous (see Section 18-5). Consequently, the null hypothesis is that the populations are identical rather than that they have identical locations. If we are also able to assume that the populations are identical except possibly for location, then the null hypothesis is that they have the same location, and this is often interpreted as referring to the median.

If the assumption and null hypothesis are true, then all the observations can be considered to be one large sample from a single population. Consequently, if we rank these observations together, we have a random permutation of n_1 objects of one kind, n_2 of another, and so on, where n_i is the number of observations drawn from the ith population. Also, all possible permutations are equally likely.

Because it is possible to enumerate all these equally likely permutations, it is, in turn, possible to generate the distribution of any proposed test criterion based on ranks. In particular, we could propose that the test criterion be the treatment mean square or sum of squares and generate its distribution. In fact, the exact distribution of the treatment sum of squares has been tabulated for some small values of n_i when all are equal, but the amount of computation becomes prohibitive for large values of n_i, whcther equal or unequal. An alternative is needed.

Since we are using a known finite population of ranks, we can compute the population mean and variance, and use this information to develop other test statistics. In particular, the ranks assigned to the sample observations from the ith population may be considered to be a random sample of n_i ranks from this finite population which has been sampled without replacement. In turn, we can take the arithmetic mean or the total of the n_i ranks and construct a variable like the $Z = (Y - \mu)/\sigma$ of Chapter 12, that is, one with zero mean and unit variance. Presumably, we will use the finite population correction associated with sampling without replacement. Next we rely on the central limit theorem and say that this variable is distributed approximately as a normal variable with zero mean and unit variance. A test of our null hypothesis will require some pooling of these treatment variables. While they are not independent, since the individual rank sums must total $n(n + 1)/2$ where $n = \sum n_i$, *Trt* SS$/\sigma^2$ for an appropriate σ^2 is still an obvious choice for a test criterion, because it could be distributed approximately as χ^2 with $t - 1$ degrees of freedom. This procedure is essentially that which Kruskal and Wallis have followed to produce their test criterion.

The Kruskal-Wallis t-sample test is conducted as follows.

1 Rank all observations together from smallest to largest. Tied observations are given the average of the ranks which would be assigned if there were no ties.
2 Sum the ranks for each sample.
3 Compute the sample value of the test criterion given as Equation 20-3-1 and compare with tabulated χ^2 values for $t - 1$ degrees of freedom from Table B-3.

$$H = \frac{12}{n(n + 1)} \sum_i \frac{r_i^2}{n_i} - 3(n + 1) \qquad (20\text{-}3\text{-}1)$$

In Equation 20-3-1, n_i is the number of observations in the ith sample, $i = 1, \cdots t$, $n = \sum n_i$; r_i is the sum of the ranks for the ith sample; and 3 and 12 are constants. H is distributed approximately as χ^2 with $t - 1$ degrees of freedom if the n_i are not too small. For $t = 2$, use Wilcoxon's test. For $t = 3$ and all combinations of the n_i's up to 5, 5, 5 a table of exact probabilities is given by Kruskal and Wallis (1952).

Illustration 20-3-1

The data of Table 20-2-1 are already ranked so that we need only the rank sums $r_1 = 211$, $r_2 = 134.5$, and $r_3 = 119.5$ to be able to compute H.

$$H = \frac{12}{30(31)} \frac{211^2 + 134.5^2 + 119.5^2}{10} - 3(31) = 6.21 \qquad 2 \text{ df}$$

Since $6.21 > 5.99$, the tabulated χ^2 value for $\alpha = .05$ and 2 degrees of freedom, the null hypothesis is rejected and we conclude that there are real differences in the locations of the three populations.

Examination of the rank sums suggests that treatments B and C have reaction times that are not very different while that for treatment A is noticeably longer. Differences among treatments will not always be as easily recognized.

A correction, given by Equation 20-3-2, may be made when ties occur, but will not usually change the value of H appreciably.

$$\text{Divisor} = 1 - \frac{\sum T}{(n-1)n(n+1)} \tag{20-3-2}$$

Here $T = (h-1)h(h+1)$ for each group of ties, and h is the number of tied observations in the group. This number is used as a divisor of H to give a corrected H. For the data of Table 20-2-1, there are four groups of $h = 2$ tied observations given ranks $2\frac{1}{2}$, $7\frac{1}{2}$, $16\frac{1}{2}$, and $27\frac{1}{2}$, five groups of $h = 3$ tied observations given ranks of 5, 14, 19, 22, and 25, and one group of $h = 4$ observations given the rank of $10\frac{1}{2}$. Hence,

$$\text{Divisor} = 1 - \frac{4[1(2)3] + 5[2(3)4] + 1[3(4)5]}{29(30)31} = .9924$$

$$\text{Corrected } H = \frac{6.21}{.9924} = 6.26$$

Corrected H is only slightly larger than H, and there is no change in the conclusion.

EXERCISES

20-3-1 Apply the Kruskal-Wallis t-sample test of the null hypothesis of no differences among population medians or among populations to the data in the Exercises referred to below. Compare your conclusion with those obtained by the F-test in the analysis of variance and the sign test of Section 20-2.

(a) Data from Exercise 19-3-1.

(b) Data from Exercise 19-3-2. Do you think, given that the assumptions for the Kruskal-Wallis test are valid, that this test has advantages over the sign test other than being more powerful?

(c) Data from Exercise 19-3-3.
(d) Data from Exercise 19-7-1.

20-3-2 It is sometimes necessary to estimate the number of plants, clover in this example, in experimental plots because counting all is too time-consuming and expensive. Randomly selected small areas are taken within each plot and the number of plants counted. The data in this exercise are the sums of the counts for three sampling units in each experimental unit of a completely random design.

Experimental Unit	Treatment				
	1	2	3	4	5
1	40	57	58	33	48
2	45	62	67	29	51
3	53	56	57	27	43
4	45	67	62	51	59

(a) What can be said for sure about the validity of the underlying assumptions if we plan to analyze the data by
(1) The sign test
(2) The Kruskal-Wallis test
(3) The analysis of variance followed by an *F*-test
(b) Analyze the data and test the null hypothesis of no differences among medians or means, as required.
(c) Compare results from the three procedures.

20-4 THE SIGN TEST FOR THE TWO-WAY CLASSIFICATION

Development of a sign test for application to data from a randomized complete block experiment with $t \geq 2$ treatments would appear to be fairly straightforward. First, the effect of blocks must be eliminated. This can be done by working from the block median for each block. Secondly, we would expect to need the number of times a treatment exceeds its block median. From this information, we could construct a $2 \times t$ table, that is, one with 2 rows and t columns, in which, if the null hypothesis of no treatment differences is true, we would expect any treatment to exceed the block median about half the time. If the null hypothesis should be false, then a "good" or "poor" treatment would give responses that tend to be more often on one side of the block medians. Consequently, the χ^2 test of homogeneity in a two-way table should be an appropriate test criterion to test the null hypothesis. This is, in fact, very close to the procedure described in this section.

For a sign test, the assumptions about the underlying distribution are minimal, namely that the distributions, whatever they may be, are identical except possibly for location. We also assume that block and treatment contributions, if any, to the median of the population from which any observation is a sample value, are independent. This assumption is not necessary if the blocks or the treatments have been chosen at random. The null hypothesis is that there are no differences among the true population medians of the treatment contributions. The proposed test is against the alternative that all medians are not identical.

We will outline the procedure and follow it immediately with an example.

The sign-test procedure for a randomized block design with t treatments and b blocks is as follows.

1 Find the median of the observations in each block.
2 Replace each observation by a + or a − according to whether it is above or not above the median of the observations in the block. If t is odd, the median of each block becomes a −.
3 Arrange the numbers of + and − signs by treatments in a $2 \times t$ table. Table 20-4-1 shows the general case.
4 Use the test criterion defined by Equation 20-4-1. It is distributed approximately as χ^2 with $t - 1$ degrees of freedom, but is slightly different from the χ^2 test criterion usually used with two-way contingency tables.

$$\chi^2 = \frac{bt(t-1)}{m(tb-m)} \sum_{i=1}^{t} \left(m_i - \frac{m}{t}\right)^2 \qquad t-1 \text{ df} \qquad (20\text{-}4\text{-}1)$$

TABLE 20-4-1 / Numbers of Observations above and Not above the Block Medians, by Treatments, in a Randomized Block Experiment

	Treatment				
Observations	1	2	...	t	Totals
Above median	m_1	m_2	...	m_t	m†
Not above median	$b - m_1$	$b - m_2$	...	$b - m_t$	$tb - m$
Number of blocks	b	b	...	b	tb

† $m = tb/2$ for t even; $m = (t-1)b/2$ for t odd because median is a −.

We now use the data of Table 19-4-1 to illustrate the procedure. These are repeated in Table 20-4-2 with signs, as required for this test, and ranks for a later test.

Illustration 20-4-1

Medians for the observations on the five treatments are found for each block. Observations larger than the median are assigned a +; there are two. Observations less than the median and the median itself are assigned a −.

TABLE 20-4-2 / Coefficients of Digestibility of Total Carbohydrates for Steers Fed Five Rations or Treatments

Ration = Treatment	Block 1†	2	3	4	5	6	r_i
1	86.5 (5)+	74.5 (3)−	68.8 (4)+	79.9 (5)+	78.2 (5)+	86.8 (5)+	27
2	78.2 (4)+	76.9 (4.5)+	67.8 (3)−	74.2 (2)−	72.5 (2)−	76.5 (4)+	$19\frac{1}{2}$
3	74.7 (3)−	72.3 (1)−	72.7 (5)+	76.3 (4)+	75.8 (4)+	76.1 (3)−	20
4	72.9 (2)−	76.9 (4.5)+	64.7 (1)−	73.2 (1)−	73.2 (3)−	73.2 (2)−	$13\frac{1}{2}$
5	70.8 (1)−	73.5 (2)−	67.2 (2)−	74.5 (3)−	71.5 (1)−	70.4 (1)−	10

† Numbers in parentheses are the ranks of treatments within blocks. The + and − signs indicate whether an observation is above or not above the block median in value.
Source: Watson et al. (1949).

Table 20-4-3 summarizes the results and is convenient when applying Equation 20-4-1 for χ^2. We find

$$\chi^2 = \frac{30(5)(4)}{12(18)}\left[\left(5 - \frac{12}{5}\right)^2 + \cdots + \left(0 - \frac{12}{5}\right)^2\right] = 8.44 \qquad 4 \text{ df}$$

Since $8.44 < 9.49$, the tabulated χ^2 for 4 degrees of freedom and $\alpha = .05$, we accept the null hypothesis of a common median; that is, we find that this hypothesis and chance together offer an adequate explanation of the data.

TABLE 20-4-3 / Numbers of Observations above and Not above the Block Medians, by Treatments, for the Data of Table 20-4-2

Observations	Ration 1	2	3	4	5	Totals
Above Median	5	3	3	1	0	12
Not above Median	1	3	3	5	6	18
Totals	6	6	6	6	6	30

EXERCISES

20-4-1 Apply the sign test to the data of Exercise 19-4-1 to test the null hypothesis of no differences among the crop contributions to the population medians involved. Compare your results with those obtained by the F-test in the analysis of variance.
20-4-2 Repeat Exercise 20-4-1 for the data of Exercise 19-4-2.
20-4-3 Repeat Exercise 20-4-1 for the data of Exercise 19-4-3, testing appropriate null hypotheses about types of plankton and the five different nets.
20-4-4 Repeat Exercise 20-4-1 for the data of Exercise 19-4-4, testing appropriate null hypotheses about drivers and days.

20-5 FRIEDMAN'S RANK TEST: RANDOMIZED BLOCKS

A rank test for data in a two-way table with more than two rows and two columns, as for randomized block experiments, should also be easy to construct. Since block effects have been assumed to be real, or at least very probable, a means of eliminating them is necessary. This could be done by ranking treatment responses within blocks. Treatment rank sums would then contain no block contributions.

A test criterion for the null hypothesis that responses attributable to treatments are all the same would likely involve a sum of squares of rank sums. Finding the true distribution might be possible. Finding an approximation to the true distribution would surely be helped by the fact that we are dealing with a finite population of ranks, and so can compute the true mean and variance and incorporate them in any test criterion. This makes χ^2 a likely candidate for the approximate distribution provided the experiment is sufficiently large. We have just described essentially what Friedman (1937) has done.

For a rank test, the assumptions are not very stringent. It is assumed only that, within any block, the observations must come from the same distribution except perhaps for location. The null hypothesis is, then, that all treatment contributions are identical.

If the assumptions and null hypothesis are both true, then all permutations of the observations are equally likely. This is sufficient information to allow us to compute the exact distribution, under the null hypothesis, of any test criterion based on ranks. For t treatments, there will be $t!$ such permutations in any block and $(t!)^b$ for the experiment. This number, $(t!)^b$, becomes large fairly rapidly as either t or b increases so that only limited tables of exact probabilities have been computed. An approximation to the exact distribution is generally used.

Friedman's rank test of the null hypothesis of no population differences attributable to treatments in a randomized block experiment is carried out as follows.

1 Rank the treatments in each block from smallest to largest. Tied observations are given the average of the ranks which would have been assigned if there had been no ties.
2 Sum the ranks for each treatment. Let these be r_i, $i = 1, \ldots, t$.

3 Compute the value of the test criterion given by Equation 20-5-1 for an experiment with b blocks.

$$\chi_r^2 = \frac{12}{bt(t+1)} \sum r_i^2 - 3b(t+1) \qquad t-1 \text{ df} \qquad (20\text{-}5\text{-}1)$$

4 Compare the value of the test statistic with critical values from the χ^2 table.

We now apply Friedman's test to the data of Table 20-4-2.

Illustration 20-5-1

Table 20-4-2 shows ranks in parentheses for treatments ranked in each block. Block 2 contains a tied pair. Rank sums for each treatment are shown in the last column of the table.

For the test statistic, we find

$$\chi^2 = \frac{12}{6(5)(5+1)}(27^2 + 19.5^2 + 20^2 + 13.5^2 + 10^2) - 3(6)(5+1)$$
$$= 11.43 \qquad 5 - 1 = 4 \text{ df}$$

Since $11.43 > 9.49$, the tabulated χ^2 for 4 degrees of freedom and $\alpha = .05$, we reject the null hypothesis that treatments contribute equally to the locations of the populations of which each observation is a random sample.

In summary, an analysis of variance was originally carried out on these data, and the F-test of the hypothesis of no treatment effects was highly significant. By Friedman's test, we find χ_r^2 significant when testing essentially the same null hypothesis. The sign test was not significant. Although this is a single experiment and cannot be expected to provide conclusive evidence, the results of the tests of significance do indicate the relative power of the test procedures to detect alternative hypotheses when the assumption of an underlying normal distribution is justified.

Friedman has given tables of the exact distribution of χ_r^2 for some pairs of small values of t and b.

EXERCISES

20-5-1 Apply Friedman's test to the data of Exercise 19-4-1. Compare your conclusion with that from each of Exercises 19-4-1 and 20-4-1.

20-5-2 Repeat Exercise 20-5-1 for the data of Exercises 19-4-2 and 20-4-2.

20-5-3 Repeat Exercise 20-5-1 for the data of Exercises 19-4-3 and 20-4-3. Tests were made of two different hypotheses.

20-5-4 Repeat Exercise 20-5-1 for the data of Exercises 19-4-4 and 20-4-4. Tests were made of two different hypotheses.

20-6 RANDOMIZATION TESTS

Randomization tests are developed under the assumption that if the null hypothesis is true, the observed data simply constitute one arrangement chosen at random from among all possible randomizations of these observations for the type of experiment used, and that all randomizations are equally likely.

For example, the data in Table 20-6-1 are the times it took each of eight men to run a mile on the same track. The times in the first column are those of men who were over 44 years of age at the time while those in the second column are those of men under 44. Most of us would assume that men in the former group would tend to require longer running times.

TABLE 20-6-1 / Times for Men to Run 1 Mile

	Over 44 Years Old	Under 44 Years Old
	8.32†	7.50
	8.93	6.17
	6.78	5.98
	7.78	7.85
Totals	31.81	27.50
Means	7.95	6.87

† Times are in minutes, to two decimal places.
Source: Linnerud (1968).

If we hypothesize that there is no reason for men who are less than 44 to have running times different from those for men over 44, then there is no reason for any one of the given numbers to have been assigned to one column rather than the other. Consequently, the assignment of the numbers to the columns has simply been a random one with the provision that there must be four in each column.

For any particular experiment, it is possible to compute the number of possible randomizations and, in fact, to tabulate them. In this tabulation, some randomizations will appear to support the null hypothesis and some will appear to deny it. Thus, if every treatment seems to have a scattering of numbers from all parts of the total set of data, we will not suspect that the null hypothesis is false; but if mostly large numbers are associated with one treatment and mostly small ones with another, we will suspect that the null hypothesis is false.

For the running times, four of eight have been assigned to column one. There are $\binom{8}{4} = 70$ ways in which this can be done, all considered to be equally likely under the null hypothesis. Table 20-6-2 shows four of the possible randomizations and their average times. Four others would be simply the same numbers under reverse column headings. In other words, tabulation of the randomizations would require us to find only a particular 70/2 = 35, say all the positive differences, since the remainder would be these with column headings reversed to give negative differences. Of the four randomizations shown, the first three can be seen to be the most extreme, whereas the fourth is a balanced one in that the means are the same although the original totals were not quite identical. If the observed set of data had been the first randomization, the null hypothesis would have been suspect, whereas if the set had been the fourth, the null hypothesis would appear to be supported.

TABLE 20-6-2 / Some Possible Randomizations of the Running-times Data, with Means

(1)		(2)	
8.93	7.50	8.93	7.78
8.32	6.78	8.32	6.78
7.85	6.17	7.85	6.17
7.78	5.98	7.50	5.98
8.22	6.61	8.15	6.68
(3)		**(4)**	
8.93	7.85	8.93	7.50
8.32	6.78	6.78	8.32
7.50	6.17	6.17	7.85
7.78	5.98	7.78	5.98
8.13	6.69	7.41	7.41

Obviously, we need a more definitive procedure than the above for deciding whether or not the null hypothesis is acceptable. From a complete tabulation of the randomizations, it is possible to compute all possible values of any test criterion that might be proposed for testing the null hypothesis, for example, t, F, or simply the difference between totals.

Table 20-6-3 is a tabulation of the differences, arranged in descending order of magnitude, between column totals. We can take this table of test-criterion values generated from the data and associate some of them with rejection and others with acceptance of the null hypothesis. The association must, of course, be a reasonable one, consistent with the alternative hypothesis. Thus, extreme values, both positive and negative, would be associated with rejection for two-tailed

TABLE 20-6-3 / Ordered Differences between Column Totals for All Possible Randomizations (70) of the Running-times Data

6.45	2.65	.71	− .93	−2.71
5.89	2.53	.55	−1.07	−2.81
5.75	2.15	.51	−1.09	−2.85
4.81	2.15	.37	−1.21	−3.09
4.45	2.15	.23	−1.45	−3.23
4.31	1.77	.15	−1.45	−3.37
3.75	1.59	.01	−1.59	−3.59
3.59	1.59	−.01	−1.59	−3.75
3.37	1.45	−.15	−1.77	−4.31
3.23	1.45	−.23	−2.15	−4.45
3.09	1.21	−.37	−2.15	−4.81
2.85	1.09	−.51	−2.15	−5.75
2.81	1.07	−.55	−2.53	−5.89
2.71	.93	−.71	−2.67	−6.45

alternatives in the present case, while one-sided alternatives would call for rejection with extreme values of the appropriate sign. Rejection probabilities can be computed exactly.

For the running-times data, let us suppose that the alternative hypothesis is that men in the older age group are slower runners. Then large positive differences in running times would be cause to deny the null hypothesis. For example, the difference in time of $6.45/4 = 1.61$ minutes per runner would be the first candidate for inclusion in the rejection region. If this were to be the only value, then the probability of rejecting the null hypothesis when true would be $\alpha = 1/70 = 0.014$. If we want α to be about 0.05, we must include 3 or 4 points for $\alpha = 0.043$ or 0.057, respectively. This would take us up to an observed difference in total times of 5.75 in the first case or 4.81 in the latter. In either case, the observed difference in totals of 4.31 minutes would not be cause for rejecting the null hypothesis. We have to conclude that, on the basis of this limited experiment with only four persons in each age group and a 5 percent rejection level, there is insufficient evidence to infer that men in the older group run more slowly than those in the younger group.

Randomization tests are widely applicable. One begins with the hypothesis that treatments have no differential effects. Under this hypothesis, all possible randomizations of the data are then to be considered as equally likely outcomes of the experiment. Of course, only one outcome has been observed. The randomizations are then tabulated or, if there are too many, a random sample of them is tabulated.

Next, a test criterion suitable for testing the null hypothesis against the alternatives is chosen. For each randomization, the test statistic is computed, and these values provide the information necessary to prepare a table of critical values for the desired α levels. The value of the test statistic for the data as observed in the experiment is then compared with the tabulated critical values, and a conclusion as to whether the null hypothesis is to be accepted or rejected is made.

It can be seen that a randomization test can require considerable computing since it is necessary to provide a table of critical values for each experiment conducted. Consequently, such a test is likely to be used only when there are facilities available for fast and inexpensive computing.

Chapter 21

LINEAR REGRESSION

21-1 INTRODUCTION

Previous chapters dealt with data consisting of observations made on a single random variable. However, bivariate observations, observations made simultaneously on two characteristics of an individual or experimental unit and recorded as an ordered pair, are also made. At least one of the pair must be an observation on a random variable for statistical techniques to apply. For example, we might observe height and weight, recording them as a pair in that order, of an individual selected at random from an adult population. Here, both characteristics are random variables. Or we might observe the weights of individuals randomly selected from populations determined by height, for example, the population of weights for all males of height 5 feet 10 inches, etc., and then choose the populations so that their heights cover a wide range. Here heights are nonrandom while weights are random.

Paired observations made on two characteristics of an experimental unit result from experiments performed in all areas of research—physics, biology, medicine, engineering, education, agriculture, and so on.

This chapter deals with some problems where the data to be analyzed are bivariate observations.

21-2 FUNCTIONAL AND STATISTICAL RELATIONS

In studying systems of weights and measurements, a student may be required to plot points consisting of paired values, say temperature on the Centigrade scale against temperature on the Fahrenheit scale. The result is a straight line with all points precisely on it. A *functional relation* exists between the variables because there is a mathematical function that assigns a value on the Centigrade scale for every value on the Fahrenheit scale, namely, $C = (\frac{5}{9})(F - 32) = (\frac{5}{9})F - \frac{160}{9}$.

For some types of data where there is an underlying functional relation, all observations will not generally be on a single line and strictly satisfy this functional relation. This is because of our inability to conduct a perfect experiment and make measurements without error. The statistical problem is to find the particular sample function, called the *regression line*, which best fits all the data.

For other sets of data, a functional relation may not exist. For example, a plot of height against weight for male college students is simply a scattering of points, where one is able to observe only a trend from lower left to upper right. Now, a fixed value of one of the variables determines a population of values of the other variable rather than a single value. This population has a mean dependent on the particular fixed value chosen. The relation between the variables is a *statistical relation.* Here, one of the statistical problems is to estimate the line through the population means. If randomness applies to both characteristics of the bivariate observation and if the roles of the two variables are interchanged, then a second and different regression line and equation result.

21-3 THE LINEAR REGRESSION OF Y ON x

In a heart study, the time it took for each of a number of men to run 1 mile was measured. Also, data were collected on the amount of loose stomach flesh, weight, age, the result of a Harvard step test, and other characteristics, all of which might have some bearing on running times.

Table 21-3-1 consists of data from the heart study for men aged 40 to 46; x = step test results and y = running times for the mile. The step test required each man to step up and down at a fixed rate for a fixed time; his pulse was counted immediately afterward and then again after a specified interval of rest. The rate at which the pulse falls off is a measure of the man's physical condition, a fast drop-off indicating good condition. Presumably there should be a statistical relationship between general physical condition and running times. Our concern will be to use the information on the step test to say something about populations of running times.

Let us look at the first observation, $(x, y) = (49, 7.30)$, to see how the data might have arisen. We agree that the step-test value of 49 has been measured without error; in other words, if the set of circumstances leading up to this measurement could be reproduced for this individual, then a repeat of the test would again yield the observation 49. Further, the value 49 determines a population of running

TABLE 21-3-1 / Heart-study Data on Joggers

	x = Step Test		y = Time in Min for Mile Run
	49		7.30
	68		11.20
	44		7.78
	36		6.83
	57		9.00
	50		7.25
	54		7.50
	68		8.25
	41		6.40
	53		7.78
Totals	520		79.29
Means	52.0		7.929
Sums of products:	x, x	x, y	y, y
(1) Unadjusted for means	28,036	4,223.74	645.2707
(2) Adjusted for means	996	100.66	16.58029

Source: Linnerud (1968).

times in that there are many men in this age group who have a step-test value of 49 but whose running times need not be 7.30 minutes. We have a single random value from this population and it is, in fact, 7.30 minutes. Since the value of x determines the population of y's, the x should be measured without error.

The variable "running time" is called a *dependent variable.* Any running-time value comes from a particular population, specified by the step-test value and, hence, is dependent on it. Since the step-test value simply determines the population of running times, step-test values need not be random and may be chosen arbitrarily if this is desired. Consequently, step test is called an *independent variable.* An independent variable is also called a *concomitant variable,* since it provides a value to accompany the random variable value. The dependent random variable is generally symbolized by Y, and the independent one by X or x according to whether it is a random or a fixed variable.

The population of running times for those persons with a step-test value of 49 has its own mean and variance. Another value of the step test will determine another population of running times, with its own mean and variance. By definition, the line passing through the means of all the running time populations is called the regression of Y (running time) on x (step-test value). Notice that nothing has been said about the shape of the regression line.

In general, the *regression of the dependent random variable Y on the independent variable x* is the line passing through the means of the populations of Y values, each population being determined by a value of x.

Having specified how the data arise and having defined regression, we now make some assumptions about the nature of our data and proceed to a solution of the problem of estimating the regression line and the variance about it.

Assume that the regression line is a straight line. As such it may be represented by an equation of the form $y = a + bx$ and illustrated by means of a graph such as Figure 21-3-1. The y intercept, where $x = 0$ and the line crosses the y axis, is a. If a is negative, then the y axis is crossed below the x axis. The slope b of the line measures the increase in y per unit increase in x. If b is negative, then the line runs downhill to the right. If we know the y intercept and the slope, the line is completely determined.

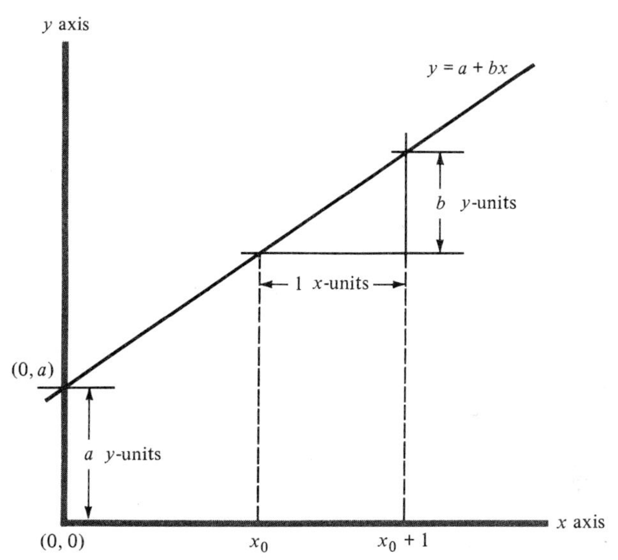

FIGURE 21-3-1 The Straight-Line Graph.

Our assumption of linearity of regression can be written as Equation 21-3-1.

$$\mu_{Y|x} = \mu_{Y.x} = \alpha + \beta x \qquad (21\text{-}3\text{-}1)$$

This equation is read as follows: "The mean of the population of Y's for a given x, namely, $\mu_{Y|x}$ or $\mu_{Y.x}$, equals a constant α plus a multiple β, of x." Thus α and β are the parameters needed to specify completely the straight line on which the population means lie. Figure 21-3-2 illustrates the situation we are describing.

A satisfactory data analysis will require an estimate of the population regression line, Equation 21-3-1, by means of a sample regression line. The estimator, which is a random variable, and the sample estimate are often written as Equation 21-3-2.

$$\hat{\mu}_{Y|x} = \hat{Y} = a + bx \qquad (21\text{-}3\text{-}2)$$

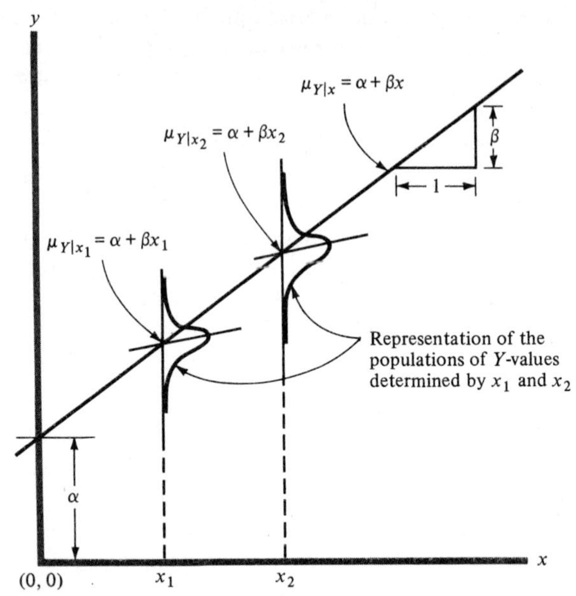

FIGURE 21-3-2 A Population Linear-Regression Line; Y Regressed on x.

Here, the parameters α and β of Equation 21-3-1 have been replaced by estimates a and b, which are sample statistics. Long-time usage has us use lowercase letters to represent both the estimators and the sample estimates. Sample lines computed from different sets of data from the same population will have slopes that differ from β and from each other, being either too steep or too shallow relative to β, and will have intercepts that are above or below α. In other words, our estimate of the regression line is subject to sampling variation in both a and b. This is further indicated by the ^ on $\mu_{Y|x}$ or on Y, whichever is used.

Another assumption about the true regression line is that the variance about it is, unlike the mean, independent of x. This assumption states that every population of y's, where a population is determined by the choice of an x, has the same variance. For data where the relation is a statistical one, this assumption is often reasonable. Solutions other than that which follows are available when the assumption of homogeneity is not met.

The *variance about regression* or the *variance of Y adjusted for x* is designated by $\sigma^2_{y \cdot x}$ when it is a population parameter and by $s^2_{y \cdot x}$ when a sample statistic. It is a measure of the variation of the individual y-values from the regression line. Figure 21-3-3 will help in understanding this measure. Here, the data of Table 21-3-1 have been plotted in a *scatter diagram*; also, on the right, the y-values have been plotted without their accompanying x-values. If we look at just the y-values, we see considerable variation, from 6.40 to 11.20 with deviations from the mean of $6.40 - 7.93 = -1.53$ to $11.20 - 7.93 = 3.27$. However, the deviations that should go into the computation of $s^2_{y \cdot x}$ are to be measured from points that estimate

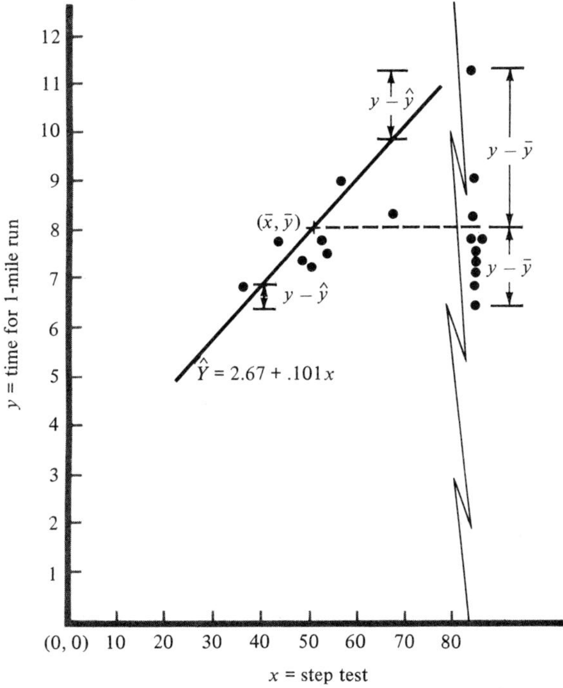

FIGURE 21-3-3 Plot of Data in Table 21-1, Showing also Regression Line and Deviations from Regression.

population means, namely those on the sample regression line. Two such deviations are shown; these are lines vertical to the x axis, since an x specifies the population from which each comes. Obviously, deviations from the regression line are smaller than those from the overall mean, $\bar{y}$. Deviations from the regression line have been *adjusted for* x and are truly *deviations from regression.* Thus, to estimate the variance about regression, we must efficiently pool the information available from all the populations sampled. Though we may have only one observation from each population, estimation of the regression line is possible because we have assumed a form for the regression equation, namely, linearity.

Estimation of the parameters α and β in Equation 21-3-1 is, by the least-squares procedure, a procedure mentioned in Section 19-5 in connection with the randomized block design. Thus in Equation 21-3-2, any two real numbers are candidates for a and b. These lead to an observed deviation from the regression line of $y_i - a - bx_i$ for each observed pair (x_i, y_i) and to the computation of the sum of squares of deviations, namely, $\sum_{i=1}^{n} (y_i - a - bx_i)^2$. From among all possible (a, b)-pairs, there is one for which the sum of squares is a minimum. This pair constitutes least-squares estimates of the parameters, and the sum of squares of the deviations is information necessary for computing the variance about regression. The sample least-squares estimates of α and β are found most easily using the tools of elementary calculus and are given by Equations 21-3-3 and 21-3-4.

$$a = \bar{y} - b\bar{x} \tag{21-3-3}$$

$$b = \frac{\sum (x_i - \bar{x})(y_i - \bar{y})}{\sum (x_i - \bar{x})^2} \tag{21-3-4}$$

For b, the above equation may be described as a definition formula in contrast to a computing formula. The numerator, when divided by the degrees of freedom, is called the *covariance* and is a measure of the extent to which x and y vary jointly. Its magnitude alone is not too helpful because it is dependent on the units of measurement. The numerator sum of products is usually computed by the right-hand side of Equation 21-3-5, a formula corresponding to that used for computing sums of squares.

$$\sum_i (x_i - \bar{x})(y_i - \bar{y}) = \sum_i x_i y_i - \frac{\sum_i x_i \sum_i y_i}{n} \tag{21-3-5}$$

Illustration 21-3-1

For the data in Table 21-3-1, we find

$$b = \frac{100.66}{996} = .10106 \text{ min/pulse unit}$$

$$a = 7.929 - .10106(52.0) = 2.67 \text{ min}$$

We may now write the equation of time to run a mile regressed on step-test value as

$$\hat{Y} = 2.67 + .101x$$

The regression equation is often conveniently expressed as in Equation 21-3-6, from which we can also see that $(\bar{x}, \bar{y})$ is a point on the regression line.

$$\begin{aligned}\hat{Y} &= \bar{y} + b(x - \bar{x}) \\ &= 7.929 + .101(x - 52.0)\end{aligned} \tag{21-3-6}$$

Now, if we wish to estimate the mean running time for a population of individuals with step-test value of 40, we find

$$\begin{aligned}\hat{Y} &= 2.67 + .101(40) && \text{by Equation 21-3-2} \\ &= 7.929 + .101(-12.0) && \text{by Equation 21-3-6} \\ &= 6.71 \text{ min}\end{aligned}$$

This value, 6.71 minutes, is our estimate of the population mean of running times of men with a step-test value of 40; the population of running times is expected to vary about this estimate. Of course, the estimate is subject to sampling variation, and another sampling and regression computation would lead to a new and presumably different estimate of this value. However, the average of the b's from all possible samplings is β. In other words, the random variable b is an unbiased estimate of β.

EXERCISES

21-3-1 Estimate the population mean running time for the population with $x = 36$, $x = 44$, $x = 52$, $x = 60$, and $x = 68$.

21-3-2 What are the deviations of the observed values from the expected, as best we can estimate them, when $x = 36$, $x = 44$, and $x = 68$? Notice that you have two values of y when $x = 68$ and, so, two deviations.

21-3-3 Bouwkamp and McCully (1972) studied the relationship between the age in years x of an asparagus bed and the percent female plant survival Y. The observations are:

x	1	2	3	5	7	8	11	12	13	15	18	19
y	58	55	56	56	48	53	42	46	52	47	49	39

The x-values are fixed and measured without error. The y-values were read from a graph and so may have some variation in addition to that associated with the original experiment.

Prepare a scatter diagram of the data. Compute the y intercept and slope of the regression line. Present the regression equation in the forms of Equations 21-3-2 and 21-3-6. Compute $\hat{Y}$ values for $x = 1$ and 19 and use these to draw the regression line on your scatter diagram. Find the deviations of the observed values from their corresponding $\hat{Y}$'s for $x = 1$, 11, and 19.

21-3-4 Baer (1956) studied the relationship between the fixed variable age x and a random variable Y that measured the mean face height minus the mean nose length in millimeters during the third decade of life.

x	19	20	21	22	23	24	25	26	27
y	.30	.35	.55	1.06	.74	1.43	1.55	1.25	1.52

x	28	29	30	31	32	33
y	1.57	2.06	1.99	2.05	1.75	2.36

Prepare a scatter diagram of the data. Compute the y intercept and slope of the regression line. Present the regression equation in the two forms suggested. Compute $\hat{Y}$ values for $x = 19$, 26, and 33. Draw the regression line on your scatter diagram. Compute the deviations of the observations from the corresponding $\hat{Y}$'s just computed.

21-3-5 Bergerud et al. (1964) studied relationships for 13 stag caribou in Newfoundland, Canada. For each animal, they obtained the weight in pounds y and the value of an independent variable x, which was the sum of the total length plus the chest girth plus the length of the hind foot, all in inches. For this experiment, the x-values are also on a random variable, namely, X.

x	130	137	134	144	146	150	156	153	155	154	158	160	166
y	165	175	200	215	280	290	305	325	355	375	375	385	390

Prepare a scatter diagram. Compute a, b, and the regression line in the two suggested forms. Draw the regression line through your plot. Compute $\hat{Y}$ values and corresponding deviations for $x = 130, 150$, and 166.

21-3-6 Levy et al. (1972) studied the body's capacity for salicyl phenolic glucuronide formation and its effects on the kinetics of salicylate elimination in man. They present the following observations on the amount x, in grams, of the latter chemical in the body and the ratio y of this amount to the excretion rate of the former per hour. Notice that both X and Y are random variables.

x	.1	.2	.2	.3	.4	.5	.6	.8	1.2	1.4
y	18	19	24	28	34	30	25	43	55	62

x	1.6	1.7	1.8	1.8	1.8	2.2	2.3	2.4
y	55	60	65	65	78	80	98	95

Prepare a scatter diagram. Compute a, b, and the regression line in the two suggested forms. Draw the regression line through your plot. Compute $\hat{Y}$-values and corresponding deviations from $x = .1, 1.2$, and 2.4.

A problem in algebra follows.

21-3-7 Prove Equation 21-3-5.

21-4 REGRESSION MODELS

Data for a regression analysis may be obtained by selecting x-values and drawing random values of Y, either one or many, or by drawing random bivariate observations.

Suppose we theorize that blood pressure is linearly dependent upon age. We can obtain data to evaluate this theory by means of a sampling scheme that calls for us to fix age in advance. Thus, we might require that a broad spectrum of ages be represented by specifying what ages the sample must contain. Here, only blood pressure is a random observation. A model to describe a regression based on such a sampling scheme is called a *fixed model*. Inferences drawn from such an experiment apply to blood pressure for the sampled ages only.

On the other hand, an agronomist, an agricultural scientist who experiments with plants, might be looking for a relation between some value of rainfall and wheat yield per acre for a specified cultivar or variety. Data are available on both variables for a number of years and at many locations. The agronomist would likely consider that the pairs of observations were random. The model in this case is a *random model*. Inferences drawn from this model can apply to populations of Y's for other than those at the observed x's.

In the case of the fixed model, we compute an equation $\hat{Y} = a + bx$ and use it to estimate mean values for populations of Y's. On occasion, we may wish to estimate an x-value. For example, if x is an insecticide dosage and Y measures the proportion killed in an experiment using a fixed number of insects at each dosage, then we choose the dosages nonrandomly by specifying the amount of dilution of the original insecticide. On the other hand, the proportion killed varies from trial to trial for the same dosage. In advertising the insecticide, the manufacturer may wish to specify the dosage that will kill 50 or 90 or some other percentage of the insects to be controlled. Since only Y was random, the appropriate equation is the same one. We simply rearrange it. Thus, we use the equation $\hat{X} = (y - a)/b$ to estimate the required dosage. Problems of this sort are of interest to many research workers; for example, see Finney (1947).

For the *random model*, neither variable is obviously the dependent one. Thus, if we arbitrarily label a bivariate observation by assigning the random variable symbol X to the first and Y to the second, then we compute $\hat{Y} = a + bx$ if we wish to estimate means of Y-populations; and we compute $\hat{X} = a' + b'y$ if we wish to estimate means of X-populations. Note that the two regression coefficients are different random variables defined by

$$b = \frac{\sum (X_i - \bar{X})(Y_i - \bar{Y})}{\sum (X_i - \bar{X})^2} \qquad \text{and} \qquad b' = \frac{\sum (X_i - \bar{X})(Y_i - \bar{Y})}{\sum (Y_i - \bar{Y})^2}$$

so that b and b' are seen *not* to be reciprocals of one another.

Regardless of whether the model is fixed or random, when Y is regressed on x, then Y must have a random component. For the regression techniques discussed here, these random components must be normally and independently distributed with zero mean and homogeneous variance, for tests of significance and confidence intervals to be valid at the tabulated significance levels and for the specified confidence coefficients.

EXERCISES

21-4-1 For the following sets of data state whether it would be appropriate to call x a random variable: the data of Exercise 21-3-3, Exercise 21-3-4, Exercise 21-3-5, and Exercise 21-3-6.

21-4-2 Do you think one would want to estimate population means for the Y variable for values of x between those given in Exercise 21-3-3? Exercise 21-3-4? Exercise 21-3-5? Exercise 21-3-6?

21-5 SOURCES OF VARIATION IN REGRESSION

We now return to the problem of estimating the variation about regression and, in so doing, will take a closer look at regression as a source of variation.

As usual, we are describing a random variable Y_i by a model that specifies an observation to be the sum of a population mean and a random component. However, the population mean is now linearly dependent upon a concomitant value of a variable x_i, not necessarily random. Equation 21-5-1 applies.

$$Y_i = \alpha + \beta x_i + \varepsilon_i \qquad (21\text{-}5\text{-}1)$$

Replacing the parameters α and β by their estimators, we write Equation 21-5-2.

$$Y_i = \bar{Y} + b(x_i - \bar{x}) + e_i \qquad (21\text{-}5\text{-}2)$$

The estimate e of the random component is also symbolized by $d_{Y.x}$, read as "the deviation of the value of the random variable Y from the regression line at a specified x," or *adjusted* for x. Equation 21-5-3 follows.

$$d_{Y.x} = (Y_i - \bar{Y}) - b(x_i - x) \tag{21-5-3}$$

These last two equations take on more meaning when related to Figure 21-5-1. Note that $d_{Y.x}$ may be either positive or negative, but, for convenience, is shown as positive in the figure; also, we assume that we have a set of data and, consequently, use lowercase letters.

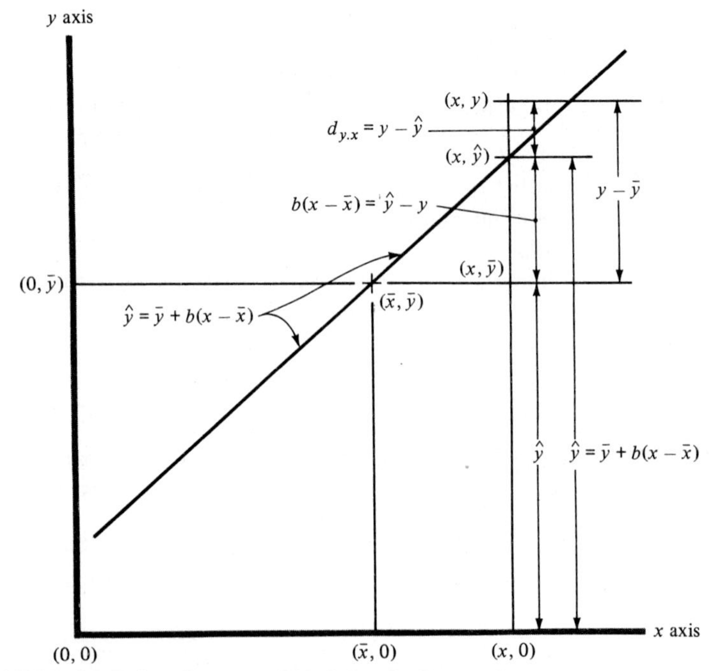

FIGURE 21-5-1 Sources of Variation in Regression.

Since the squared deviations are to be the basis of an estimate of $\sigma^2_{y.x}$, we square the expression for $d_{Y.x}$, sum the squared deviations, substitute the definition of b given in Equation 21-3-4, and, after some algebraic manipulation, obtain Equation 21-5-4. This is our estimator of the sum of squares of the deviations from the regression line.

$$\sum d^2_{Y.x} = \sum (Y_i - \bar{Y})^2 - \frac{[\sum (x_i - \bar{x})(Y_i - \bar{Y})]^2}{\sum (x_i - \bar{x})^2} \tag{21-5-4}$$

Equation 21-5-4 shows the sum of squares about regression to be the sum of squares of deviations from the mean, less an additional component. This latter component is called the *sum of squares attributable to regression* or the *regression sum of squares.* It has a single degree of freedom. Since the first term on the right has $n - 1$ degrees of freedom, $\sum d_{Y.x}^2$ has $(n - 1) - 1 = n - 2$ degrees of freedom.

The estimated *variation about regression* $s_{y.x}^2$ is found by dividing the sum of squares about regression by the degrees of freedom to give Equation 21-5-5.

$$s_{y.x}^2 = \frac{\sum d_{Y.x}^2}{n - 2}$$

$$= \frac{\sum (Y_i - \bar{Y})^2 - [\sum (x_i - \bar{x})(Y_i - \bar{Y})]^2 / \sum (x_i - \bar{x})^2}{n - 2} \quad (21\text{-}5\text{-}5)$$

The subscript states that we are talking about variation in the random variable and that any contribution attributable to variation in x has been removed. If we had used the | symbol instead of the ., we would be suggesting, at least, that variation in the Y-population for the specified x depends upon its value. We have, however, assumed that the variance is homogeneous over all x and, so, independent of it. This sample variance is an unbiased estimate of the population variance about regression $\sigma_{y.x}^2$.

Illustration 21-5-1

For the running-time data, we have the following results:

$$\sum (y_i - \bar{y})^2 = 16.5803 \qquad 10 - 1 = 9 \text{ df}$$

$$\frac{[\sum (x_i - \bar{x})(y_i - \bar{y})]^2}{\sum (x_i - \bar{x})^2} = \frac{(100.66)^2}{996} = 10.1731 \qquad 1 \text{ df}$$

$$\text{Difference} = 6.4072 \qquad 9 - 1 = 8 \text{ df}$$

$$s_{y.x}^2 = \frac{6.4072}{8} = .8009 \qquad 8 \text{ df}$$

The standard deviation or *standard error of estimate* $s_{y.x}$ is found by taking the square root of $s_{y.x}^2$. It does not give an unbiased estimate of $\sigma_{y.x}$.

For the data, $s_{y.x} = .895$ minutes.

In the preceding development, we sought $\sum d_{Y.x}^2$ directly, ending with Equation 21-5-4, which shows it as the difference between two components. As an alternative, we can begin with Equation 21-5-2, square each side, add all observations, and so obtain Equation 21-5-6.

$$\sum Y_i^2 = n\bar{Y}^2 + b^2 \sum (x_i - \bar{x})^2 + \sum d_{Y.x}^2 \quad (21\text{-}5\text{-}6)$$

This is really the same algebraic result as in Equation 21-5-4 but, in this form, is seen to be an algebraic identity that suggests an analysis-of-variance presentation.

Equation 21-5-7 is easily proved.

$$b^2 \sum (x_i - \bar{x})^2 = \frac{[\sum (x_i - \bar{x})(Y_i - \bar{Y})]^2}{\sum (x_i - \bar{x})^2} \qquad (21\text{-}5\text{-}7)$$

The analysis-of-variance presentation of the regression computations based on Equations 21-5-6 and 21-5-7 is given in Table 21-5-1.

TABLE 21-5-1 / Definition and Computing Formulas for an Analysis of Variance with Regression

Source	df	Sums of Squares: Definition		Sums of Squares: Computing	Mean Square
Linear regression	1	$\frac{[\sum (x_i - \bar{x})(y_i - \bar{y})]^2}{\sum (x_i - \bar{x})^2}$	=	$\frac{[\sum x_i y_i - (\sum x_i)(\sum y_i)/n]^2}{\sum x_i^2 - (\sum x_i)^2/n}$	Same
Residual	$n - 2$	$\sum d_{y.x}^2$	=	Subtraction	$s_{y.x}^2$
Total	$n - 1$	$\sum (y_i - \bar{y})^2$	=	$\sum y_i^2 - (\sum y_i)^2/n$	

EXERCISE

21-5-1 For the data at the end of Section 21-3, compute the sum of squares attributable to regression and the residual sum of squares or sum of squares about regression, $s_{y.x}^2$ and $s_{y.x}$. Present your results in analysis-of-variance form. The data are those for:
(a) Exercise 21-3-3. (b) Exercise 21-3-4.
(c) Exercise 21-3-5. (d) Exercise 21-3-6.

21-6 CONFIDENCE INTERVALS AND TESTS OF HYPOTHESES

The sample linear-regression equation provides point estimates of population means. An interval estimate is, however, a more desirable one in that it specifies a range, about the point estimate, within which the population mean will lie for a fixed percentage of such interval estimates. To compute an interval estimate, a standard deviation applicable to the point estimate is required.

At first glance, it might appear that $s_{y.x}$ is the standard deviation of any $\hat{Y}$. Figure 21-6-1 shows that this is not so. First, relate the figure to the regression equation in the form of Equation 21-3-6. The $\bar{Y}$ is obviously subject to sampling variation. The dotted lines indicate an appropriate interval. This sort of variation alone, as indicated by parallel lines, is to be expected only if β is known. If β is to be estimated, as in our case, then we must allow for sampling variation in b as well. For b alone, variation is shown by the dashed line. Finally, to obtain a standard deviation, sum the variances of $\bar{Y}$ and b, since they can be shown to be independent, and take the square root.

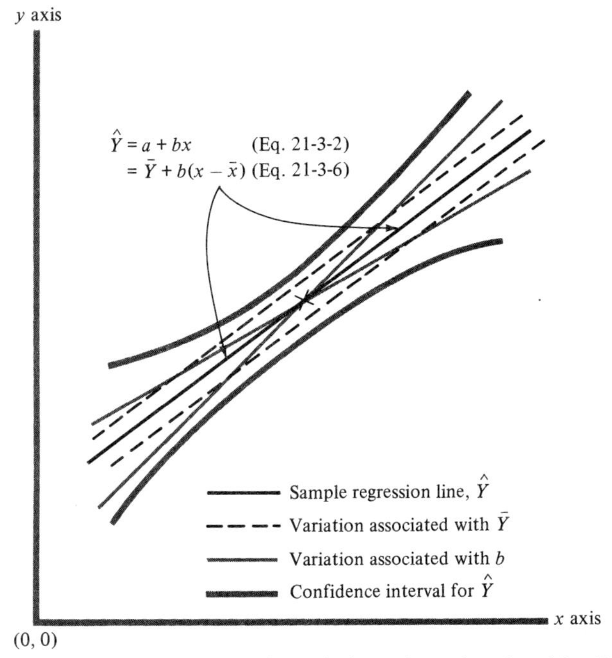

FIGURE 21-6-1 Illustration of Variation Associated with Estimation by Regression Equation.

First consider the variance of $\bar{Y}$ by supposing that, in repeated sampling, we use the same set of x's; that is, we have a fixed model. Then $\bar{x}$ remains constant but $\bar{Y}$ and b vary. From Figure 21-6-1, it is seen that variation in b has no effect on $\bar{Y}$, the only Y-value for which this is so. Consequently, Equation 21-6-1 holds for the population variance.

$$\sigma_{\bar{y}}^2 = \frac{\sigma_{y \cdot x}^2}{n} \tag{21-6-1}$$

When $\sigma_{y \cdot x}^2$ is unknown, we use $s_{y \cdot x}^2$ and Equation 21-6-2.

$$s_{\bar{y}}^2 = \frac{s_{y \cdot x}^2}{n} \tag{21-6-2}$$

For the variance of b, we have Equation 21-6-3.

$$\sigma_b^2 = \frac{\sigma_{y \cdot x}^2}{\sum (x_i - \bar{x})^2} \tag{21-6-3}$$

In terms of our sample values, we have the corresponding Equation 21-6-4.

$$s_b^2 = \frac{s_{y \cdot x}^2}{\sum (x_i - \bar{x})^2} \tag{21-6-4}$$

Finally, by using Equations 21-6-1 and 21-6-3, we are led to Equation 21-6-5.

$$\sigma_{\hat{Y}}^2 = \frac{\sigma_{y \cdot x}^2}{n} + \frac{(x - \bar{x})^2 \sigma_{y \cdot x}^2}{\sum (x_i - \bar{x})^2}$$
$$= \left[\frac{1}{n} + \frac{(x - \bar{x})^2}{\sum (x_i - \bar{x})^2}\right] \sigma_{y \cdot x}^2 \qquad (21\text{-}6\text{-}5)$$

In turn, Equations 21-6-2 and 21-6-4 give us Equation 21-6-6.

$$s_{\hat{Y}}^2 = s_{y \cdot x}^2 \left[\frac{1}{n} + \frac{(x - \bar{x})^2}{\sum (x_i - \bar{x})^2}\right] \qquad (21\text{-}6\text{-}6)$$

The unsubscripted x in Equations 21-6-5 and 21-6-6 defines the population of y's for which a mean is being estimated. This x does not have to be an observed x. Note also that the variance increases as x moves away from $\bar{x}$, as indicated by $(x - \bar{x})^2$. This is also apparent from Figure 21-6-1.

We are now in a position to construct confidence intervals for the means of populations of Y-values or to test hypotheses that specify what these parameters might be.

First, let us consider $\bar{Y}$ itself. This is the value of $\hat{Y}$ when $x = \bar{x}$. Its sample variance is given by $s_{\bar{y}}^2 = s_{y \cdot x}^2/n$. To compute the $(1 - \alpha)100$ percent confidence interval for the mean of the population of y's for which $x = \bar{x}$, use Equation 21-6-7.

$$\text{CI}(\mu_{Y|\bar{x}}) = \bar{Y} \pm \frac{t_{\alpha/2} s_{y \cdot x}}{\sqrt{n}} \qquad \text{for } t = \text{tabulated } t \text{ for } n - 2 \text{ df} \qquad (21\text{-}6\text{-}7)$$

Illustration 21-6-1

Suppose we want a 95 percent confidence interval for $\mu_{\hat{Y}|\bar{x}}$, the mean number of minutes for a 1-mile run for joggers with a step-test value of 52.0, the same as $\bar{x}$. We refer again to the data of Table 21-3-1 with $\bar{y} = 7.929$ and $s_{y \cdot x}^2 = .8009$ with $n - 2 = 8$ degrees of freedom, and compute as follows:

$$\text{CI}(\mu_{Y|52}) = 7.929 \pm 2.306 \sqrt{\frac{.8009}{10}}$$
$$= 7.929 \pm 2.306(.283)$$
$$= 7.28 \text{ to } 8.58 \text{ min}$$

To test the null hypothesis that $\mu_{Y|\bar{x}} = \mu_0$, compute the sample t of Equation 21-6-9 and compare with the appropriate tabulated t.

$$t = \frac{\bar{Y} - \mu_0}{s_{y \cdot x}/\sqrt{n}} \qquad (21\text{-}6\text{-}8)$$

Now consider confidence-interval estimation for and tests of hypotheses concerning β. The variance of b is given by Equation 21-6-4. With this, we can construct a confidence interval with confidence coefficient $1 - \alpha$ using Equation 21-6-9 or test the null hypothesis that $\beta = \beta_0$ using Equation 21-6-10.

$$\text{CI}(\beta) = b \pm t_{\alpha/2}\sqrt{\frac{s_{y.x}^2}{\sum (x_i - \bar{x})^2}} \tag{21-6-9}$$

where t = tabulated t for $n - 2$ degrees of freedom.

$$t = \frac{b - \beta_0}{\sqrt{s_{y.x}^2/\sum (x_i - \bar{x})^2}} \tag{21-6-10}$$

Illustration 21-6-2

The variance and standard deviation of b for the heart-study data follow.

$$s_b{}^2 = \frac{s_{y.x}^2}{\sum (x_i - \bar{x})^2} = \frac{.800895}{996} = .0008041$$

$$s_b = .0284 \text{ min}$$

A 90 percent confidence interval for β is given by application of Equation 21-6-9.

$$90\% \ \text{CI}(\beta) = .101 \pm 1.86(.0284) = .048 \text{ to } .154 \text{ min/pulse unit}$$

This confidence interval does not contain $\beta = 0$ so we know that a two-tailed test of $H_0{:}\beta = 0$, at a significance level of $\alpha = .10$, would lead us to reject $H_0{:}\beta = 0$. Also, using Equation 21-6-10, we have

$$t = \frac{.101}{.0284} = 3.56$$

This sample t is compared with the tabulated value of 1.86, used in constructing the confidence interval, for a test against two-tailed alternatives. The sample t is much larger than the tabulated t, and we reject the hypothesis that $\beta = 0$.

The same test can be made within the analysis of variance of Table 21-5-1. If there is no linear regression or contribution to the variance of Y that is attributable to variation in x, then the mean square for linear regression with 1 degree of freedom will be of the same order of magnitude, except for random variation, as $s^2_{y.x}$. Hence

$$F = \frac{\text{regression mean square}}{s^2_{y.x}}$$

tests the null hypothesis of no linear contribution attributable to x, that is, that $\beta = 0$. The critical value of F for 1 and 8 degrees of freedom and $\alpha = .10$ is found, in Table B-7, to be 3.46. Notice that this is the square of the tabulated t used earlier.

Our illustrations up to this point have been concerned with the parts of an estimate of a population mean, that is, with $\bar{Y}$ and b. Equation 21-6-6 gave the general case of a variance applicable to a population of y's defined by any x. From Equation 21-6-6, we develop Equation 21-6-11 to construct a confidence interval with coefficient $1 - \alpha$ about a population mean, and Equation 21-6-12 to test the null hypothesis that its location is at μ_0.

$$\text{CI}(\mu_{Y|x}) = \bar{Y} + b(x - \bar{x}) \pm t\sqrt{s^2_{y.x}\left[\frac{1}{n} + \frac{(x - \bar{x})^2}{\sum (x_i - \bar{x})^2}\right]} \qquad \text{(21-6-11)}$$

where t = tabulated t for $\alpha/2$ and $n - 2$ degrees of freedom.

$$t = \frac{\bar{Y} + b(x - \bar{x}) - \mu_0}{\sqrt{s^2_{y.x}[1/n + (x - \bar{x})^2/\sum (x_i - \bar{x})^2]}} \qquad \text{(21-6-12)}$$

Illustration 21-6-3

Suppose we set a confidence interval on the mean of the population of y's for which $x = 40$ in the heart-study data. In Section 21-3, we computed the estimate as 6.71 minutes. Using Equation 21-6-11, we construct the following 99 percent confidence interval for the population mean.

$$\text{CI}(\mu_{Y|40}) = 6.71 \pm 3.355\sqrt{.8009\left[\frac{1}{10} + \frac{(-12)^2}{996}\right]}$$

$$= 5.23 \text{ to } 8.19$$

Notice that the length of this confidence interval is almost twice as long as that for $x = \bar{x} = 52$. This is in part due to changing from $1 - \alpha = .95$ in the former case to $1 - \alpha = .99$ at present. Also note that $(-12)^2/996$ is approximately 1/7, and this means that the variance is more than doubled so that the standard deviation is about $1\frac{1}{2}$ times as large as formerly.

The t-test of Equation 21-6-12 allows us to test the null hypothesis that the regression line goes through the origin, the point (0, 0). By setting $\mu_0 = 0$ and $x_0 = 0$ in the equation, we have an appropriate t-test criterion.

EXERCISES

21-6-1 For the heart-study data, compute sample t by Equation 21-6-8 to test the null hypothesis that $\mu_{Y|\bar{x}} = 7$ minutes. How could you reach the same conclusion by using the confidence interval computed in this section? Can you use Equation 21-6-12 and get the same result?

21-6-2 For the same data, compute sample t by Equation 21-6-12 to test the null hypothesis that $\mu_{Y|40} = 6$ minutes. Can you use the confidence interval computed in this section for $\mu_{Y|40}$ to reach the same conclusion?

21-6-3 Use the analysis of variance to test the null hypothesis that $\beta = 0$. Compare this sample F with sample t^2 for the same null hypothesis.

21-6-4 Compute 95 and 99 percent confidence intervals for the means of populations of y's where $x = 36$ and $x = 67$. Compare the lengths of your 95 percent confidence intervals with that computed in Illustration 21-6-1 for $x = 52$.

21-6-5 For the data of Exercise 21-3-3, compute the standard deviation of $\bar{Y}$, b, and $\hat{Y} = \hat{\mu}_{Y|x}$ for $x = 1$, 11, and 19. Compute 95 percent confidence intervals for $\mu_{Y|x}$ at $x = 1$, 11, and 19. Test the null hypothesis that $\beta = 0$ using a t-test and an F-test. For the F-test, use the analysis-of-variance presentation of your computations.

21-6-6 For the data of Exercise 21-3-4, compute the standard deviation of $\bar{Y}$, b, and $\hat{Y} = \hat{\mu}_{Y|x}$ for $x = 19$, 26, and 33. Compute 90 percent confidence intervals for $\mu_{Y|x}$ and for β at each of these x's.

21-6-7 For the data of Exercise 21-3-5, compute the standard deviation of $\bar{Y}$, b, and $\hat{Y} = \hat{\mu}_{Y|x}$ for $x = 130$, 150, and 166. Compute 99 percent confidence intervals for $\mu_{Y|x}$ at the three x-values. Compare their lengths. Test the null hypothesis that $\mu_{Y|x} = 300$ for $x = 150$. Test the null hypothesis that $\mu_{Y|x} = 400$ at $x = 166$.

21-6-8 For the data of Exercise 21-3-6, compute the standard deviation of $\bar{Y}$, b, and $\hat{Y} = \hat{\mu}_{Y|x}$ for $x = .1$, 1.2, and 2.4. Compute 95 percent confidence intervals for each $\mu_{Y|x}$ and for the common β at the three x-values.

21-7 CONFIDENCE INTERVAL VERSUS CONFIDENCE BAND

Our first experience with estimation was to draw a random sample and construct a confidence interval for the population mean. The confidence coefficient was a probability that applied before the sampling was carried out, or was a long-run average that applied to the number of correct statements made if we always said the confidence interval contained the parameter. Clearly, there was only one mean to be estimated, and the concept of "in repeated sampling" or "taking an expectation" applied to the process of sampling and the computation of a single confidence interval.

In regression, we draw a sample but then may wish to compute a confidence interval for a single mean, for several means, or for the complete regression line. For the procedure given by Equation 21-6-11, the confidence coefficient applies when we draw a sample and construct a single interval. If we should construct a second interval, then the confidence coefficient applies again but we must forget, so to speak, about the first interval. Basically, we have been concerned with a comparisonwise error rate and corresponding confidence coefficient. We cannot then say that in $(1 - \alpha)100$ percent of samplings where two confidence intervals are estimated both intervals will contain the appropriate population means simultaneously. As a matter of fact, we cannot multiply probabilities to give a confidence coefficient of $(1 - \alpha)^2$ for the joint statement because the two events, where an event begins with the sampling and concludes with the confidence-interval computation, are dependent on the same sampling. Now we need a simultaneous inference procedure and an experimentwise error rate.

Suppose we require an equation to be used in repeated calibrations, as is often required in chemical analysis. In this case, we need a *confidence band*, a pair of lines like the solid curved ones of Figure 21-6-1, but with a coefficient that applies to the total line. The solution to this problem is given by Equation 21-7-1.

$$\text{Confidence band} = \bar{Y} + b(x - \bar{x}) \pm \left\{2Fs_{y.x}^2\left[\frac{1}{n} + \frac{(x - \bar{x})^2}{\sum (x_i - \bar{x})^2}\right]\right\}^{1/2} \tag{21-7-1}$$

where F is tabulated F for 2 and $n - 2$ degrees of freedom and the specified α.

Computations are carried out for as many x's as desired, and the confidence band and coefficient apply to the whole line.

A confidence band, as opposed to a confidence interval, carries a cost with it in that individual intervals within the band are longer than those associated with estimation of single intervals. The ratio of the two lengths is seen to be $\sqrt{2F}/t$. If a confidence band made sense for our heart data, this ratio would be $\sqrt{2(4.46)}/2.306 = 1.3$, and a confidence interval from the band would be about 1.3 times as long as the single interval estimate.

21-8 CORRELATION AND REGRESSION

In any discussion of regression, the term correlation is almost certain to be introduced. When one wishes simply to measure the degree of relationship between two variables, the problem is one of correlation. When one wishes to use a relationship between two variables to improve the estimation of population means for one of the variables, then the problem is one of regression.

As a measure of the linear relation between two random variables, we have the linear *correlation coefficient* or *product-moment correlation coefficient* given by Equation 21-8-1.

$$r = \frac{\sum (X_i - \bar{X})(Y_i - \bar{Y})}{\sqrt{\sum (X_i - \bar{X})^2 \sum (Y_i - \bar{Y})^2}} \tag{21-8-1}$$

Here we are considering r, although not capitalized, as a random variable defined in terms of random observations obtained from a bivariate distribution. It is seen to be symmetric in X and Y, and these symbols are not to be interpreted as those for a dependent and an independent variable. Numerically, $-1 \leq r \leq 1$, where values of -1 and $+1$ call for a perfect linear relation, allowing nothing for random errors.

If the numerator and denominator of Equation 21-8-1 are both divided by the degrees of freedom, $n - 1$, it may be written as Equation 21-8-2.

$$r = \frac{\text{cov}(XY)}{s_X s_Y} \tag{21-8-2}$$

where $\text{cov}(XY)$ is the covariance, mentioned earlier in connection with Equations 21-3-4 and 21-3-5, and s_X and s_Y are the standard deviations of X and Y.

Standardized variables, first discussed in terms of observations in Section 4-3, have zero means and unit variances. In a sample situation, standardization is easily accomplished by measuring the observations as deviations from the sample mean, in which case the new mean is 0, and by dividing each deviation by the sample standard deviation, in which case the standard deviation of the new set becomes unity. Thus, $(X_i - \bar{X})/s_X$ is standardized.

The covariance between two standardized variables is a linear correlation coefficient (see Equation 21-8-3).

$$r = \frac{\sum [(X_i - \bar{X})/s_X][(Y_i - \bar{Y})/s_Y]}{n - 1} \tag{21-8-3}$$

From Equations 21-8-2 and 21-8-3, it is clear that r is independent of the units of measurement.

By squaring Equation 21-8-1, then rearranging the expression somewhat, it is easily shown that Equation 21-8-4 holds.

$$\begin{aligned} r^2 &= \frac{[\sum (X_i - \bar{X})(Y_i - \bar{Y})]^2 / \sum (X_i - \bar{X})^2}{\sum (Y_i - \bar{Y})^2} \\ &= \frac{\text{regression SS } (Y \text{ adjusted for } x)}{\text{total SS}(Y)} \end{aligned} \tag{21-8-4}$$

Now r^2 is seen to be the proportion of the total sum of squares attributable to regression if it is appropriate to view the problem as one on regression. In any case, we often compute r^2 in order to have a measure of this proportion when our observations are not random bivariate ones; in such cases, r^2 should not be called the square of a correlation coefficient. Instead, it is called the *coefficient of determination.*

The sample linear correlation coefficient r estimates the corresponding parameter ρ, Greek rho. Like r, ρ must lie between -1 and $+1$. Due to the limited range of r, it is fairly easy to see that its distribution will be symmetric only when $\rho = 0$. This lack of symmetry causes a real problem when it is necessary to test hypotheses that state that ρ is a value other than zero. However, when the null hypothesis is that $\rho = 0$, it is equivalent to the null hypothesis $H_0: \beta = 0$, and may be tested by t; see Equation 21-6-1. The test criterion may also be written as in Equation 21-8-5.

$$t = \frac{r}{\sqrt{(1 - r^2)/(n - 2)}} \qquad n - 2 \text{ df} \tag{21-8-5}$$

EXERCISES

21-8-1 Compute r for the following sets of data:
(a) Exercise 21-3-3 (b) Exercise 21-3-4 (c) Exercise 21-3-5 (d) Exercise 21-3-6

What computational alternative do you have to Equation 21-8-1? For which of the four sets of data would you be prepared to call r a correlation coefficient? When you do not call r a correlation coefficient, what alternatives do you have? Test the null hypothesis that the corresponding $\rho = 0$.

21-8-2 Frey and Watson (1950) studied the relationships among several constituents of the oat kernel, using 16 cultivars. Data on three constituents, namely, niacin y, thiamin x, and riboflavin z, are given in the accompanying table. Compute r and test the null hypothesis that $\rho = 0$ for the following variables:
(a) y and x (b) y and z (c) x and z

y	9.6	8.6	11.2	8.7	10.2	8.4	4.4	7.6	10.6	11.0
x	9.2	9.7	8.1	5.5	7.7	5.6	5.7	5.4	7.5	7.9
z	1.25	1.11	1.63	1.20	1.66	1.17	1.09	1.05	1.63	1.54

y	9.8	9.4	10.4	7.3	10.8	11.7
x	6.6	5.9	8.2	4.6	8.7	9.7
z	1.71	1.22	1.87	1.10	1.23	1.77

Compute r^2 for each of the three cases above. What can you do with this piece of information?

Two algebraic problems follow.

21-8-3 Beginning with the definition of r, develop Equation 21-8-4. Check carefully for the formulas necessary to justify each step.

21-8-4 Consider Equation 21-6-10 and a test where $\beta = 0$. Show that Equation 21-8-5 can be derived directly from Equation 21-6-10 under the null hypothesis that $\beta = 0$.

21-9 NONPARAMETRIC TECHNIQUES FOR REGRESSION AND CORRELATION

When it is known that the assumptions associated with a certain statistical technique are not satisfied, then several alternatives are available. One of these is the use of nonparametric or distribution-free statistics where the assumptions are a minimum. We now look at two such procedures.

A correlation coefficient based on minimum assumptions is given by *Spearman's rank correlation coefficient*. For this statistic, the y-observations are ranked from smallest to largest and then the x-observations are ranked separately in the same manner. One next computes a correlation coefficient using ranks. Ranks are now the basic observations and tables of the test criterion can be constructed and made available.

We ask nothing more of the original observations than that each set be drawn from a continuous distribution function although our measurement does not actually need to be more precise than a ranking. If our problem had been one of ordering a number of cakes according to coarseness, presumably an actual measurement could be made but would not necessarily provide a more accurate ranking than a visual one.

Spearman's rank correlation coefficient is computed by Equation 21-9-1.

$$r_s = 1 - \frac{6\sum (x_i - y_i)^2}{n^3 - n} \qquad (21\text{-}9\text{-}1)$$

The same result will be obtained if we compute a correlation coefficient, using ranks, by Equation 21-8-1. Note that $\sum (x_i - y_i) = 0$, always; this provides a check.

Spearman's rank correlation coefficient lies in the interval $(-1, 1)$. The value $r_s = -1$ is found when the two sets of ranks run opposite to one another, that is, when 1 is opposite n, 2 opposite $n - 1$, and so on. In this case, we find $6\sum (x_i - y_i)^2/(n^3 - n) = 2$.

On occasion, more than one individual may receive the same rank. When this occurs, the rank assigned is the average of the ranks that would have been assigned if these individuals were not tied. For a small number of ties, no adjustment to r_s need be made. If the proportion is large, an adjustment should be made; for example, see Siegel (1956).

Suggested critical values of Spearman's r_s are given in Table B-8. These have been computed using ranks, and so involve a discrete variable. Consequently, we are not ordinarily able to find critical values corresponding exactly to the customarily used values of .10, .05, and .01. Exact probabilities corresponding to the tabulated critical values are given in Table B-8. Unfortunately, the table goes no further than $n = 9$.

Student's t-test may be used as an alternative to the above test procedure, particularly when tables of the exact distribution are not adequate. The appropriate computation is given by Equation 21-9-2, which makes use of r_s.

$$t = r_s \sqrt{\frac{n - 2}{1 - r_s^2}} \qquad n - 2 \text{ df} \qquad (21\text{-}9\text{-}2)$$

Note that this is Equation 21-8-5 again.

This approximation appears to be reasonably satisfactory beyond $n = 9$. Computed values of t tend to have associated probabilities which are somewhat smaller than those given by the exact distribution of r_s.

Illustration 21-9-1

A class of graduate students was given a test early in the semester. Grades and ranks for this test are given as x and x', respectively. Further tests and a course exam were given, and a final grade, as a percentage of a total possible score, arrived at for each student.

Final grades and ranks are given as y and y', respectively, in Table 21-9-1.

The instructor now wishes to know if there is a correlation between first and final grades. If there is a relation, she would like to have a measure of how strong it is. Spearman's rank correlation coefficient provides such a measure.

TABLE 21-9-1 / Grade Data from a Statistics Course

Grade on First Test x	Rank x'	Final Grade y	Rank y'	$x' - y'$	$(x' - y')^2$
78	7	73	6	1	1
88	5	64	9	−4	16
98	2	80	4	−2	4
100	1	82	3	−2	4
84	6	66	8	−2	4
64	10	67	7	3	9
74	8	91	1	7	49
66	9	61	10	−1	1
96	3	83	2	1	1
92	4	76	5	−1	1
Sums				0	90

$$r_s = 1 - \frac{6\sum (x' - y')^2}{n^3 - n} = 1 - \frac{6(90)}{10^3 - 10} = .455$$

We compute $r_s = .455$. To complete our test of significance, we now require tables of critical values. Unfortunately, such values are not available in Table B-8 for the exact distribution when $n = 10$. Consequently, we will rely on the approximation given by Equation 21-9-2.

$$t = \frac{.455\sqrt{8}}{\sqrt{1 - (.455)^2}} = 1.445 \qquad 8 \text{ df}$$

A one-tailed test is appropriate since the alternative would reasonably call for a positive relation. From Table B-2, we find

$$P(t > 1.397) = .10 \qquad \text{and} \qquad P(t > 1.860) = .05$$

Hence, the probability, under the null hypothesis, of finding a random value of t larger than that observed is not much less than .10. We conclude that the null hypothesis and chance offer an adequate explanation of the data.

An alternative to a parametric test of $H_0: \beta = 0$ in regression, such as t in Equation 21-6-10, is a *randomization test*. If we hypothesize that $\beta = 0$, then there is no reason to expect anything more than random pairing of the sample x's and y's; that is, all possible pairings are equally likely. We, then, construct all possible pairings or a large random set of them and generate a distribution of b's, or some other appropriate test statistic, as was done for differences between two sample means in Section 20-6. From this empirical distribution, critical values are obtained and the sample statistic, say b, for the data paired as observed is compared with them. One then concludes whether or not the null hypothesis is to be accepted.

To obtain the pairings for producing the distribution of the sample statistic for the randomization test, it is convenient to order the y's and simply randomize the x's. There will, of course, be $n!$ randomizations. The number $n!$ becomes large very fast and is already well over 3 million for $n = 10$. Consequently, the availability of high-speed computing equipment is essential for this randomization test, and the full set of randomizations is likely to be randomly sampled rather than enumerated in its entirety.

EXERCISES

21-9-1 Compute Spearman's rank correlation coefficient for the data of Exercise 21-3-5. Test the null hypothesis that the corresponding population parameter is 0. Compare your result with that obtained in Exercise 21-8-1*c* when testing $\rho = 0$.

21-9-2 Compute Spearman's rank correlation coefficient for the data of Exercise 21-8-2, all three cases. Test the null hypothesis that the corresponding population parameter is 0. Compare your results with those obtained when testing the null hypothesis that the linear correlation coefficient $\rho = 0$.

21-10 MULTIPLE LINEAR REGRESSION

In Section 21-3, it was suggested that one's time to run a mile might be affected by several variables. Many situations exist where an outcome measured on one variable may be dependent on several others. Analyses of such data are multiple-linear-regression analyses.

Construction of a multiple-linear-regression equation calls for the solution of simultaneous linear equations. As the number of independent variables increases, the arithmetic complexity of the solution increases at an even more rapid rate. Consequently, we leave this topic for texts more suited to the problem; see, for example, Steel and Torrie (1960, chapter 14).

Appendix A

A-1 / Greek Alphabet (Letter and Name)

Α	α	Alpha	Ν	ν	Nu
Β	β	Beta	Ξ	ξ	Xi
Γ	γ	Gamma	Ο	ο	Omicron
Δ	δ	Delta	Π	π	Pi
Ε	ε	Epsilon	Ρ	ρ	Rho
Ζ	ζ	Zeta	Σ	σ	Sigma
Η	η	Eta	Τ	τ	Tau
Θ	θ	Theta	Υ	υ	Upsilon
Ι	ι	Iota	Φ	φ	Phi
Κ	κ	Kappa	Χ	χ	Chi
Λ	λ	Lambda	Ψ	ψ	Psi
Μ	μ	Mu	Ω	ω	Omega

A-2 / Symbols and Notation

$W, X, Y, \ldots$ — Capital letters are used, in general, to designate random variables and so are concerned with events that have not, but could, occur. Often used with subscripts.

$w, x, y, \ldots$ — Lowercase letters are used, in general, to designate sample values, ones that have occurred or been specified. Often subscripted.

$\neq$ — "Is not equal to." For example, $7 \neq 9$.

$\approx$ — "Is approximately equal to."

$<$ — "Less than." For example, $3 < 6$.

$\leq$ — "Less than or equal to."

$>$ — "Greater than."

$\geq$ — "Greater than or equal to."

$|\cdots|$ — Absolute value. For example, $|y|$ represents the magnitude of an observation regardless of whether it is positive or negative.

1[1]5 — Going by steps of 1 from 1 to 5, that is, 1, 2, 3, 4, 5. Other endpoints and step sizes may be used.

$\sum$ — A summation sign to replace multiple + signs. For example, $\sum y_i = y_1 + y_2 + \cdots + y_n$.

n, N — Used as sample and finite population sizes. Often subscripted.

S — Sample space; for example $S = \{\text{HH, HT, TH, TT}\}$ for two tossed coins. Also, the standard deviation in a finite population sampled without replacement. Also, a success.

E — An event or occurrence; a point in a sample space. Often used with subscript. For example, E_1 = HH for two tossed coins, E_2 = HT, and so on. Also, an expectation or population average; an average over an experiment repeated infinitely often under the same apparent conditions.

$\not{E}$ — The event "not-$\not{E}$." For example, if E represents a head in the toss of a coin, then E means a head was not observed.

$\bar{E}$ — The complement of E; that is, all points not in E.

$\varnothing$ — The empty set.

$\in$ — "Is an element of." For example HH $\in$ (HH, HT, TH, TT).

$\notin$ — "Is not an element of."

$\cup$ — Union. An "and/or" symbol. For example, $A \cup B$ means either A or B or both; if A refers to black cards and B to honors, then $A \cup B$ includes *all* black cards and *all* honors.

$\cap$ — Intersection. Strictly an "and" symbol. For example, $A \cap B$ means both A and B; if A and B are as just defined, then $A \cap B$ must mean *only* black honors.

A-2 / Symbols and Notation (*continued*)

$\cdots \mid \cdots$	"Given." For example, $A \mid B$ might mean the event of observing four heads in four tosses, A, *given* that there is at least one head, B. Clearly this would restrict the sample space involved since it cannot include four tails.
P	Probability. For example, $P(S) = .25$ says that the probability of a success is 1/4. Also used with a subscript to indicate a percentile; for example, 90 percent of sample values are less than P_{90}.
$n!$	A factorial number; $n! = n(n-1)\cdots(2)(1)$.
${}_nP_r$	Number of permutations when drawing r from n different objects; ${}_nP_r = n!/(n-r)!$
$\binom{n}{r} = {}_nC_r$	Number of combinations of n things taken r at a time; ${}_nC_r = n!/r!\,(n-r)!$
Greek alphabet	Used to symbolize population parameters, constants rather than variables.
Roman alphabet	Used to symbolize sample values and statistics, variables (caps), and observed values (lowercase).
$\hat{}$	"Hat." Used over a letter, usually Greek, to indicate an estimator; for example, $\hat{\mu} = \bar{Y}$ for the mean.
$Y_{(1)}, \ldots, Y_{(n)}$	Order statistics, as indicated by parens.
$y_{(1)}, \ldots, y_{(n)}$	The sample values ordered from smallest to largest.
df	Degrees of freedom.
SS	Sum of squares; generally of deviations from their mean.
MS	Mean square; a sum of squares divided by its degrees of freedom.
CV	Coefficient of variation; $(s/\bar{y})100$ percent.
H_0	A null hypothesis, one for which a test criterion can be computed.
H_1	An alternative hypothesis; usually includes many alternatives to H_0.
α	Level of significance, error rate, size of test. The probability of rejecting a true null hypothesis.
$1 - \alpha$	Confidence coefficient.
β	Probability of accepting a null hypothesis when an alternative is true.
$1 - \beta$	Power of a test.
z, t, χ^2, F	Commonly used test criteria developed from the normal distribution. Also, z is used to designate a standardized score.
r	Correlation coefficient. Equation 21-8-1.
r_s	Spearman's rank correlation coefficient. Equation 21-9-1.
π, e	Numerical constants; approximately 3.14159 and 2.71828, respectively.

Appendix B

LIST OF TABLES

TABLE B-1 / Probability of a Random Value of $Z = (X - \mu)/\sigma$ Being Greater Than the Values Tabulated in the Margins

z	.00	.01	.02	.03	.04	.05	.06	.07	.08	.09	z
.0	.5000	.4960	.4920	.4880	.4840	.4801	.4761	.4721	.4681	.4641	.0
.1	.4602	.4562	.4522	.4483	.4443	.4404	.4364	.4325	.4286	.4247	.1
.2	.4207	.4168	.4129	.4090	.4052	.4013	.3974	.3936	.3897	.3859	.2
.3	.3821	.3783	.3745	.3707	.3669	.3632	.3594	.3557	.3520	.3483	.3
.4	.3446	.3409	.3372	.3336	.3300	.3264	.3228	.3192	.3156	.3121	.4
.5	.3085	.3050	.3015	.2981	.2946	.2912	.2877	.2843	.2810	.2776	.5
.6	.2743	.2709	.2676	.2643	.2611	.2578	.2546	.2514	.2483	.2451	.6
.7	.2420	.2389	.2358	.2327	.2296	.2266	.2236	.2206	.2177	.2148	.7
.8	.2119	.2090	.2061	.2033	.2005	.1977	.1949	.1922	.1894	.1867	.8
.9	.1841	.1814	.1788	.1762	.1736	.1711	.1685	.1660	.1635	.1611	.9
1.0	.1587	.1562	.1539	.1515	.1492	.1469	.1446	.1423	.1401	.1379	1.0
1.1	.1357	.1335	.1314	.1292	.1271	.1251	.1230	.1210	.1190	.1170	1.1
1.2	.1151	.1131	.1112	.1093	.1075	.1056	.1038	.1020	.1003	.0985	1.2
1.3	.0968	.0951	.0934	.0918	.0901	.0885	.0869	.0853	.0838	.0823	1.3
1.4	.0808	.0793	.0778	.0764	.0749	.0735	.0721	.0708	.0694	.0681	1.4
1.5	.0668	.0655	.0643	.0630	.0618	.0606	.0594	.0582	.0571	.0559	1.5
1.6	.0548	.0537	.0526	.0516	.0505	.0495	.0485	.0475	.0465	.0455	1.6
1.7	.0446	.0436	.0427	.0418	.0409	.0401	.0392	.0384	.0375	.0367	1.7
1.8	.0359	.0351	.0344	.0336	.0329	.0322	.0314	.0307	.0301	.0294	1.8
1.9	.0287	.0281	.0274	.0268	.0262	.0256	.0250	.0244	.0239	.0233	1.9
2.0	.0228	.0222	.0217	.0212	.0207	.0202	.0197	.0192	.0188	.0183	2.0
2.1	.0179	.0174	.0170	.0166	.0162	.0158	.0154	.0150	.0146	.0143	2.1
2.2	.0139	.0136	.0132	.0129	.0125	.0122	.0119	.0116	.0113	.0110	2.2
2.3	.0107	.0104	.0102	.0099	.0096	.0094	.0091	.0089	.0087	.0084	2.3
2.4	.0082	.0080	.0078	.0075	.0073	.0071	.0069	.0068	.0066	.0064	2.4
2.5	.0062	.0060	.0059	.0057	.0055	.0054	.0052	.0051	.0049	.0048	2.5
2.6	.0047	.0045	.0044	.0043	.0041	.0040	.0039	.0038	.0037	.0036	2.6
2.7	.0035	.0034	.0033	.0032	.0031	.0030	.0029	.0028	.0027	.0026	2.7
2.8	.0026	.0025	.0024	.0023	.0023	.0022	.0021	.0021	.0020	.0019	2.8
2.9	.0019	.0018	.0018	.0017	.0016	.0016	.0015	.0015	.0014	.0014	2.9
3.0	.0013	.0013	.0013	.0012	.0012	.0011	.0011	.0011	.0010	.0010	3.0
3.1	.0010	.0009	.0009	.0009	.0008	.0008	.0008	.0008	.0007	.0007	3.1
3.2	.0007	.0007	.0006	.0006	.0006	.0006	.0006	.0005	.0005	.0005	3.2
3.3	.0005	.0005	.0005	.0004	.0004	.0004	.0004	.0004	.0004	.0003	3.3
3.4	.0003	.0003	.0003	.0003	.0003	.0003	.0003	.0003	.0003	.0002	3.4
3.6	.0002	.0002	.0001	.0001	.0001	.0001	.0001	.0001	.0001	.0001	3.6
3.9	.0000										3.9

TABLE B-2 / Values of Student's *t*

df	Probability of a Numerically Larger Value of t, Sign Ignored									df
	0.5	0.4	0.3	0.2	0.1	0.05	0.02	0.01	0.001	
1	1.000	1.376	1.963	3.078	6.314	12.706	31.821	63.657	636.619	1
2	.816	1.061	1.386	1.886	2.920	4.303	6.965	9.925	31.598	2
3	.765	.978	1.250	1.638	2.353	3.182	4.541	5.841	12.941	3
4	.741	.941	1.190	1.533	2.132	2.776	3.747	4.604	8.610	4
5	.727	.920	1.156	1.476	2.015	2.571	3.365	4.032	6.859	5
6	.718	.906	1.134	1.440	1.943	2.447	3.143	3.707	5.959	6
7	.711	.896	1.119	1.415	1.895	2.365	2.998	3.499	5.405	7
8	.706	.889	1.108	1.397	1.860	2.306	2.896	3.355	5.041	8
9	.703	.883	1.100	1.383	1.833	2.262	2.821	3.250	4.781	9
10	.700	.879	1.093	1.372	1.812	2.228	2.764	3.169	4.587	10
11	.697	.876	1.088	1.363	1.796	2.201	2.718	3.106	4.437	11
12	.695	.873	1.083	1.356	1.782	2.179	2.681	3.055	4.318	12
13	.694	.870	1.079	1.350	1.771	2.160	2.650	3.012	4.221	13
14	.692	.868	1.076	1.345	1.761	2.145	2.624	2.977	4.140	14
15	.691	.866	1.074	1.341	1.753	2.131	2.602	2.947	4.073	15
16	.690	.865	1.071	1.337	1.746	2.120	2.583	2.921	4.015	16
17	.689	.863	1.069	1.333	1.740	2.110	2.567	2.898	3.965	17
18	.688	.862	1.067	1.330	1.734	2.101	2.552	2.878	3.922	18
19	.688	.861	1.066	1.328	1.729	2.093	2.539	2.861	3.883	19
20	.687	.860	1.064	1.325	1.725	2.086	2.528	2.845	3.850	20
21	.686	.859	1.063	1.323	1.721	2.080	2.518	2.831	3.819	21
22	.686	.858	1.061	1.321	1.717	2.074	2.508	2.819	3.792	22
23	.685	.858	1.060	1.319	1.714	2.069	2.500	2.807	3.767	23
24	.685	.857	1.059	1.318	1.711	2.064	2.492	2.797	3.745	24
25	.684	.856	1.058	1.316	1.708	2.060	2.485	2.787	3.725	25
26	.684	.856	1.058	1.315	1.706	2.056	2.479	2.779	3.707	26
27	.684	.855	1.057	1.314	1.703	2.052	2.473	2.771	3.690	27
28	.683	.855	1.056	1.313	1.701	2.048	2.467	2.763	3.674	28
29	.683	.854	1.055	1.311	1.699	2.045	2.462	2.756	3.659	29
30	.683	.854	1.055	1.310	1.697	2.042	2.457	2.750	3.646	30
40	.681	.851	1.050	1.303	1.684	2.021	2.423	2.704	3.551	40
60	.679	.848	1.046	1.296	1.671	2.000	2.390	2.660	3.460	60
120	.677	.845	1.041	1.289	1.658	1.980	2.358	2.617	3.373	120
∞	.674	.842	1.036	1.282	1.645	1.960	2.326	2.576	3.291	∞
	0.25	0.2	0.15	0.1	0.05	0.025	0.01	0.005	0.0005	
df	Probability of a Larger Positive Value of t, Sign Considered									df

Source: This table is abridged from Table III of Fisher and Yates, "Statistical Tables for Biological, Agricultural, and Medical Research," published by Longman Group Ltd., London (previously published by Oliver and Boyd, Edinburgh), and by permission of the authors and publishers.

TABLE B-3 / Values of χ^2

df	Probability of a Larger Value of χ^2													df
	.995	.990	.975	.950	.900	.750	.500	.250	.100	.050	.025	.010	.005	
1	$.0^4393$	$.0^3157$	$.0^3982$	$.0^2393$	.0158	.102	.455	1.32	2.71	3.84	5.02	6.63	7.88	1
2	.0100	.0201	.0506	.103	.211	.575	1.39	2.77	4.61	5.99	7.38	9.21	10.6	2
3	.0717	.115	.216	.352	.584	1.21	2.37	4.11	6.25	7.81	9.35	11.3	12.8	3
4	.207	.297	.484	.711	1.06	1.92	3.36	5.39	7.78	9.49	11.1	13.3	14.9	4
5	.412	.554	.831	1.15	1.61	2.67	4.35	6.63	9.24	11.1	12.8	15.1	16.7	5
6	.676	.872	1.24	1.64	2.20	3.45	5.35	7.84	10.6	12.6	14.4	16.8	18.5	6
7	.989	1.24	1.69	2.17	2.83	4.25	6.35	9.04	12.0	14.1	16.0	18.5	20.3	7
8	1.34	1.65	2.18	2.73	3.49	5.07	7.34	10.2	13.4	15.5	17.5	20.1	22.0	8
9	1.73	2.09	2.70	3.33	4.17	5.90	8.34	11.4	14.7	16.9	19.0	21.7	23.6	9
10	2.16	2.56	3.25	3.94	4.87	6.74	9.34	12.5	16.0	18.3	20.5	23.2	25.2	10
11	2.60	3.05	3.82	4.57	5.58	7.58	10.3	13.7	17.3	19.7	21.9	24.7	26.8	11
12	3.07	3.57	4.40	5.23	6.30	8.44	11.3	14.8	18.5	21.0	23.3	26.2	28.3	12
13	3.57	4.11	5.01	5.89	7.04	9.30	12.3	16.0	19.8	22.4	24.7	27.7	29.8	13
14	4.07	4.66	5.63	6.57	7.79	10.2	13.3	17.1	21.1	23.7	26.1	29.1	31.3	14
15	4.60	5.23	6.26	7.26	8.55	11.0	14.3	18.2	22.3	25.0	27.5	30.6	32.8	15
16	5.14	5.81	6.91	7.96	9.31	11.9	15.3	19.4	23.5	26.3	28.8	32.0	34.3	16
17	5.70	6.41	7.56	8.67	10.1	12.8	16.3	20.5	24.8	27.6	30.2	33.4	35.7	17
18	6.26	7.01	8.23	9.39	10.9	13.7	17.3	21.6	26.0	28.9	31.5	34.8	37.2	18
19	6.84	7.63	8.91	10.1	11.7	14.6	18.3	22.7	27.2	30.1	32.9	36.2	38.6	19
20	7.43	8.26	9.59	10.9	12.4	15.5	19.3	23.8	28.4	31.4	34.2	37.6	40.0	20
21	8.03	8.90	10.3	11.6	13.2	16.3	20.3	24.9	29.6	32.7	35.5	38.9	41.4	21
22	8.64	9.54	11.0	12.3	14.0	17.2	21.3	26.0	30.8	33.9	36.8	40.3	42.8	22
23	9.26	10.2	11.7	13.1	14.8	18.1	22.3	27.1	32.0	35.2	38.1	41.6	44.2	23
24	9.89	10.9	12.4	13.8	15.7	19.0	23.3	28.2	33.2	36.4	39.4	43.0	45.6	24
25	10.5	11.5	13.1	14.6	16.5	19.9	24.3	29.3	34.4	37.7	40.6	44.3	46.9	25
26	11.2	12.2	13.8	15.4	17.3	20.8	25.3	30.4	35.6	38.9	41.9	45.6	48.3	26
27	11.8	12.9	14.6	16.2	18.1	21.7	26.3	31.5	36.7	40.1	43.2	47.0	49.6	27
28	12.5	13.6	15.3	16.9	18.9	22.7	27.3	32.6	37.9	41.3	44.5	48.3	51.0	28
29	13.1	14.3	16.0	17.7	19.8	23.6	28.3	33.7	39.1	42.6	45.7	49.6	52.3	29
30	13.8	15.0	16.8	18.5	20.6	24.5	29.3	34.8	40.3	43.8	47.0	50.9	53.7	30
40	20.7	22.2	24.4	26.5	29.1	33.7	39.3	45.6	51.8	55.8	59.3	63.7	66.8	40
50	28.0	29.7	32.4	34.8	37.7	42.9	49.3	56.3	63.2	67.5	71.4	76.2	79.5	50
60	35.5	37.5	40.5	43.2	46.5	52.3	59.3	67.0	74.4	79.1	83.3	88.4	92.0	60
100	67.3	70.1	74.2	77.9	82.4	90.1	99.3	109.1	118.5	124.3	129.6	135.8	140.2	100

Source: This table is abridged from Table 8, Percentage Points of the χ^2 distribution, "Biometrika Tables for Statisticians," 3d ed., vol. 1, E. S. Pearson and H. O. Hartley (eds.), Cambridge University Press, London, 1965, with permission of the editors and the Biometrika trustees.

TABLE B-4 / Values of $c_{\alpha,n}$, Critical Values for the Sign-test Count (Confidence coefficient = $1 - \alpha$; two-sided rejection rate = α)

n	α 0.10	0.05	0.01	n	α 0.10	0.05	0.01
...	...	...	...	26	8.3	7.5	6.0
...	...	...	...	27	8.8	7.9	6.3
...	...	...	...	28	9.2	8.3	6.7
...	...	...	...	29	9.6	8.7	7.1
...	...	...	...	30	10.1	9.2	7.5
6	0.5	0.2	...	31	10.5	9.6	7.9
7	0.9	0.5	...	32	10.9	10.0	8.3
8	1.2	0.8	0.1	33	11.3	10.4	8.7
9	1.6	1.1	0.4	34	11.7	10.8	9.1
10	2.0	1.5	0.6	35	12.2	11.2	9.5
11	2.3	1.8	0.9	36	12.6	11.6	9.9
12	2.7	2.2	1.2	37	13.0	12.1	10.3
13	3.1	2.5	1.5	38	13.4	12.5	10.7
14	3.5	2.9	1.9	39	13.9	12.9	11.1
15	3.9	3.2	2.2	40	14.3	13.4	11.5
16	4.3	3.6	2.5	41	14.8	13.8	11.9
17	4.6	3.9	2.9	42	15.2	14.2	12.3
18	5.0	4.3	3.2	43	15.6	14.7	12.7
19	5.4	4.7	3.6	44	16.1	15.1	13.1
20	5.9	5.1	3.9	45	16.5	15.5	13.5
21	6.3	5.5	4.3	46	16.9	15.9	13.9
22	6.7	5.8	4.6	47	17.4	16.4	14.3
23	7.1	6.4	5.0	48	17.8	16.8	14.7
24	7.5	6.7	5.4	49	18.3	17.2	15.1
25	7.9	7.1	5.7	50	18.7	17.6	15.5

Source: Reproduced with permission from an unpublished table of J. W. Tukey, later published by Thomas E. Kurtz, "Basic Statistics," Prentice-Hall, Englewood Cliffs, N.J., 1963.

TABLE B-5 / Values of $s_{(\alpha,n)}$, Critical Values for the Signed Rank-sum Test (Confidence coefficient = $1 - \alpha$; two-sided rejection rate = α)

n	α 0.10	0.05	0.01	n
6	2.1	0.6	...	6
7	3.7	2.1	...	7
8	5.7	3.7	0.3	8
9	8.1	5.7	1.6	9
10	10.8	8.1	3.1	10
11	13.9	10.8	5.1	11
12	17.5	13.8	7.3	12
13	21.4	17.2	9.8	13
14	25.7	21.1	12.7	14
15	30.4	25.3	15.9	15
16	35.6	29.9	19.5	16
17	41.2	34.9	23.4	17
18	47.2	40.3	27.7	18
19	53.6	46.1	32.4	19
20	60.4	52.3	37.5	20
21	67.6	58.9	42.9	21
22	75.3	66.0	48.7	22
23	83.9	73.4	54.9	23
24	91.9	81.3	61.5	24
25	100.9	89.5	68.5	25

Source: Abridged with permission from J. W. Tukey, *Memorandum Report* 17, 1949, Statistical Research Group, Princeton University, and Thomas E. Kurtz, "Basic Statistics," Prentice-Hall, Englewood Cliffs, N.J., 1963.

TABLE B-6 / Critical Values for Wilcoxon's Rank-Sum Test (Tabulated values of T such that smaller values occur by chance with stated probability)†

n_2 = Larger n	P	n_1 = Smaller n												
		3	4	5	6	7	8	9	10	11	12	13	14	15
4	.05	6	11											
	.01	5	9											
5	.05	6	12	18										
	.01	5	10	15										
6	.05	7	12	19	26									
	.01	5	10	16	23									
7	.05	7	13	20	28	37								
	.01	6	11	17	24	33								
8	.05	8	14	21	29	39	49							
	.01	6	11	18	25	34	44							
9	.05	8	15	22	31	41	51	63						
	.01	6	12	19	27	36	46	57						
10	.05	9	16	24	32	43	54	66	79					
	.01	6	12	19	28	37	47	59	71					
11	.05	10	17	25	34	45	56	68	82	96				
	.01	7	13	20	29	39	49	61	74	88				
12	.05	10	17	26	36	46	58	71	85	100	116			
	.01	7	13	21	30	40	51	63	76	91	106			
13	.05	11	18	27	37	48	61	74	88	103	120	137		
	.01	7	14	22	31	42	53	65	79	94	109	126		
14	.05	11	19	29	39	50	63	77	91	107	124	141	160	
	.01	8	15	23	32	43	55	68	82	97	113	130	148	
15	.05	12	20	30	41	52	65	79	94	110	128	146	165	185
	.01	8	15	24	34	45	57	70	84	100	116	133	152	171

† Probabilities are for two-tailed tests. For one-tailed tests, the above probabilities become .025 and .005.
Source: Abridged from F. Wilcoxon and R. A. Wilcox, "Some Rapid Approximate Statistical Procedures," American Cyanamid Company, Stamford., Conn. 1964, with permission of the authors and the American Cyanamid Company.

TABLE B-7 / Values of F

Denominator df	Probability of a Larger F	Numerator df										Probability of a Larger F	Denominator df
		1	2	3	4	5	6	7	8	9	10		
1	.100	39.86	49.50	53.39	55.83	57.24	58.20	58.91	59.44	59.86	60.19	.100	1
	.050	161.4	199.5	215.7	224.6	230.2	234.0	236.8	238.9	240.5	241.9	.050	
	.025	647.8	799.5	864.2	899.6	921.8	937.1	948.2	956.7	963.3	968.6	.025	
	.010	4052	4999.5	5403	5625	5764	5859	5928	5982	6022	6056	.010	
	.005	16211	20000	21615	22500	23056	23437	23715	23925	24091	24224	.005	
2	.100	8.53	9.00	9.16	9.24	9.29	9.33	9.35	9.37	9.38	9.39	.100	2
	.050	18.51	19.00	19.16	19.25	19.30	19.33	19.35	19.37	19.38	19.40	.050	
	.025	38.51	39.00	39.17	39.25	39.30	39.33	39.36	39.37	39.39	39.40	.025	
	.010	98.50	99.00	99.17	99.25	99.30	99.33	99.36	99.37	99.39	99.40	.010	
	.005	198.5	199.0	199.2	199.2	199.3	199.3	199.4	199.4	199.4	199.4	.005	
3	.100	5.54	5.46	5.39	5.34	5.31	5.28	5.27	5.25	5.24	5.23	.100	3
	.050	10.13	9.55	9.28	9.12	9.01	8.94	8.89	8.85	8.81	8.79	.050	
	.025	17.44	16.04	15.44	15.10	14.88	14.73	14.62	14.54	14.47	14.42	.025	
	.010	34.12	30.82	29.46	28.71	28.24	27.91	27.67	27.49	27.35	27.23	.010	
	.005	55.55	49.80	47.47	46.19	45.39	44.84	44.43	44.13	43.88	43.69	.005	
4	.100	4.54	4.32	4.19	4.11	4.05	4.01	3.98	3.95	3.94	3.92	.100	4
	.050	7.71	6.94	6.59	6.39	6.26	6.16	6.09	6.04	6.00	5.96	.050	
	.025	12.22	10.65	9.98	9.60	9.36	9.20	9.07	8.98	8.90	8.84	.025	
	.010	21.20	18.00	16.69	15.98	15.52	15.21	14.98	14.80	14.66	14.55	.010	
	.005	31.33	26.28	24.26	23.15	22.46	21.97	21.62	21.35	21.14	20.97	.005	
5	.100	4.06	3.78	3.62	3.52	3.45	3.40	3.37	3.34	3.32	3.30	.100	5
	.050	6.61	5.79	5.41	5.19	5.05	4.95	4.88	4.82	4.77	4.74	.050	
	.025	10.01	8.43	7.76	7.39	7.15	6.98	6.85	6.76	6.68	6.62	.025	
	.010	16.26	13.27	12.06	11.39	10.97	10.67	10.46	10.29	10.16	10.05	.010	
	.005	22.78	18.31	16.53	15.56	14.94	14.51	14.20	13.96	13.77	13.62	.005	
6	.100	3.78	3.46	3.29	3.18	3.11	3.05	3.01	2.98	2.96	2.94	.100	6
	.050	5.99	5.14	4.76	4.53	4.39	4.28	4.21	4.15	4.10	4.06	.050	
	.025	8.81	7.26	6.60	6.23	5.99	5.82	5.70	5.60	5.52	5.46	.025	
	.010	13.75	10.92	9.78	9.15	8.75	8.47	8.26	8.10	7.98	7.87	.010	
	.005	18.63	14.54	12.92	12.03	11.46	11.07	10.79	10.57	10.39	10.25	.005	

7	.100	3.59	3.26	3.07	2.96	2.88	2.83	2.78	2.75	2.72	2.70	.100	7
	.050	5.59	4.74	4.35	4.12	3.97	3.87	3.79	3.73	3.68	3.64	.050	
	.025	8.07	6.54	5.89	5.52	5.29	5.12	4.99	4.90	4.82	4.76	.025	
	.010	12.25	9.55	8.45	7.85	7.46	7.19	6.99	6.84	6.72	6.62	.010	
	.005	16.24	12.40	10.88	10.05	9.52	9.16	8.89	8.68	8.51	8.38	.005	
8	.100	3.46	3.11	2.92	2.81	2.73	2.67	2.62	2.59	2.56	2.54	.100	8
	.050	5.32	4.46	4.07	3.84	3.69	3.58	3.50	3.44	3.39	3.35	.050	
	.025	7.57	6.06	5.42	5.05	4.82	4.65	4.53	4.43	4.36	4.30	.025	
	.010	11.26	8.65	7.59	7.01	6.63	6.37	6.18	6.03	5.91	5.81	.010	
	.005	14.69	11.04	9.60	8.81	8.30	7.95	7.69	7.50	7.34	7.21	.005	
9	.100	3.36	3.01	2.81	2.69	2.61	2.55	2.51	2.47	2.44	2.42	.100	9
	.050	5.12	4.26	3.86	3.63	3.48	3.37	3.29	3.23	3.18	3.14	.050	
	.025	7.21	5.71	5.08	4.72	4.48	4.32	4.20	4.10	4.03	3.96	.025	
	.010	10.56	8.02	6.99	6.42	6.06	5.80	5.61	5.47	5.35	5.26	.010	
	.005	13.61	10.11	8.72	7.96	7.47	7.13	6.88	6.69	6.54	6.42	.005	
10	.100	3.29	2.92	2.73	2.61	2.52	2.46	2.41	2.38	2.35	2.32	.100	10
	.050	4.96	4.10	3.71	3.48	3.33	3.22	3.14	3.07	3.02	2.98	.050	
	.025	6.94	5.46	4.83	4.47	4.24	4.07	3.95	3.85	3.78	3.72	.025	
	.010	10.04	7.56	6.55	5.99	5.64	5.39	5.20	5.06	4.94	4.85	.010	
	.005	12.83	9.43	8.08	7.34	6.87	6.54	6.30	6.12	5.97	5.85	.005	
11	.100	3.23	2.86	2.66	2.54	2.45	2.39	2.34	2.30	2.27	2.25	.100	11
	.050	4.84	3.98	3.59	3.36	3.20	3.09	3.01	2.95	2.90	2.85	.050	
	.025	6.72	5.26	4.63	4.28	4.04	3.88	3.76	3.66	3.59	3.53	.025	
	.010	9.65	7.21	6.22	5.67	5.32	5.07	4.89	4.74	4.63	4.54	.010	
	.005	12.23	8.91	7.60	6.88	6.42	6.10	5.86	5.68	5.54	5.42	.005	
12	.100	3.18	2.81	2.61	2.48	2.39	2.33	2.28	2.24	2.21	2.19	.100	12
	.050	4.75	3.89	3.49	3.26	3.11	3.00	2.91	2.85	2.80	2.75	.050	
	.025	6.55	5.10	4.47	4.12	3.89	3.73	3.61	3.51	3.44	3.37	.025	
	.010	9.33	6.93	5.95	5.41	5.06	4.82	4.64	4.50	4.39	4.30	.010	
	.005	11.75	8.51	7.23	6.52	6.07	5.76	5.52	5.35	5.20	5.09	.005	
13	.100	3.14	2.76	2.56	2.43	2.35	2.28	2.23	2.20	2.16	2.14	.100	13
	.050	4.67	3.81	3.41	3.18	3.03	2.92	2.83	2.77	2.71	2.67	.050	
	.025	6.41	4.97	4.35	4.00	3.77	3.60	3.48	3.39	3.31	3.25	.025	
	.010	9.07	6.70	5.74	5.21	4.86	4.62	4.44	4.30	4.19	4.10	.010	
	.005	11.37	8.19	6.93	6.23	5.79	5.48	5.25	5.08	4.94	4.82	.005	

Values of F (continued)

Denominator df	Probability of a Larger F	Numerator df 1	2	3	4	5	6	7	8	9	10	Probability of a Larger F	Denominator df
14	.100	3.10	2.73	2.52	2.39	2.31	2.24	2.19	2.15	2.12	2.10	.100	14
	.050	4.60	3.74	3.34	3.11	2.96	2.85	2.76	2.70	2.65	2.60	.050	
	.025	6.30	4.86	4.24	3.89	3.66	3.50	3.38	3.29	3.21	3.15	.025	
	.010	8.86	6.51	5.56	5.04	4.69	4.46	4.28	4.14	4.03	3.94	.010	
	.005	11.06	7.92	6.68	6.00	5.56	5.26	5.03	4.86	4.72	4.60	.005	
15	.100	3.07	2.70	2.49	2.36	2.27	2.21	2.16	2.12	2.09	2.06	.100	15
	.050	4.54	3.68	3.29	3.06	2.90	2.79	2.71	2.64	2.59	2.54	.050	
	.025	6.20	4.77	4.15	3.80	3.58	3.41	3.29	3.20	3.12	3.06	.025	
	.010	8.68	6.36	5.42	4.89	4.56	4.32	4.14	4.00	3.89	3.80	.010	
	.005	10.80	7.70	6.48	5.80	5.37	5.07	4.85	4.67	4.54	4.42	.005	
16	.100	3.05	2.67	2.46	2.33	2.24	2.18	2.13	2.09	2.06	2.03	.100	16
	.050	4.49	3.63	3.24	3.01	2.85	2.74	2.66	2.59	2.54	2.49	.050	
	.025	6.12	4.69	4.08	3.73	3.50	3.34	3.22	3.12	3.05	2.99	.025	
	.010	8.53	6.23	5.29	4.77	4.44	4.20	4.03	3.89	3.78	3.69	.010	
	.005	10.58	7.51	6.30	5.64	5.21	4.91	4.69	4.52	4.38	4.27	.005	
17	.100	3.03	2.64	2.44	2.31	2.22	2.15	2.10	2.06	2.03	2.00	.100	17
	.050	4.45	3.59	3.20	2.96	2.81	2.70	2.61	2.55	2.49	2.45	.050	
	.025	6.04	4.62	4.01	3.66	3.44	3.28	3.16	3.06	2.98	2.92	.025	
	.010	8.40	6.11	5.18	4.67	4.34	4.10	3.93	3.79	3.68	3.59	.010	
	.005	10.38	7.35	6.16	5.50	5.07	4.78	4.56	4.39	4.25	4.14	.005	
18	.100	3.01	2.62	2.42	2.29	2.20	2.13	2.08	2.04	2.00	1.98	.100	18
	.050	4.41	3.55	3.16	2.93	2.77	2.66	2.58	2.51	2.46	2.41	.050	
	.025	5.98	4.56	3.95	3.61	3.38	3.22	3.10	3.01	2.93	2.87	.025	
	.010	8.29	6.01	5.09	4.58	4.25	4.01	3.84	3.71	3.60	3.51	.010	
	.005	10.22	7.21	6.03	5.37	4.96	4.66	4.44	4.28	4.14	4.03	.005	
19	.100	2.99	2.61	2.40	2.27	2.18	2.11	2.06	2.02	1.98	1.96	.100	19
	.050	4.38	3.52	3.13	2.90	2.74	2.63	2.54	2.48	2.42	2.38	.050	
	.025	5.92	4.51	3.90	3.56	3.33	3.17	3.05	2.96	2.88	2.82	.025	
	.010	8.18	5.93	5.01	4.50	4.17	3.94	3.77	3.63	3.52	3.43	.010	
	.005	10.07	7.09	5.92	5.27	4.85	4.56	4.34	4.18	4.04	3.93	.005	

20	.100	2.97	2.59	2.38	2.25	2.16	2.09	2.04	2.00	1.96	1.94	.100	20
	.050	4.35	3.49	3.10	2.87	2.71	2.60	2.51	2.45	2.39	2.35	.050	
	.025	5.87	4.46	3.86	3.51	3.29	3.13	3.01	2.91	2.84	2.77	.025	
	.010	8.10	5.85	4.94	4.43	4.10	3.87	3.70	3.56	3.46	3.37	.010	
	.005	9.94	6.99	5.82	5.17	4.76	4.47	4.26	4.09	3.96	3.85	.005	
30	.100	2.88	2.49	2.28	2.14	2.05	1.98	1.93	1.88	1.85	1.82	.100	30
	.050	4.17	3.32	2.92	2.69	2.53	2.42	2.33	2.27	2.21	2.16	.050	
	.025	5.57	4.18	3.59	3.25	3.03	2.87	2.75	2.65	2.57	2.51	.025	
	.010	7.56	5.39	4.51	4.02	3.70	3.47	3.30	3.17	3.07	2.98	.010	
	.005	9.18	6.35	5.24	4.62	4.23	3.95	3.74	3.58	3.45	3.34	.005	
40	.100	2.84	2.44	2.23	2.09	2.00	1.93	1.87	1.83	1.79	1.76	.100	40
	.050	4.08	3.23	2.84	2.61	2.45	2.34	2.25	2.18	2.12	2.08	.050	
	.025	5.42	4.05	3.46	3.13	2.90	2.74	2.62	2.53	2.45	2.39	.025	
	.010	7.31	5.18	4.31	3.83	3.51	3.29	3.12	2.99	2.89	2.80	.010	
	.005	8.83	6.07	4.98	4.37	3.99	3.71	3.51	3.35	3.22	3.12	.005	
60	.100	2.79	2.39	2.18	2.04	1.95	1.87	1.82	1.77	1.74	1.71	.100	60
	.050	4.00	3.15	2.76	2.53	2.37	2.25	2.17	2.10	2.04	1.99	.050	
	.025	5.29	3.93	3.34	3.01	2.79	2.63	2.51	2.41	2.33	2.27	.025	
	.010	7.08	4.98	4.13	3.65	3.34	3.12	2.95	2.82	2.72	2.63	.010	
	.005	8.49	5.79	4.73	4.14	3.76	3.49	3.29	3.13	3.01	2.90	.005	
120	.100	2.75	2.35	2.13	1.99	1.90	1.82	1.77	1.72	1.68	1.65	.100	120
	.050	3.92	3.07	2.68	2.45	2.29	2.17	2.09	2.02	1.96	1.91	.050	
	.025	5.15	3.80	3.23	2.89	2.67	2.52	2.39	2.30	2.22	2.16	.025	
	.010	6.85	4.79	3.95	3.48	3.17	2.96	2.79	2.66	2.56	2.47	.010	
	.005	8.18	5.54	4.50	3.92	3.55	3.28	3.09	2.93	2.81	2.71	.005	
∞	.100	2.71	2.30	2.08	1.94	1.85	1.77	1.72	1.67	1.63	1.60	.100	∞
	.050	3.84	3.00	2.60	2.37	2.21	2.10	2.01	1.94	1.88	1.83	.050	
	.025	5.02	3.69	3.12	2.79	2.57	2.41	2.29	2.19	2.11	2.05	.025	
	.010	6.63	4.61	3.78	3.32	3.02	2.80	2.64	2.51	2.41	2.32	.010	
	.005	7.88	5.30	4.28	3.72	3.35	3.09	2.90	2.74	2.62	2.52	.005	

Source: This table is abridged from Table 18, Percentage Points of the *F*-Distribution (Variance Ratio), "Biometrika Tables for Statisticians," 3d ed., vol. 1, E. S. Pearson and H. O. Hartley (eds.), Cambridge University Press, London, 1965, with permission of the editors and the Biometrika trustees.

TABLE B-8 / Suggested Critical Values of Spearman's r_s

n	One-tailed Alternatives		Two-tailed Alternatives	
	Critical Value	α†	Critical Value	α†
	...	...	...	...
4	1.	.0417	...	...
	...	...	1.	.0833
	1.	.0083	1.	.0167
5	.9	.0417		
	.7	.1167	.9	.0833
	.9429	.0083	.9429	.0167
6	.7714	.0514	.8286	.0583
	.6571	.0875	.7714	.1028
	.8571	.0119	.8929	.0123
7	.6786	.0548	.7857	.0480
	.5714	.1000	.6786	.1095
	.8095	.0109	.8571	.0107
8	.6429	.0481	.7381	.0458
	.5238	.0983	.6429	.0962
	.7667	.0107	.8167	.0108
9	.6000	.0484	.6833	.0503
	.4833	.0969	.6000	.0968

† α is the exact probability of finding an r_s as extreme, or more so, as the tabulated value. Probabilities are chosen to give a rough correspondence to signfiicance levels of .01, .05, and .10.
Source: Courtesy of Dr. James H. Goodnight, Department of Statistics, North Carolina State University, Raleigh, N.C.

References

1. "Automobile Facts and Figures," p. 36, Percentage of Householders Owning Automobiles, 1960, Automobile Manufacturers Association Inc., Detroit, 1963.
2. "Automobile Facts and Figures," p. 42, Age of Licensed Drivers in the United States, 1962, Automobile Manufacturers Association Inc., Detroit, 1963.
3. "Automobile Facts and Figures," p. 22, Number of Passenger Cars in the United States by Age Groups, 1964, Automobile Manufacturers Association Inc., Detroit, 1965.
4. Baer, M. J.: Dimensional Changes in the Human Head and Face in the Third Decade of Life, *Am. J. Phys. Anthropol.*, **14:**557–575 (1956).
5. Barker, M. G., and D. J. VanRest: Some Effects of Variability of Individual Drivers on Measures of Tractor Performance, *J. Agric. Eng. Res.*, **7:**208–213 (1962).
6. Bergerud, A. T., A. Butt, H. L. Russell, and H. Whalen: Immobilization of Newfoundland Caribou and Moose with Succinylcholine Chloride and Cap-chur Equipment, *J. Wildlife Manage.*, **28;**49–53 (1964).
7. Bernard, R. L., and B. B. Singh: Inheritance of Pubescence Types in Soybeans, *Crop Sci.*, **9:**192–197 (1969).
8. Bouwkamp, J. C., and J. E. McCully: Competition and Survival in Female Plants of *Asparagus officinalis L., J. Am. Soc. Hortic. Sci.*, **97:**74–76 (1972).
9. Brumby, P. J.: The Grazing Behavior of Dairy Cattle in Relation to Milk Production, *N. Z. J. Agric. Res.*, **2:**797–807 (1959).
10. Clagett, C. O., T. E. Stoa, H. J. Klosterman, A. F. Kingsley, and W. W. Sisler: The Soil Depleting Power of the Flax Crop Compared with That of Hard Red Spring Wheat, Oats, and Barley, *N.D. Agric. Exp. Stn. Bull.*, **378:**1–26 (1952).
11. Chu, F. C. H., M. Paulshock, S. L. Wallenstein, R. W. Houde, and R. P. Phillips: A Controlled Clinical Study of Metopimazine and Perphenazine in Treatment of Radiation Nausea and Vomiting, *Clin. Pharmacol. Therap.*, **10:**800–809 (1969).
12. Cochran, W. G.: The χ^2 Test of Goodness of Fit, *Ann. Math. Stat.*, **23:**315–345 (1952).
13. Cochran, W. G.: "Sampling Techniques," Wiley, New York, 1965.
14. Cochran, W. G., and G. M. Cox: "Experimental Designs," 2d ed., Wiley, New York, 1957.
15. Ditman, K. S., G. G. Crawford, E. W. Forgy, H. Moskowitz, and C. MacAndrew: A Controlled Experiment on the Use of Court Probation for Drunk Arrests, *Am. J. Psychiatry*, **124:**160–163 (1967).
16. Dixon, W. J., and F. J. Massey, Jr.: "Introduction to Statistical Analysis," 3d ed., McGraw-Hill, New York, 1969.
17. Eblinger, W. H.: Wisconsin during Mid-century, Sources of Gross Farm Income United States and Wisconsin, Ten Year Average, 1944–1953, "Wisconsin Crop and Livestock Report," Wisconsin State Department of Agriculture cooperating with U.S. Department of Agriculture, *U.S. Dept. Agric. Bull.*, **331:**41 (1955).
18. Edwards, G. J., W. L. Thompson, J. R. King, and P. J. Jutras: Optical Determination of Spray Coverage, *Trans. Am. Soc. Agric. Eng.*, **4:**206–207 (1961).
19. Engel, M., G. Marsden, and S. W. Pollock: Child Work and Social Class, *Psychiatry*, **34:**140–155 (1971).

20. Evans, A. S., K. A. Shepard, and V. A. Richards: A B O Blood Groups and Viral Diseases, *Yale J. Biol. Med.*, **45:**81–92 (1972).
21. Finney, D. J.: "Probit Analysis," Cambridge University Press, London, 1947.
22. Fisher, R. A., and F. Yates: "Statistical Tables for Biological, Agricultural, and Medical Research," 6th ed., Longman Group Ltd., London, 1963. (Previously published by Oliver and Boyd, Edinburgh.)
23. Frey, K. J., and G. I. Watson: Chemical Studies in Oats, I. Thiamine, Niacin, Riboflavin, and Pantothenic Acid, *Agron. J.*, **42:**434–436 (1950).
24. Friedman, M.: The Use of Ranks to Avoid the Assumption of Normality Implicit in the Analysis of Variance, *J. Am. Stat. Assoc.*, **32:**675–701 (1937).
25. Goldstein, A., and S. Kaizer: Psychotropic Effects of Caffein in Man, III. A Questionnaire Survey of Coffee Drinking and Its Effects in a Group of Housewives, *Clin. Pharmacol. Therap.*, **10:**477–488 (1969).
26. Gorbet, D. W., and D. E. Weibel, Inheritance and Genetic Relationships of Six Endosperm Types in Sorghum, *Crop Sci.*, **12:**378–382 (1972).
27. Hale, James B, Unpublished data concerning acre yield in muskrats in Horicon Marsh, Wildlife Research Section, Wisconsin Department of Natural Resources.
28. Hendricks, W. A., and J. C. Scholl: Techniques in Measuring Joint Relationships, *N.C. Agric. Exp. Stn. Tech. Bull.* 74, table 3, 1943.
29. Hopkins, J. W., and J. Biely: Variation in Weight of Some Internal Organs in the Domestic Fowl, *Can. J. Res.*, **12:**651–656 (1935).
30. Kirkpatrick, C. M., and H. E. Moses: The Effect of Streptomycin against Spontaneous Quail Disease in Bob Whites, *J. Wildlife Manag.*, **17:**24–28 (1953).
31. Kriz, L. B., and C. H. Raths: Connections in Precast Concrete Structures, *J. Prestressed Concr. Inst.*, **10:**16–61 (1965).
32. Kruskal, W. H., and W. A. Wallis: Use of Ranks in One-criterion Variance Analysis, *J. Am. Stat. Assoc.*, **47:**583–621 (1952).
33. Kurtz, T. E.: "Basic Statistics," Prentice-Hall, Englewood Cliffs, N.J., 1963.
34. Larsson, Y. A. A., C. G. Goran, M. D. Sterky, K. E. K. Ekengren, and T. G. H. O. Moller: Physical Fitness and the Influence of Training in Diabetic Adolescent Girls, *Diabetes*, **11:**109–117 (1962).
35. Lawrence, D. R., R. Moulton, and M. D. Rosenheim: Problems of Short-term Comparison of Antihypertensive Drugs, *Clin. Pharmacol. Ther.*, **1:**617–623 (1960).
36. Levy, G., T. Tsuchiya, and L. P. Amsel: Limited Capacity for Salicyl Phenolic Glucuronide Formation and Its Effect on the Kinetics of Salicylate Elimination in Man, *Clin. Pharmacol. Ther.*, **13:**258–268 (1972).
37. Lieberman, G. J., and D. B. Owen: "Tables of the Hypergeometric Probability Distribution," Stanford University Press, Stanford, Calif., 1961.
38. "Life Insurance Fact Book," Types of Ordinary Life Insurance in Force in the United States, 1962, Expressed as a Percentage, Institute of Life Insurance, New York, 1964.
39. "Life Insurance Fact Book," Types of Life Insurance in Force in the United States, 1963, by Amount and Number of Policies, Institute of Life Insurance, New York, 1965.
40. Linnerud, Ardell: Data collected for a study concerning uric acid, cholesterol, and fitness variables in professional men, 1968.
41. Mann, H. B., and D. R. Whitney: On a Test of Whether One of Two Variables is Stochastically Larger than the Other, *Ann. Math. Stat.*, **18:**50–60 (1947).

42. McMullen, L.: Foreword, "Student's Collected Papers," E. S. Pearson and John Wishart (eds.), *Biometrika* Office, University College, London, 1947.
43. Mendel, G.: "Versuche über Pflanzen Hybriden," 1866; English translation from the Harvard University Press, Cambridge, Mass., 1948.
44. Molina, E. C.: "Poisson's Exponential Binomial Limit," D. Van Nostrand, Princeton, N.J., 1949.
45. Nichols, R. E.: Unpublished data on apparent surface tension in dynes of rumen fluid before and after feeding, Department of Veterinary Science, University of Wisconsin, Madison, Wis.
46. Olson, E. C., and R. L. Miller: "Morphological Integration," University of Chicago Press, Chicago, 1958.
47. Patton, R. F.: Unpublished data on age of parent tree and reaction of grafts to blister rust, Department of Plant Pathology, University of Wisconsin, Madison, Wis.
48. Pearse, G. E.: On Corrections for the Moment Coefficients of Frequency Distributions When There Are Infinite Ordinates at One or Both of the Terminals, *Biometrika*, **20a**:314–355 (1928).
49. Pearson, E. S.: Karl Pearson, An Appreciation of Some Aspects of His Life and Work, Part I, 1857–1906, *Biometrika*, **28**:193–257 (1936).
50. Pearson, E. S.: Karl Pearson, An Appreciation of Some Aspects of His Life and Work, Part II, 1906-1936, *Biometrika*, **29**:161–248 (1938).
51. Pearson, E. S., and H. O. Hartley (eds.): "Biometrika Tables for Statisticians," 3d ed., vol. 1, Cambridge University Press, London, 1966.
52. Peterson, H. L.: Pollination and Seed Setting in Lucerne, "Royal Veterinary and Agriculture Yearbook," pp. 138–169, Copenhagen, Denmark, 1954.
53. Pinkerton, P.: The Influence of Sociopathology in Childhood Asthma, in "Recent Research In Psychosomatics," 8th Europ. Conference on Psychosomatics Research, Knokke, 1970 (Karger, Basel, 1970); *Psychother. Psychosom*, **18**:231–238 (1970).
54. Siegel, S.: "Nonparametric Statistics," McGraw Hill, New York, 1956.
55. Smith, D. C., and E. L. Nielsen: Comparison of Clonal Isolations of *Poa pratensis L.*, *Agron. J.*, **43**:214–218 (1951).
56. Sofranko, A. J., and M. F. Nolan: Early Life Experience and Adult Sport Participation, *J. Leisure Res.*, **4**:6–18 (1971).
57. Steel, R. G. D., and J. H. Torrie: "Principles and Procedures of Statistics," McGraw Hill, New York, 1960.
58. Street and Smith, Age of the National League Pitchers in 1965, "Baseball Yearbook," pp. 74–83, Condé Nast, Greenwich, Conn., 1965.
59. "Survey of Current Business," Monthly Commodity Price Indices for 1964, Bureau of Labor Statistics, U.S. Department of Commerce, 1965.
60. "Tables of the Binomial Probability Distribution," Applied Mathematics Series 6, National Bureau of Standards, 1950.
61. Tanikawa, M.: Changes in the Heart Rate after Procaine, *Japanese J. Pharm.*, **3**:112–117 (1954).
62. Tanur, Judith M., et al.: "Statistics: A Guide to the Unknown," Holden-Day, San Francisco, 1972.
63. Ten Ween, J. H., and T. E. W. Feltkamp: Studies in Drug Induced *Lupers erythomatosus* in Mice, *Clin. Exp. Immunol.*, **11**:265–276 (1972).
64. Thomas, G. P., and D. G. Podmore: Studies in Forestry Pathology, XI, Decay in

Blackcottonwood in the Middle Fraser Region, British Columbia, *Can. J. Bot.*, **31**:675–692 (1953).
65. Walker, Helen M.: The Contributions of Karl Pearson, *J. Am. Stat. Assoc.*, **53**:11–27 (1958).
66. Watson, G. J., J. W. Kennedy, W. M. Davidson, C. H. Robinson, and G. W. Muir: Digestibility Studies with Ruminants, XIII, The Effect of the Plane of Nutrition on the Digestibility of Linseed Oil Meal, *Sci. Agric.*, **29**:263–272 (1949).
67. Wax, J., F. W. Ellis, and A. J. Lehman: Absorption and Distribution of Isopropyl Alcohol, *J. Pharmacol. Exp. Ther.*, **97**:229–237 (1949); Williams and Wilkins, Baltimore.
68. Wernimont, G.: Quality Control in the Chemical Industry, I, Statistical Quality Control in the Chemical Laboratory, *Indust. Qual. Cont.*, **11**:5–11 (1947).
69. Wilcoxon, F.: Individual Comparisons by Ranking Methods, *Biom. Bull.*, **1**:80–83 (1945).
70. Wilcoxon, F., and R. A. Wilcox: "Some Rapid Approximate Statistical Procedures," 2d ed., American Cyanamid Co., Pearl River, N.Y., 1964.
71. Winsor, C. P., and G. P. Clarke: A Statistical Study of Variation in the Catch of Plankton Nets, *J. Marine Res.*, **3**:1–34 (1940).

Answers to Selected Problems

CHAPTER 2

2-2-1 See Table A-1.

TABLE A1

Variable	Qualitative	Quantitative	Continuous	Discrete
Hair color	x			x
Incomes		x		x†
Lengths		x	x	
Accidents		x		x
Weights		x	x	
Dow-Jones		x	x‡	
Fish caught		x		x

† Arguable but such salaries usually change by increments of several hundred dollars.
‡ Customarily given to two decimal points.

2-3-1 4, 5, 9, 16, 17, 21, 32, 43, 51, 81.
2-3-3 Need: 1, table headings; 2, four column headings; 3, three column totals.

Column 1: Number of heads. Values 0, 1, . . . , 10.
Column 2: Frequency. Found by experimentation. Has total of 150. Maximum value will probably be at 4, 5, or 6 and frequency not likely much over 40.

Column 3: Relative frequency found as (frequency/150). Should total to 1.
Column 4: Percentage frequency found by multiplying column 3 by 100. Should total to 100.

2-3-5 See Table A2. The information lost in this summary includes the fact that there were no pitchers of age 40 to 43, inclusive.

TABLE A2

Age	Frequency	Percentage
18–20	20	11.0
21–23	41	22.7
24–26	45	24.9
27–29	32	17.7
30–32	20	11.0
33–35	9	5.0
36–38	12	6.6
39–41	1	0.6
42–44	1	0.6

2-3-7 See Table A3.

TABLE A3

Count	Frequency	Percentage
2–3	4	2.0
4–5	20	10.0
6–7	12	6.0
8–9	30	15.0
10–11	35	17.5
12–13	36	18.0
14–15	22	11.0
16–17	23	11.5
18–19	9	4.5
20–21	8	4.0
22–23	0	0.0
24–25	1	0.5

2-4-1 See Table A4.

TABLE A4

Type	Amount	Number
Ordinary	206°	122°
Group	113°	61°
Industrial	20°	113°
Credit	21°	64°

2-4-3 See Table A5.

TABLE A5

Year	1	2	3	4	5	6	7	8	9
1953–54	143°	...	49°	103°	11°	...	...	13°	41°
1973–74	85°	18°	65°	53°	10°	31°	36°	10°	52°

2-4-5 Plot number versus age, using rectangles on unit intervals. For ages 12 to 14 and 14 to 16, height should be halved. For ogives, frequencies must be accumulated.
2-4-7 Center rectangles on given integers.
2-4-9 For range of 300–1,650, intervals of 100 would lead to 14 classes. However, classes from 1,000 on up have none or very few observations. They could be combined in a single class. In this case, it would be well to reduce the height so that the area of this class corresponds to the relative frequency. If an open class is used, for 1,100 or greater, there is no way to make the area correspond to the relative frequency.

CHAPTER 3

3-3-1 (a) $y_1 + y_2 + y_3 + y_4 + y_5 + y_6$; (b) $y_1 + y_2 + \cdots + y_n$; (c) $y_2 + y_3 + y_4 + y_5$; (d) $(y_1 - 8) + (y_2 - 8) + (y_3 - 8)$; (e) $(x_1 + y_1) + (x_2 + y_2) + (x_3 + y_3) + (x_4 + y_4)$; (f) $K + K + \cdots + K$ (show K n times); (g) $w_1Y_1 + w_2Y_2 + \cdots + w_nY_n$; (h) $(y_1 - 3)^2 + (y_2 - 3)^2 + (y_3 - 3)^2 + (y_4 - 3)^2$; (i) $(y_1{}^2 - 3^2) + (y_2{}^2 - 3^2) + (y_3{}^2 - 3^2) + (y_4{}^2 - 3^2)$.
3-3-3 (a) Let $y_1 = 1$ and $y_2 = 2$; $\sum y_i{}^2 = 1 + 4 = 5$ but $(\sum y_i)^2 = (1 + 2)^2 = 9$. (b) Let $x_1 = 1$ and $x_2 = 2$, $y_1 = 3$ and $y_2 = 5$. Then $\sum x_iy_i = 1(3) + 2(5) = 13$ but $(\sum x_i)(\sum y_i) = 3(8) = 24$.
3-6-1 $\bar{y} = 76.36$, $Q_2 = 77$.
3-6-3 $\bar{y} = 147.62$, $Q_2 = 140$.
3-6-5 $\bar{y} = 85.23$, $Q_2 = 84.98$.
3-6-7 Meat, poultry, and fish (one value missing): $\bar{y} = 98.73$, $Q_2 = 98.9$; dairy products: $\bar{y} = 104.65$, $Q_2 = 104.55$; fruits and vegetables: $\bar{y} = 115.32$, $Q_2 = 114.8$.

3-6-9 $\bar{y} = 2.46$, $Q_2 = 2$.
3-6-11 Q_2 is in the 5–6 class. Use 1,586.5/5,586 to get $Q_2 = 5.3$ years.
3-6-13 $\bar{y} = 26.25$, $Q_2 = 25$. (If we assume true ages to be uniformly distributed in this class, then $Q_2 = 25.96$). For groupings 18–20, 21–23, . . . , midpoints are perhaps best located at 19.5, 22.5, etc. In this case, $\bar{y} = 26.78$ and Q_2 is (29.5/45)ths of the way into the 24–26 interval, which is really a three-unit interval. Hence $Q_2 = 25.97$.
3-6-15 $\bar{y} = 11.57$, $Q_2 = 11$. With grouping, $\bar{y} = 11.52$. Consider intervals to be 1.5 to 3.5, 3.5 to 5.5, etc. Q_2 will be in 10–11 interval and $Q_2 = 9.5 + (34/35)(2) = 11.44$.
3-7-1 $P_{25} = 26.1$, $P_{75} = 32.2$ (directly from data), midrange = 29.15.
3-7-3 $P_{25} = 67.25$, $P_{75} = 71.05$ (directly from data with interpolation since 25 and 75 percent of 50 arc not integers), midrange = 69.15.
3-7-5 $P_{25} = 516.5$, $P_{75} = 809.5$ (directly from data), midrange = 663.
3-8-1 (a) 6; (b) 8; next values: 162 and 32.
3-8-3 (a) 72/13 = 5.54;(b) 150.86.

CHAPTER 4

4-3-1 $s^2 = 75.32$, $s = 8.68$. See Table A6. Conversion formula $[(y - 76.36)/8.68]10 + 70$. 56 becomes 46.5, 75 becomes 68.4, 93 becomes 89.2.

TABLE A6

Interval	67.68–85.04	59.00–93.72	50.32–102.40
Count	18	24	25
Percent	72	96	100
Chebyshev	0	75	89
Normal	67	95	99

4-3-3 Heart-rate data, Exercise 3-6-3: $s^2 = 754.09$, $s = 27.46$. Girls before, Exercise 3-6-4: $s^2 = 636.36$, $s = 25.23$. Girls after, Exercise 3-6-4: $s^2 = 819.65$, $s = 28.63$. Bronze-copper, Exercise 3-6-5: $s^2 = .4833$, $s = .6952$. Gas bills, Exercise 3-6-6: $s^2 = 103.46$, $s = 10.17$. Meat, poultry, and fish, Exercise 3-6-7: $s^2 = 2.18$, $s = 1.48$. Dairy products, Exercise 3-6-7: $s^2 = .3082$, $s = .5551$. Fruits and vegetables, Exercise 3-6-7: $s^2 = 10.62$, $s = 3.26$.
4-3-5 With grouping: $\bar{y} = 78.40$, $s^2 = 117.84$, and $s = 10.86$. Without grouping: $\bar{y} = 78.45$, $s^2 = 117.87$, and $s = 10.86$.
4-3-7 For baseball data of Exercise 3-6-13, use intervals 18–20, 21–23, etc., which cover ages 18 through 20, etc., so that midpoints are 19.5, 22.5, and so on. Then $\bar{y} = 26.78$, $s^2 = 26.71$, $s = 5.17$, and the intervals are 21.61 to 31.95, 16.44 to 37.12, and 11.27 to 42.29. Intervals include 121 or 67 percent, 177 or 98 percent, and 180 or 99 percent.

For high-card point-count data of Exercise 3-6-15, counts are at points and not throughout an interval. Thus, for intervals 2–3, 4–5, etc., use midpoints of 2.5, 4.5, and so on. Now $\bar{y} = 11.52$, $s^2 = 19.98$, $s = 4.47$, and the intervals are 7.05 to 15.99, 2.58 to 20.46, and 0 to 24.93. There is no point in having a negative endpoint for an interval.

4-4-1 See Table A7. No apparent relationship between sample size and range. A relationship between sample size and range/s is a possibility.

TABLE A7

Sample Size	Exercise	Range	Standard Deviation (s)	Range/s
10	3-6-5	1.99	.70	2.84
11	3-6-4 (1)	75.0	25.23	2.97
	3-6-4 (2)	117.0	28.63	4.09
	3-6-7 (1)	4.8	1.48	3.24
12	3-6.6	25.94	10.17	2.55
	3-6-7 (2)	1.7	.56	3.04
	3-6-7 (3)	10.6	3.26	3.25
13	3-6-3	93.0	27.46	3.39
16	3-6-2	.50	.15	3.33
25	3-6-1	37.0	8.68	4.26

4-5-1 See Table A8.

TABLE A8

Exercise	Interquartile Range	Semi-interquartile Range	Quartile Variation	Midrange / Range
2-3-4	6.1	3.05	10.5	1.43
2-3-5	6.	3.	11.5	1.19
2-3-6	3.8	1.9	2.7	6.05
2-3-7	7.	3.5	30.4	.59
2-3-8	293.	146.5	22.1	.72

4-5-3 See Table A9.

TABLE A9

Exercise	3-6-1	3-6-2	3-6-3	3-6-4	3-6-4	3-6-5	3-6-6	3-6-7	3-6-7	3-6-7
CV	11.5	3.22	18.60	13.85	27.43	.82	68.39	1.49	.53	2.83
Midrange/Range	5.31	9.40	1.63	2.38	1.77	43.0	.67	20.63	61.62	11.04

CHAPTER 5

5-3-1 1, 2, 3, 4, 5, and 6. Same. Sample space. 1, 3, 5. 2, 4, 6. 3, 6. Yes; for example, the event of an odd number and the event of a number divisible by 3 have the point 3 in

common. Yes; for example, the event of an even number and the event of an odd number have no point in common.
5-3-3 $S = \{2, 3, 4, 5, 6, 7, 8\}$. 7 elements. No. Yes. (1, 1) goes into 2; (1, 2) and (2, 1) into 3; (1, 3), (2, 2) and (3, 1) into 4; (1, 4), (2, 3), (3, 2) and (4, 1) into 5; (2, 4), (3, 3) and (4, 2) into 6; (3, 4) and (4, 3) into 7; (4, 4) into 8.
5-3-5 $S = \{$MMMM, MMMF, MMFM, . . . , FMMM, MMFF, . . . , FFMM, MFFF, . . . , FFFM, FFFF$\}$. $2^4 = 16$. 15. 5. No. 8.
5-4-1 $S = \{$H1, . . . , H6, T1, . . . , T6$\}$. $P(E_i) = 1/12$, $i = 1, \ldots, 12$. $P[(a)] = 1/4$. $P[(b)] = 1/4$. $P[(c)] = 1/2$. $P[(d)] = 1/2$.
5-4-3 $P[(a)] = 8/130$. $P[(b)] = 8/130$. $P[(c)] = 4/130$. $P[(d)] = 20/130$. $P[(e)] = 28/130$. $P[(f)] = 15/130$. $P[(g)] = 34/130$. $P[(h)] = 33/130$. $P[(i)] = 91/130$.
5-5-1 (a) Each point might be represented by student name or number, and the letter(s) required to specify the language(s) taken.
(b) Start with circle. About half the circle $[(80 + 10)/190]$ will go into the two parts representing all of R and part of S, separately. Next to S, *both* F and S will carry you past halfway in the circle. Next to *both* F and S, nearly half the remainder goes to only F. Then one-third of what still remains goes into *both* F and G, leaving the remainder for only G.
(c) No. They are not disjoint.
(d) (1) 10, (2) 45, (3) 100, (4) 70, (5) 180, (6) 145, (7) 90, (8) 120, (9) 55, (10) 135, (11) 0, (12) 190, (13) 150, (14) 40, (15) 20, (16) 170, (17) 100, (18) 90, (19) 15, (20) 175, (21) 180, (22) 10, (23) Neither F nor G nor $S = R$, (24) Those taking F and G and S = empty set = $\varnothing$, (25) Those not taking all of F and G and S = sample space.
(e) $P = 1/190$.
(f) 70/190, 45/190, 10/190, 100/190, 100/190, 80/190, 150/190, 55/190, 145/190, 110/190, 155/190, 15/190, 0, 20/190, 0, 0, 0, 0.
(g) $\overline{S \cup G \cup R}$ (= F only), $\overline{F \cup S \cup R}$ (= G only), $\overline{S \cup F \cup G}$ (= R only), $\overline{R \cup G \cup F}$ (= S only), $F \cap S$ and $F \cap G$.
5-5-3 (a) (1) $A \cup B = \bar{C}$, (2) $\bar{C} \cap F$, (3) $\bar{C} \cap F$, (4) $\overline{A \cup B}$, (5) $(A \cup B) \cap (E \cup F) = \bar{C} \cap \bar{D}$, (6) $B \cap E$, (7) $D \cap (B \cup C)$; (b) (1) 210/260, (2) 65/260, (3) 15/260, (4) 50/260, (5) 137/260, (6) 14/260, (7) 32/260; (c) A, B, C. Also, D, E, F.

CHAPTER 6

6-2-1 (a) 1/2. Yes. (b) 1/4. (c) 1/2.
6-2-3 (a) MMM, MMF, MFM, FMM, MFF, FMF, FFM, FFF; (b) 1/8, (c) 6/8, (d) 7/8.
6-2-5 (a) 1296, (b) 1, (c) 4, (d) 10, (e) 20, (f) 35, (g) 70, (h) $70/1{,}296 \approx .054$. (1) Personal opinion. Approximately 1 time in 20 is generally considered small.
6-3-1 (a) 1/2, yes; (b) 2/3, no; (c) 1/3, no; (d) 2/3, no.
6-3-3 (a) 3/7, no; (b) 1/4, yes; (c) 1/4, no.
6-3-5 (a) 7/12, (b) 3/7.
6-3-7 (a) .04, (b) .36.

CHAPTER 7

7-2-1 $5! = 120$
7-2-3 210; 343; 7; 126
7-2-5 $(20!)/(7!)(13!) = 77{,}520$
7-2-7 120, 480
7-2-9 $13!$, $(13!)^4$
7-3-1 210, 5,040

7-3-3 151, 800 **7-3-5** 1/32, 10/32 **7-3-7** 1,024, 56/1,024 = .055
7-3-9 125

CHAPTER 8

8-2-1 (a)

y	0	1
P	1/2	1/2

; (b)

y	0	1	2
P	1/4	1/2	1/4

; (c)

y	2	3	4
P	9/16	3/16	4/16

8-2-3 (a) S = {HHHH, HHHT, . . . , TTTT}; 16 points, each with $P = 1/16$;

(b)

y	4	3	2	1	0
P	1/16	4/16	6/16	4/16	1/16

8-2-5

y	2	3	4	5	6
P	1/15	2/15	3/15	4/15	5/15

8-3-1 (a) 5, (b) 1/2, (c) 1/5, (d) 2/5, (e) 2/5, (f) 3/5, (g) 1/10, (h) 7/10, (i) 3/5, (j) 3/5, (k) 1/5.
8-3-3 1/6. No.

CHAPTER 9

9-2-1 216,

y	0	1	2	3
P	125/216	75/216	15/216	1/216

, $1 - (1/216)^4$

9-2-3

y	0	1	2	3	4	5	6	7	8	9	10
P	$\frac{1}{4^{10}}$	$\frac{10(3)}{4^{10}}$	$\frac{45(3^2)}{4^{10}}$	$\frac{120(3^3)}{4^{10}}$	$\frac{210(3^4)}{4^{10}}$	$\frac{252(3^5)}{4^{10}}$	$\frac{210(3^6)}{4^{10}}$	$\frac{120(3^7)}{4^{10}}$	$\frac{45(3^8)}{4^{10}}$	$\frac{10(3^9)}{4^{10}}$	$\frac{3^{10}}{4^{10}}$

;
sum P's under 5, 6, 7, 8, 9, and 10.
9-2-5 $[9^{20} + 20(9^{19}) + 190(9^{18})]/10^{20}$.
9-2-7 .90112.
9-3-1 P(no student) = 11/522 = P(no faculty). P(one student) = 25/174. P(no more than 3 faculty) = 218/261.
9-3-3 2/1,131; 100/3,393; 2,584/23,751; 8,075/23,751.
9-3-5 .2817; .7183; .7513; .7513. .4096; .5904; .8192; .8192.
9-4-1 Presumably satisfactory. $P(0) = .0224$, $P(0, 1) = .1075$, $P(y > 5) = .1834$.
9-4-3 $P(10) = .1240$, $P(0) = .000045$, $P(y \leq 5) = .066495$.
9-4-5 .472, .354, .133, .033, .006, .001, .001. .470, .358, .133, .033, .006, .001, .000.

CHAPTER 10

10-3-1 (a) .1587, (b) .0228, (c) .0013, (d) .0505, (e) .2514, (f) .0250, (g) .0051.
10-3-3 (a) .6170, (b) .3174, (c) .2006, (d) .1010, (e) .0512, (f) .0198.
10-3-5 (a) 0, (b) $.25 < z_\alpha < .26$, (c) $.52 < z_\alpha < .53$, (d) $.67 < z_\alpha < .68$, (e) Approx. .84, (f) Approx. 1.28, (g) 1.645, (h) 2.33.
10-3-7 (a) Approx. 2.33, (b) 1.645, (c) Approx. 1.28, (d) $.67 < z < .68$, (e) 0, (f) $-.68 < z < -.67$, (g) Approx. -1.28, (h) -1.645, (i) Approx. -2.33.
10-3-9 (a) 2.575, (b) Approx. 2.33, (c) 1.96, (d) 1.645, (e) Approx. 1.15, (f) $.67 < z < .68$.

10-4-1 (a) .0475, (b) .9332, (c) .0475, (d) Approx. .3372, (e) Approx. .6628, (f) .8944.
10-4-3 (a) (1) 6.5, (2) 4.6, (3) 1.9; (b) (1) 7.2, (2) 5.5, (3) 4.6; (c) (1) 3.9, (2) 4.6, (3) 6.5.

CHAPTER 11

11-2-1 1/2, 4, .2384.

11-2-3

y	1	2	3	4
P	1/4	1/4	1/4	1/4

, $E = 2.5$

11-2-5 7
11-4-1 2.5, $P = 0$ since one can't answer half a question.
11-4-3 $E = 12.5$ cents. Play.
11-5-1 3/16.
11-5-3 (a) 1/2, (b) .715.
11-5-5 3.5, 35/12.
11-5-7 5/12.
11-5-9 15/8.

CHAPTER 12

12-2-1 All distributions normal. See Table A10. Variance decreasing faster than standard deviation.

TABLE A10

n	μ_Y	${\sigma_Y}^2$	σ_Y
10	μ	$\sigma^2/10 = .1\sigma^2$	$.3162\sigma$
20	μ	$\sigma^2/20 = .05\sigma^2$	$.2236\sigma$
30	μ	$\sigma^2/30 = .033\sigma^2$	$.1826\sigma$
40	μ	$\sigma^2/40 = .025\sigma^2$	$.1581\sigma$
50	μ	$\sigma^2/50 = .02\sigma^2$	$.1414\sigma$

12-2-3 3.5; 35/24; no.
12-3-1 See Table A11.

TABLE A11

n	μ	σ^2	Points Required
1	2.5	1.25	4
2	2.5	.625	7
3	2.5	.4167	10
4	2.5	.3125	13

12-4-1 See Table A12.

TABLE A12

n	$\hat{p}$, $P(\hat{p})$	μ	σ^2
1	0, 1 1/2, 1/2	1/2	1/4
2	0, 1/2, 1 1/4, 1/2, 1/4	1/2	1/8
3	0, 1/3, 2/3, 1 1/8, 3/8, 3/8, 1/8	1/2	1/12
4	0, 1/4, 1/2, 3/4, 1 1/16, 4/16, 6/16, 4/16, 1/16	1/2	1/16
5	0, 1/5, 2/5, 3/5, 4/5, 1 1/32, 5/32, 10/32, 10/32, 5/32, 1/32	1/2	1/20
6	0, 1/6, 2/6, 3/6, 4/6, 5/6, 1 1/64, 6/64, 15/64, 20/64, 15/64, 6/64, 1/64	1/2	1/24

12-4-3

y	15	14	13	12	11	10	9	8
P_N	.0000	.0000	.0003	.0016	.0070	.0237	.0611	.1208
y	7	6	5	4	3	2	1	0
P_N	.1814	.2082	.1814	.1208	.0611	.0237	.0070	.0019

12-4-5 $P(y \leq 50) = .5398 = P(30 \leq y \leq 50)$.
12-4-7 .8016; .9880.
12-4-9 .0175.
12-4-11 31.

CHAPTER 13

13-2-1 $.95^{20} = .3585$, .3774, binomial.
13-4-1 (a) .55, (b) .549995, (c) .550006.
13-4-3 Jones: (.549, .651), (.539, .661). Smith: (.349, .451), (.339, .461). Complementary relation on unit interval. Finite.
13-4-5 (.687, .713), (.683, .717).
13-4-7 (0, .063), (0, .074).
13-5-1 (72.77, 79.88), (4.66, 4.82), (84.73, 85.73), (8.41, 21.34), (97.74, 99.72), (104.10, 105.00), (114.15, 116.45).
13-5-3 (71.51, 81.21), (4.63, 4.85), (84.52, 85.94), (5.75, 23.99).
13-5-5 (−.0414, .0614).
13-5-7 $2t_{\alpha/2}s/\sqrt{n}$. Yes, because s does.
13-7-1 Minimum of 896 for $p = .3$ or .7 to maximum of 1,067 for $p = .5$.

13-7-3 456 if a very large number of fuses are involved.
13-7-5 20.
13-8-1 For $1 - \alpha = .90$: (347.74, 1,615.14) for girls before, (447.90, 2,080.34) for girls after. For $1 - \alpha = .95$: (45.88, 145.79) for exams, (.0127, .0557) for *Hyopsodus*, (388.37, 2,056.61) for heart data. For $1 - \alpha = .99$: (.1843, 2.5142) for bronze.
13-8-3 62 but not dependent on σ^2, as can be seen from formula.

CHAPTER 14

14-2-1 Grade data: (73, 83), $1 - \alpha = .957$ (approx.). *Hyopsodus* data: (4.65, 4.87), $1 - \alpha = .979$ (approx.). Bronze data: (84.54, 86.12), $1 - \alpha = .979$ (approx.).
14-2-3 (73, 82.6), (4.67, 4.86), (84.63, 85.92).
14-3-1 (73, 80.25), (4.65, 4.83), (84.72, 85.72).
14-3-3 Arrange table with three data types across the top; sign test, adjusted sign test, rank sum test, and normal procedures in the left column; and intervals and/or lengths of intervals in body of table. (See grade data and note what ties do to left endpoint.)

CHAPTER 15

15-1-1 See Table A13.

TABLE A13

	Best Machine is	
Machine Chosen Is	*A*	*B*
A	Correct	Incorrect
B	Incorrect	Correct

15-2-1 See Table A14.

TABLE A14

	Decision	
Truth	H_0 accepted; salesperson believed; buy.	H_0 rejected; salesperson not believed; don't buy.
H_0; salesperson spoke truth.	Good decision; driving safe car.	Poor decision; driving old car; repairs needed.
H_1; salesperson spoke falsely.	Poor decision; possible accident due to steering failure.	Good decision; avoid accident due to steering failure unless age-related.

15-2-3 (a) .167, .180, .382 for $p = .4$; .678, .678. .879 for $p = .2$. (b) .833, .820, .618 for $p = .4$; .322, .322, .121 for $p = .2$. (c) .055, .109, .172.
15-2-5 (a) Reject H_0 for observed 20:0, $\cdots$, 14:6; accept H_0 for 15:5, $\cdots$, 0:20; $\alpha = .058$; $1 - \beta = .786$. (b) Reject H_0 for observed 0:20, $\cdots$, 11:9; accept H_0 for 12:8, $\cdots$, 20:0; $\alpha = .41$; $1 - \beta = .748$. (c) .942, .058; .786, .214.
15-3-1 If coin falls 7:13, $\cdots$, 13:7, accept H_0: $p = .5$. Otherwise reject H_0. True $\alpha =$ .115.
15-3-3 The value of p is not constant from trial to trial.
15-4-1 Accept H_0. $P(|Z| > .52) = .6030$. Results seem comparable.
15-4-3 Accept H_0. $P(|Z| > .50) = .6170$.
15-4-5 For data from Exercise 15-5-3: (a) accept H_0, (b) accept H_0. For data from Exercise 15-5-4: (a) reject H_0, (b) accept H_0, (c) reject H_0. For data from Exercise 15-5-5: (a) reject H_0, (b) reject H_0.
15-5-1 $\chi^2 = .267$; accept H_0. $.75 > P(\chi^2 > .267) > .50$.
15-5-3 $\chi^2 = .479$; accept H_0. $\chi^2 = 3.084$; accept H_0.
15-5-5 $z = 2$, $\chi^2 = 4$, reject H_0. $z = 6.32$, $\chi^2 = 40$, reject H_0. Reject in both cases, tests agree, large samples make it easier to detect alternatives when true. Observe n in

$$\frac{\hat{p} - p}{\sqrt{p(1 - p)/n}} = \frac{\sqrt{n}(p - \hat{p})}{\sqrt{p(1 - p)}}$$

15-5-7 (a) $\chi^2 = 1.18$ with 2 df, accept H_0. Nothing obvious. Problem is to define "more extreme than observed" simply. (b) $\chi^2 = .550$ with 2 df, accept H_0, P is somewhat larger than .75. (c) $\chi^2 = 1.101$ with 2 df, accept H_0, $.750 > P > .500$. (d) $\chi^2 = 1.382$ with 2 df, accept H_0, P is very close to .50.

CHAPTER 16

We now follow the convention of using * and ** to indicate that values of the test criterion are beyond those at the 5 percent and 1 percent levels of significance, respectively.
16-2-1 Yes. $\chi^2 = 10.10$**, 1 df.
16-2-3 Effective, $\chi^2 = 110$** with 1 df. Realistic alternative would have been $H_1 : P(A \mid T) > P(A \mid U)$.
16-2-5 $\chi^2 = .5625$ with 1 df. No relation apparent. Would expect treatment to improve situation though we would not need to test this since observed proportions are seen to be counter to this alternative. $\chi^2 = 8.95$**, 1 df; yes, but a one-sided alternative would seem suitable since new pasture might not be infected with parasites.
16-2-7 $\chi^2 = 8.11$**, 1 df. Reject H_0.
16-3-1 $\chi^2 = 1.51$, 1 df. Do not reject H_0. $\chi^2 = 4.31$*, 1 df. Reject H_0. Evidence of a relation. Conclusion changed but same people used.
16-3-3 $\chi^2 = .08$, 1 df. Do not reject H_0.
16-3-5 $\chi^2 = 4.96$*, 1 df. Yes. $H_1 : p_{11}/p_{12} \neq p_{21}/p_{22}$. Compute $P(L \mid M) = .264$ and $P(L \mid F) = .416$ rather than $P(L)$
16-4-1 $\chi^2 = 8.89$**, 1 df. Reduced. $P(+ \mid \text{treated}) = 25/28$ whereas $P(+ \mid \text{placebo}) = 16/28$. If we use a one-sided alternative, then $P(\chi^2 > 8.89) < 1/2(.005)$.
16-4-3 $\chi^2 = .67$, 1 df. $H_0{:}p(+ \mid 4 \text{ mg P}) = p(+ \mid 2.5 \text{ mg M})$. H_0 is not denied regardless

of alternative. Without prior knowledge concerning drugs, alternatives will be two-sided. $\chi^2 = .11$, 1 df. H_0 is not denied.
16-5-1 Appears to be completely random. $\chi^2 = 5.58$ with 3 df. Do not reject. $H_0: p_{ij} = p_{i.}p_{.j}$ where $p_{i.} = p_{i1} + p_{i2}$, all i, and $p_{.j} = \sum_i p_{ij}, j = 1, 2$. $H_1: p_{ij} \neq p_{i.}p_{.j}$.
16-5-3 $\chi^2 = 20.11^{**}$ with 2 df. Association; not independent.
16-5-5 $\chi^2 = 4.48$ with 4 df. Evidence is not against homogeneity. $H_0: p_{11} = p_{21} = p_{31}$, $p_{12} = p_{22} = p_{32}$, $p_{13} = p_{23} = p_{33}$, simultaneously.

CHAPTER 17

17-2-1 $t = 4.41^{**}$, df $= 9$, reject H_0, $P(|t| > 4.41)$ is near .001.
17-2-3 $t = -3.19^{**}$, df $= 199$, $P(t < -3.19) = .0007$ from normal tables, reject H_0. Population is of mileages based on a hypothetical population of tires made and used under the existing conditions.
17-2-5 $t = 5.79^{**}$ with 19 df. Reject manufacturers claim. One-sided alternative.
17-4-1 $t = 2.09$ with 18 df. Do not reject H_0 if a purist; $t_{.05} = 2.101$, so evidence is rather marginal.
17-4-3 $t = 1.304$ with 17 df. Do not reject H_0.
17-4-5 $t = 1.453$ with 48 df. Do not reject H_0. If teaching is on a class, rather than an individual, basis, then there is no true replication of treatments.
17-5-1 $t = 3.572^{**}$ with 7 df. Reject H_0. CI(95 percent) = (45.48, 223.77), (41.8, 283.8), and (48.25, 225.35).
17-5-3 $t = .600$ with 11 df. Do not reject H_0. CI(90 percent) = (−.498, 1.075), and (−.559, .366).
17-5-5 $t = 2.830^{**}$ with 9 df. Reject H_0. $P(t > 2.830)$ is very close to .01.
17-7-1 $F = 1.64$ with 4 and 9 df; $F = 1.15$ with 8 and 9 df; $F = 1.42$ with 19 and 24 df; $F = 1.27$ with 24 and 24 df. In no case would one reject the null hypothesis.

CHAPTER 18

18-2-1 $\chi^2 = 6.4^*$ with 1 df. Compare $P(\chi^2 > 6.63) = .01$.
18-2-3 $\chi^2 = 6.25^*$ with 1 df. Compare $P(\chi^2 > 6.63) = .01$.
18-3-1 Sum $= +2^{**}$. Critical value $= 3.1$ for $\alpha = .01$. Reject H_0.
18-3-3 Sum $= -14^{**}$. Critical value is 19.5 for $\alpha = .01$.
18-4-1 $\chi^2 = 5.56^*$ with 1 df; $P(\chi^2 > 5.02) = .025$. P(3 tables) $= .0267 + .0033 + .0000 = .0300$, which is for one-sided alternatives.
18-4-3 P's are .000017, .000936, .014034, .084204, .235770, and .330079. Except for last, each appears twice.
18-5-1 Sum $= 79^*$. Critical value $= 79$ for $\alpha = .05$.
18-6-1 $\chi^2 = 8^{**}$ with 1 df. $P(\chi^2 > 7.88) = .005$. Binomial $P = 2(1/2)^8 = .0078$.
18-6-3 $\chi^2 = .33$ with 1 df. Do not reject H_0. $.750 > P(\chi^2 > .33) > .500$. Binomial $P = .774414$.
18-6-5 $\chi^2 = 3.6$ with 1 df. Do not reject H_0. $\chi^2_{.05} = 3.84$. Binomial $P = .1094$.
18-7-1 Sum $= 0^{**}$; critical value $= .3$ for $\alpha = .01$.
18-7-3 Sum $= 35$. Do not reject H_0.

CHAPTER 19

19-3-1 See Table A15. CV = 7.56 percent, $s_{\bar{y}}$ = 5.06 grams, s(mean difference) = 7.16 grams. Assumptions: random components are normally and independently distributed with zero mean and common variance. Were they fed and housed together? Were they fed *ad libitum* or a fixed ration?

TABLE A15

Source	df	MS	F
Rations	3	7363.08	35.92**
Error	28	205.01†	

† 1,365.5 + 1,333.875 + 1,010.875 + 2,030.

19-3-3 See Table A16. $F_{.10}$ = 2.33. Do not reject H_0. CV = 14.54 percent. $s_{\bar{y}}$ = 1.94 seconds, s(mean difference) = 2.74 seconds. Were the students trained individually or simultaneously?

TABLE A16

Source	df	MS	F
Groups	3	60.42	2.29
Error	24	26.37	

19-4-1 See Table A17. Reject the null hypothesis.

TABLE A17

Source	df	MS	F
Blocks	5	.0009275	
Crops	3	.0129486	15.08**
Error	15	.0008586	

Crop	O	F	B	W
Means	.1783	.2550	.2717	.2800

Reject H_0: μ(oats) = μ(flax). Result-guided. Since a t-test calls for a difference between two means that has been determined by a random process, our choice of adjacent pairs probably leads to lower differences, on the average. Consequently, we have an under-estimate. CV = 11.90 percent, $s_{\bar{y}}$ = .0120 percent, s(mean difference) = .0169 percent.

19-4-3 Sampling scheme. See Table A18. Reject H_0 for nets and for types. $s_{\bar{y}}(\text{net}) = .0436$; $s(\text{mean difference}) = .0616$. $s_{\bar{y}}(\text{type}) = .0390$; $s(\text{mean difference}) = .0551$.

TABLE A18

Source	df	MS	F
Nets	4	.0284	3.73*
Types	3	3.0145	396.**
Error	12	.0076	

19-5-1 Treatments fixed. Estimate differences. Assume blocks random. Generalize. Measure a part.
19-5-3 Type probably fixed. Nets designed or chosen to be fixed. No. Not a mixed model.
19-7-1 See Table A19. Do not reject H_0. CV = 15.31 percent. $s_{\bar{y}_2} = 14.19$ minutes, $s_{\bar{y}_3} = 10.99$ minutes, and $s_{\bar{y}_2 - \bar{y}_3} = 17.95$ minutes.

TABLE A19

Source	df	MS	F
Groups	2	2,518.2415	2.08
Error	21	1,208.594†	

† 9,394.875 with 7 df, 9,596 with 5 df, and 6,389.6 with 9 df, add to SSE.

CHAPTER 20

20-2-1 $\chi^2 = 14.53^{**}$ with 3 df; $\chi^2_{.005} = 12.8$.
20-2-3 $\chi^2 = 4.846$ with 3 df; $.25 > P(\chi^2 > 4.846) > .10$.
20-3-1 (a) $\chi^2 = 23.43^{**}$ with 3 df; (b) $\chi^2 = 6.29$ with 3 df; $\chi^2_{.10} = 6.25$; (c) $\chi^2 = 5.47$ with 3 df; $\chi^2_{.10} = 6.25$; (d) $\chi^2 = 3.84$ with 2 df; $\chi^2_{.10} = 4.61$.
20-4-1 $\chi^2 = 9^*$ with 3 df; $\chi^2_{.025} = 9.35$.
20-4-3 $\chi^2 = 12^{**}$ with 3 df, for types; $\chi^2_{.005} = 12.8$. $\chi^2 = 4.33$ with 4 df, for nets; $.5 > P(\chi^2 > 4.33) > .25$.
20-5-1 $\chi^2_r = 11.4^{**}$ with 3 df; $\chi^2_{.005} = 12.8$.
20-5-3 $\chi^2_r = 15^{**}$ with 3 df, for types; $\chi^2_{.005} = 12.8$. $\chi^2_r = 9.4$ with 4 df, for nets; $\chi^2_{.05} = 9.49$.

CHAPTER 21

21-3-1 6.31, 7.11, 7.93, 8.73, and 9.54 minutes.
21-3-3 $a = 57.34$, $b = -.7639$. $\hat{Y} = 57.34 - .7639x = 50.08 - .7639(x - 9.5)$. $\hat{y}_1 = 56.5765$, $\hat{y}_{11} = 48.9375$, $\hat{y}_{19} = 42.8263$. Deviation = 1.4235, −6.9375, −3.8263.

21-3-5 $a = -794.08$, $b = 7.287$; $\hat{Y} = -794.08 + 7.287x = 295 + 7.287(x - 149.46)$; $\hat{y} = 153.2$, 298.9, and 415.5. Deviation $= 11.8, -8.9, -25.5$.
21-4-1 Probably: fixed, fixed, random (must be), random.
21-5-1 See Table A20.

TABLE A20

Regression SS	$s^2_{Y.x}$, df	$s_{Y.x}$
(a) 241.02	14.79, 10	3.8%
(b) 5.23	.05275, 13	.230 mm
(c) 73,231.04	751.72, 11	27.4 lb
(d) 10,544.65	38.95, 16	6.2

21-6-1 $t = 3.28^*$ with 8 df. Reject H_0. CI will not contain $\mu_{Y \mid \bar{x}} = 7$ minutes. Equation 21-6-12 satisfactory because $(x - \bar{x}) = 0$.
21-6-3 $F = 12.70^{**}$ with 1, 8 df. Reject $H_0\!:\beta = 0$; $F_{.01} = 11.26$. $t^2 = 3.56^2 = 12.67$.
21-6-5 $s_Y = 1.2161$; $s_b = .1892$; $s_{\hat{Y}} = 1.9544$, 1.1459, and 2.1129. $\text{CI}(\mu_{\hat{Y}}) = (52.22, 60.93)$, (46.38, 51.49), and (38.12, 47.53). $t = 4.038^{**}$ with 10 df; $F = 16.296^{**}$ with 1, 10 df; reject $H_0\!:\beta = 0$.
21-6-7 $s_Y = 7.6043$, $s_b = .7383$; $s_{\hat{Y}} = 16.2559$, 7.6147, and 14.3841. Accept both H_0's.
21-8-1 (a) $-.7872$, (b) .9402, (c) .9479, (d) .9717. $r = \pm\sqrt{\text{regn SS/total SS}}$, but would have to check covariance for sign. Correlation coefficient for (c) and (d). Call r^2 the coefficient of determination. In all cases, reject H_0.
21-9-1 $r_s = .9492$; $t = 10.00^{**}$ with 13 df; reject H_0. $P(t > 10.00) < .001$.

INDEX